ENGINEERING METALLURGY

Part I

APPLIED PHYSICAL METALLURGY

RAYMOND A. HIGGINS

B.Sc. (Birm.), F.I.M.

Senior Lecturer in Metallurgy, The College of Commerce and Technology, West Bromwich; Examiner in Metallurgy to the Institution of Production Engineers, the Union of Lancashire and Cheshire Institutes, the Union of Educational Institutes, and the City and Guilds of London Institutes; formerly Chief Metallurgist, Messrs. Aston Chain and Hook Co., Ltd., Birmingham

THE ENGLISH UNIVERSITIES PRESS LTD

Paperback edition ISBN 0 340 17673 3
Boards edition ISBN 0 340 17672 5

First edition 1957
Reprinted 1960, 1961, 1964, 1965, 1967
Second edition 1968
Reprinted (with revisions) 1970
Third edition 1971
Fourth edition 1973

The English Universities Press Ltd
St Paul's House, Warwick Lane, London EC4P 4AH

Printed in Great Britain by
Richard Clay (The Chaucer Press), Ltd, Bungay, Suffolk.

PREFACE TO THE FIRST EDITION

THIS text-book constitutes Part I of "Engineering Metallurgy" and is intended primarily for students taking metallurgy as an examination subject for a Higher National Certificate in Mechanical or Production Engineering. The author hopes that it may also prove useful to undergraduates studying metallurgy as an ancillary subject in an Engineering Degree course. To students for whom metallurgy is a principal subject the book can offer a helpful approach to certain sections of the work in preparation for the Higher National Certificate and the City and Guilds Final Certificates in Metallurgy.

Comprehensive tables covering most of the alloys of importance to engineers are given in the appropriate chapters. In these tables an attempt has been made to relate British Standard Specifications to many commercially produced alloys. The author hopes that these tables will remain of use when the reader, no longer a student, finds it necessary to choose alloys for specific engineering purposes.

A generation ago much of a student's time was spent in dealing with the principles of extraction metallurgy. The widened scope of applied physical metallurgy has, however, in recent years, established prior claims upon the time available. Hence the brief survey in Chapter II of the production of iron and steel has to serve as a sufficient introduction to the methods of extraction metallurgy in general.

In the main, then, this book deals with the microstructural and mechanical properties of metals and alloys. Processes such as heat-treatment, surface hardening and welding are dealt with from the theoretical as well as the practical aspect. The author trusts that the treatment in Chapter I of the basic principles of chemistry will enable readers to follow the study of engineering metallurgy without being at a disadvantage if they have not previously studied chemistry as an independent subject.

It has been considered desirable to provide a basis for practical metallography; hence details of laboratory techniques are dealt with in Chapter X.

At the end of each chapter will be found a selection of questions and exercises. Many of these have been taken from Higher National Certificate examination papers, and the author is greatly indebted to those authorities who have given permission for such questions to be used.

Although "Engineering Metallurgy" Part II (Metallurgical Process Technology) is strictly speaking a sequel to the present volume, it may on occasion be read with advantage as a companion book—particularly when the approved syllabus for the Engineering Higher National Certificate is more than usually ambitious, or when direct contacts of students with metallurgical processes are limited by local circumstances.

In both Parts sections are numbered on the decimal system. In Part II frequent references to appropriate sections of Part I make it easy for the reader to look up the metallurgical principles governing any particular process under study.

Except where otherwise stated, the photomicrographs in this book are the work of the author or his students.

The author wishes to record his thanks to his wife for considerable help in producing the line diagrams; and to his friends J. H. Parry, Esq., F.I.M., of the School of Technology, Ipswich, and A. N. Wyers, Esq., A.I.M., of the Chance Technical College, for reading the original MS. and making many helpful suggestions. He also wishes to record his appreciation of the generous assistance given to him at all stages in the production of this book by W. E. Fisher, Esq., O.B.E., D.Sc.

The author wishes to acknowledge the considerable help given by those connected with various industrial organisations, but in particular W. E. Bardgett, Esq., B.Sc.; F.I.M. (Messrs. United Steel Companies Ltd., Sheffield); J. F. Hinsley, Esq. (Messrs. Edgar Allen and Co. Ltd., Sheffield); Dr. J. R. Rait (Messrs. Hadfields Ltd., Sheffield); Dr. R. T. Parker and Dr. A. N. Turner (Messrs. Aluminium Laboratories Ltd., Banbury); Messrs. Samuel Osborn & Co. Ltd., and Prof. Dr. Fritz Gabler and Messrs. C. Reichert of Vienna.

R. A. HIGGINS

Department of Science,
The Technical College,
West Bromwich, Staffs.

PREFACE TO THE FOURTH EDITION

DURING the fifteen years which have passed since this volume was first published, substantial progress in metallurgy, as in other sciences, has taken place; consequently considerable revision of the subject matter has been necessary with each subsequent edition. These comments apply particularly to those topics like the dislocation theory which have become established as corner-stones of metallurgical dogma. Similarly, Chapter XXII dealing with metallic corrosion has been largely re-written to bring it into line with modern convention.

In recent years many of the metals specifications adopted by the British Standards Institution have been completely revised; in particular, the old "En system" for wrought steels has at last disappeared. Accordingly, all the tables of alloys have been brought up to date, not only in respect of BS numbers but also in terms of trade names and code numbers. Mergers and "take-overs" in the steel industry have led to the re-naming of some products, and their new designations have been incorporated in the revised tables.

One had hoped that by now some uniformity in the use of SI units would have been achieved. With the adoption of "pure SI" in mind, the unit of stress MNm^{-2} was used in the first "metric edition" of this book. Currently, however, the unit N/mm^2 seems to be the more popular, and one can see the reason for this in that it provides a value of a magnitude which is easier to appreciate. Accordingly, we have changed to the unit N/mm^2 in this edition, though of course stresses expressed in this unit are numerically identical to those expressed in MNm^{-2}. Throughout, "negative indices" have been replaced by the "solidus", e.g. N/mm^2 rather than Nmm^{-2}, since this notation seems to be preferred by all but the "pure scientists".

The selections of examination questions which occur at the end of each chapter were compiled *before* recent trends towards the adoption of SI units and terminology. Consequently many of the questions are set in "imperial units" and contain some pre-metric nomenclature. This point is mentioned in fairness to those Colleges and Universities who very kindly provided the questions used.

The author wishes to thank all those colleagues and friends who, over the years, have shown an interest in this book and have made helpful comments as to how the subject matter could be improved. In particular he wishes to thank Dr. A. E. W. Smith, lately Associate Professor, R.M.C.S., Shrivenham, for the very detailed analysis made some time ago and which has been extremely useful in preparing the present edition.

R. A. HIGGINS

The Division of Materials Technology,
  The College of Commerce and Technology,
    West Bromwich, Staffs.

NOTES ON THE "METRIC" (SI) EDITION

"SI" is the abbreviation by which the Système International d'Unités (International System of Units) is known universally. It can be regarded as the logical development of the MKS (metre, kilogram, second) system of units which has been used for a considerable time by scientists and which grew into the MKSA (metre, kilogram, second, ampere) system of Professor Giorgi.

Since 1875 the Conférence Générale des Poids et Mesures (CGPM) has been entrusted with the international co-ordination of metric matters, and the SI units used in this book are those which have been accepted by this body. Although final agreement has not been reached between the various scientific and industrial organisations concerned regarding the use of certain SI units, it was felt that both with respect to notation and the majority of units employed, this book could, with advantage, "go metric" at this stage. Accordingly, SI units have been used throughout, though the former "imperial" equivalents have also been included in some of the tables.

It is recommended that the following conventions be adopted when using the SI notation:

(1) Abbreviations are given in singular form and without the use of full stop, e.g. kg *not* kg. or kgs.

(2) Although the use of the solidus is permissible, e.g. N/mm^2, it is desirable to avoid its use where possible and to write $N\ mm^{-2}$.*

(3) Multiples or submultiples of simple or derived units can be used. As far as possible these should differ from the base unit in steps of 10^3 and the multiples 10^2, 10, 10^{-1} and 10^{-2} should only be used in special circumstances.

(4) Where the multiple or submultiple prefix immediately precedes the symbol for a unit having a power different from unity, the prefix should also be incorporated within the power, e.g. $mm^2 = 10^{-6}$ m^2 and *not* 10^{-3} m^2.

The following base units and derived units are used in this book:

---

* In spite of this the solidus is now more commonly used than "negative indices", and so we have returned to the use of the solidus in this book.

Base Units

Length	metre	m
Mass	kilogram	kg
Time	second	s
Electric current	ampere	A
Thermodynamic temperature*	kelvin	K

Derived Units

Force	newton	N	$N = kg\,m/s^2$
Work, energy and quantity of heat	joule	J	$J = N\,m$
Power	watt	W	$W = J/s$
Electric charge	coulomb	C	$C = A\,s$
Electric potential	volt	V	$V = W/A$
Electric resistance	ohm	Ω	$\Omega = V/A$
Frequency	hertz	Hz	$Hz = 1/s$
Magnetic flux	weber	Wb	$Wb = V\,s$
Magnetic flux density	tesla	T	$T = Wb/m^2$
Resistivity			Ωm
Magnetic field strength			A/m
Pressure, stress			N/m^2

Multiples and Submultiples of Coherent Units

	Multiplication factor	Prefix	Symbol
Preferred	10^{12}	tera	T
	10^{9}	giga	G
	10^{6}	mega	M
	10^{3}	kilo	k
	10^{-3}	milli	m
	10^{-6}	micro	μ
	10^{-9}	nano	n
	10^{-12}	pico	p
	10^{-15}	femto	f
	10^{-18}	atto	a
	10^{2}	hecto	h
	10	deca	da
	10^{-1}	deci	d
	10^{-2}	centi	c

* For normal temperature measurement the Celsius scale has been retained at this stage ($K = {}^{\circ}C + 273.15$).

Conversion Factors ("Imperial" to SI)

Unit	Symbol	Definition in SI
inch	in	25·4 mm
foot	ft	0·304 8 m
square inch	in²	645·16 mm²
pound	lb	0·453 592 kg
ton	ton	1 016·05 kg
		= 1·016 05 Mg = 1·016 05 tonne
ton force	tonf	9·964 02 kN

One of the most controversial points regarding the use of SI at present concerns the units involving pressure and stress. This is of paramount importance in metallurgy, since it governs the choice of units for proof stress and tensile strength. The following metric conversions to replace "tons force per square inch" are possible:

$$1 \text{ tonf/in}^2 = 1·574\ 9 \text{ kgf/mm}^2 \text{ (former metric units)}$$
$$= 1·544\ 43 \text{ hbar (non-coherent)}$$
$$= 1·544\ 43 \text{ kN/cm}^2 \text{ (coherent)}$$
$$= 15·444\ 3 \text{ N/mm}^2 \text{ (coherent)}$$
$$= 15·444\ 3 \text{ MN m}^{-2} \text{ (pure SI)}$$
$$= 15·444\ 3 \text{ MPa (proposed new derived unit)}$$

The proposed new unit to denote pressure and stress, the Pascal (Pa), is already used in France and is equivalent to N/m². In this book stress is quoted in terms of N/mm², so that whichever units are used (other than former metric units), conversion is simple and involves only movement of the decimal point. The British Standards Institution used, "on an interim basis", the property of tensile strength expressed in hectobars (1 hbar = 10 MN/m² = 10 N/mm²), but now appears to have forsaken them in favour of N/mm².

Since most values quoted in tables and elsewhere in this book are "representative", the conversions have similarly been "rounded off". Thus, whilst 1 ton = 1 016·05 kg, for most purposes, it has been sufficient to use the approximate value of 1 000 kg; or 1 tonne ("metric ton") as being equivalent to 1 "imperial" ton. Similar approximations have been made in conversions of other units where appropriate.

R.A.H.

CONTENTS

CHAPTER I

CHEMISTRY, METALLURGY AND THE ENGINEER

1.10. Until comparatively recent years the term "metallurgy" denoted little but the applied science of the extraction of metals from their ores and subsequent refining of the crude metals produced. Any further treatment covered by the term was usually of a relatively simple nature and restricted to the shaping of those metals by straightforward mechanical working processes, together with simple heat-treatment in the case of steels.

During the present century the scope of metallurgical science has expanded enormously, so that the subject can now be studied under the following headings:

(*a*) Extraction metallurgy, which is that branch already referred to, dealing with the winning of metals from the Earth's crust.

(*b*) Process metallurgy, which deals with the alloying, shaping and subsequent heat-treatment of metals.

(*c*) Physical metallurgy, that section concerned mainly with the crystal structure of metals and alloys and its effect on their physical and mechanical properties.

The contacts which the producer and the user of metals have with metallurgical science may be quite diverse. The practising metallurgist may be dealing with any, if not all, of the sections mentioned above. The engineer, on the contrary, will be concerned chiefly with the third section, namely, applied physical metallurgy, and with the second section only in so far as processes such as heat-treatment, die-casting, welding and press-work are involved.

This book, then, will deal mainly with physical metallurgy, since behind it lies all the fundamental "know how" of the science. A few of the more important manufacturing processes will also be considered briefly. The extraction of iron will be dealt with mainly to satisfy the student's curiosity regarding the methods used and also to furnish some information with respect to the quality of the finished product.

Unlike the alchemist of the Middle Ages, the modern research metallurgist does not spend his time in fruitless attempts to turn base metals into gold—a useless metal from a metallurgical point of view when all is said and done. Instead he has of recent years given to humanity an ever-growing range of useful alloys. Whilst many of these alloys are put to purposes of destruction, we must not forget that others have contributed to the material progress of mankind and to his domestic comfort.

As a result of such research a somewhat bewildering array of compositions confronts the engineer in his search for an alloy suitable for his needs. Fortunately, the most useful alloys have been classified, and rigid specifications laid down for them, by such active bodies as the British Standards Institution (BSI); the Directorate of Technical Development (DTD); and, in America, by the American Society for Testing Materials (ASTM). Hence, forearmed with the necessary metallurgical knowledge, the engineer is able to select with confidence an alloy suited to his needs and to quote the relevant specification numbers when the time comes to convert the blue-print into reality.

But before we proceed to a detailed study of metallurgical science, it will assist us if we consider, quite briefly, the chemical nature and the behaviour of the substances concerned.

Atoms, Elements and Compounds

1.20. Not so very long ago, in our early chemistry lessons, we used to say that the atom was the smallest unit of which matter was composed and was indivisible. Now, it is not quite so simple as that, and the chemist no longer regards the atom as being in the nature of an indestructible little billiard-ball which is held by some mysterious force of attraction to its neighbours. Other, more simple, particles are now known to science, and whilst from the point of view of our metallurgical studies as engineers it will often suffice if we retain the billiard-ball concept of the atom, some knowledge of its architecture is becoming increasingly necessary to technologists in the modern world. Consequently, the principles of atomic structure will be discussed later in this chapter.

Atoms possess varying numbers of electrons (negatively charged particles), protons (positively charged particles) and neutrons (which carry no charge). The ultimate nature of the substance depends upon the number of each type of particle which its individual atoms contain. Chemical properties are related to the numbers of electrons and protons present and in this respect there are altogether ninety-two basically different types of atom which occur naturally. Of late the scientists have succeeded in building up a series of new ones. These are the atoms which have been in the news so much of recent years because they can be made to disintegrate with an enormous release of energy.

1.21. When two or more atoms, either of one type or of different types, are joined together chemically, the unit which is produced is called a molecule. In a pure substance all the molecules are alike. If these like molecules are built of atoms all of which are identical the substance is a chemical **element**. Thus the salient property of a chemical element is that it cannot be split up into simpler substances whether by mechanical or chemical means. Most of the elements are chemically reactive, so that we find very few of them in their elemental state in the Earth's crust—oxygen and nitrogen **mixed** together in the atmosphere are the most common, whilst a few of the metals, such as copper, gold and silver, also occur uncombined. Typical molecules occurring naturally contain atoms of two or more kinds.

1.22. Thus most of the substances we encounter are either chemical **compounds** or **mixtures**. The difference between the two is that a compound is formed when there is an association of the electrons at the surface of two or more different atoms so that the atoms are held together, whilst a mixture is formed by mechanically "entangling" the two substances together. In this latter state they can easily be separated. For example, the powdered element sulphur can be mixed with iron filings and easily separated again by means of a magnet, but if the mixture is gently heated a chemical reaction takes place and a compound called iron sulphide is formed. This is different in appearance from either of the parent elements, and its decomposition into the parent elements, sulphur and iron, is now a more difficult matter which can be accomplished only by chemical means.

1.23. Chemical elements can be represented by a sort of short-hand cipher or **symbol** which is usually an abbreviation of either the English or Latin name, e.g. O stands for oxygen whilst Fe stands for "ferrum", the Latin equivalent of "iron". Ordinarily, a symbol written thus refers to a single atom of the element, whilst two atoms (constituting a molecule) would be indicated so: O_2.

1.24. Table 1.1 gives a list of some of the more important elements with which we are likely to come into contact in our metallurgical studies. The term "atomic weight"* mentioned in this table is really a measure of the relative mass of the atom and should in no way be confused with the relative density of the element. The latter value will depend upon how closely the atoms, whether simple or complex, are packed together. Since atoms are very small particles (the mass of the hydrogen atom is $1·673 \times 10^{-27}$ kg), it would be inconvenient to use such small values in everyday chemical calculations. Consequently, since the hydrogen atom was known to be the smallest, its relative

---

* This term is a relic of the past, and now that we make precise distinction between weight and mass it should be called "relative atomic mass".

TABLE I.I

Element	Symbol	Atomic weight (C = 12·0000)	Relative Density (Specific Gravity)	Melting point (° C)	Properties and Uses
Aluminium .	Al	26·98	2·7	659·7	The most widely used of the light metals.
Antimony .	Sb	121·75	6·6	630·5	A brittle, crystalline metal which, however, is used in bearings and type.
Argon . .	Ar	39·948	$1·78 \times 10^{-3}$	−189·4	An inert gas present in small amounts in the atmosphere. Used in "argon-arc" welding.
Arsenic . .	As	74·92	5·7	814	A black crystalline element—used to harden copper at elevated temperatures.
Barium .	Ba	137·34	3·5	850	Its compounds are useful because of their fluorescent properties.
Beryllium .	Be	9·012	1·8	1 285	A light metal which is used to strengthen copper. Also used un-alloyed in atomic-energy plant.
Bismuth .	Bi	208·98	9·8	271·3	A metal similar to antimony in many ways—used in the manufacture of fusible (low-melting-point) alloys.
Boron . .	B	10·811	2·3	2 300	Known chiefly in the form of its compound, "borax".
Cadmium .	Cd	112·4	8·6	320·9	Used for plating some metals and alloys and for strengthening copper telephone wires.
Calcium .	Ca	40·08	1·5	845	A very reactive metal met chiefly in the form of its oxide, "quicklime".
Carbon . .	C	12·011	2·2	—	The basis of all fuels and organic substances and an essential ingredient of steel.
Cerium .	Ce	140·12	6·9	640	A "rare-earth" metal. Used as an "inoculant" in cast iron, and in the manufacture of lighter flints.
Chlorine .	Cl	35·45	$3·2 \times 10^{-3}$	−103	A poisonous reactive gas, used in the de-gasification of light alloys.
Chromium .	Cr	51·996	7·1	1 890	A metal which resists corrosion—hence it is used for plating and in stainless steels and other corrosion-resistant alloys.
Cobalt .	Co	58·933	8·9	1 495	Used chiefly in permanent magnets and in high-speed steel.
Copper .	Cu	63·54	8·9	1 083	A metal of high electrical conductivity which is used widely in the electrical industries and in alloys such as bronzes and brasses.
Gold . .	Au	196·967	19·3	1 063	Of little use in engineering, but mainly as a system of exchange and in jewellery.
Helium . .	He	4·0026	$0·16 \times 10^{-3}$	Below −272	A light non-reactive gas present in small amounts in the atmosphere.
Hydrogen .	H	1·00797	$0·09 \times 10^{-3}$	−259	The lightest element and a constituent of most gaseous fuels.
Iridium .	Ir	192·2	22·4	2 454	A heavy precious metal similar to platinum.
Iron . .	Fe	55·847	7·9	1 535	A fairly soft white metal when pure, but rarely used thus in engineering.
Lead . .	Pb	207·69	11·3	327·4	Not the densest of metals, as the metaphor "as heavy as lead" suggests.
Magnesium .	Mg	24·312	1·7	651	Used along with aluminium in the lightest of alloys.
Manganese .	Mn	54·938	7·2	1 260	Similar in many ways to iron and widely used in steel as a deoxidant.
Mercury .	Hg	200·59	13·6	−38·8	The only liquid metal at normal temperatures—known as "quicksilver".

Table 1.1 (continued)

Element	Symbol	Atomic weight (C = 12·0000)	Relative Density (Specific Gravity)	Melting point(° C)	Properties and Uses
Molybdenum	Mo	95·94	10·2	2 620	A heavy metal used in alloy steels.
Nickel . .	Ni	58·71	8·9	1 458	An adaptable metal used in a wide variety of ferrous and non-ferrous alloys.
Niobium .	Nb	92·906	8·6	1 950	Used in steels and, un-alloyed, in atomic-energy plant. Formerly called "Columbium" in the United States.
Nitrogen .	N	14·0067	$1·16 \times 10^{-3}$	−210	Comprises about ⅘ of the atmosphere. Can be made to dissolve in the surface of steel and so harden it.
Osmium .	Os	190·2	22·5	2 700	The densest element and a rare white metal like platinum.
Oxygen . .	O	15·999	$1·32 \times 10^{-3}$	−218	Combined with other elements it comprises nearly 50% of the Earth's crust and 20% of the Earth's atmosphere in the uncombined state.
Palladium .	Pd	106·4	12·0	1 555	Another platinum-group metal.
Phosphorus .	P	30·9738	1·8	44	A reactive element; in steel it is a deleterious impurity, but in some bronzes it is an essential addition.
Platinum .	Pt	195·09	21·4	1 773	Precious white metal used in jewellery and in scientific apparatus because of its high corrosion resistance.
Potassium .	K	39·102	0·86	62·3	A very reactive metal which explodes on contact with water.
Rhodium .	Rh	102·905	12·4	1 985	A platinum-group metal used in the manufacture of thermocouple wires.
Selenium .	Se	78·96	4·8	220	Mainly useful in the manufacture of photo-electric cells.
Silicon . .	Si	28·086	2·4	1 427	Known mainly as its oxide, silica (sand, quartz, etc.), but also, in the elemental form, in cast irons, some special steels and non-ferrous alloys.
Silver . .	Ag	107·87	10·5	960	Widely used for jewellery and decorative work. Has the highest electrical conductivity of any metal— used for electrical contacts.
Sodium .	Na	22·9898	0·97	97·5	A metal like potassium. Used in the treatment of some of the light alloys.
Strontium .	Sr	87·62	2·6	772	Its compounds produce the red flames in fireworks. An isotope "Strontium 90" present in radioactive "fall out".
Sulphur .	S	32·064	2·1	113	Present in many metallic ores—the steelmaker's greatest enemy.
Tantalum .	Ta	180·948	16·6	3 207	Sometimes used in the manufacture of super-hard cutting tools of the "sintered-carbide" type.
Tellurium .	Te	127·6	6·2	452	Used in small amounts to strengthen lead.
Thallium .	Tl	204·37	11·85	303	A soft heavy metal forming poisonous compounds.
Thorium .	Th	232·038	11·2	1 850	A rare metal—0·75% added to tungsten filaments (gives improved electron emission).
Tin . .	Sn	118·69	7·3	231·9	A widely used but rather expensive metal. "Tin cans" carry only a very thin coating of tin on mild steel.
Titanium .	Ti	47·9	4·5	1 725	Small additions are made to steels and aluminium alloys to improve their properties. Used in the un-alloyed form in the aircraft industry.

TABLE 1.1 *(continued)*

Element	Sym-bol	Atomic weight (C = 12·0000)	Relative Density (Specific Gravity)	Melting point (° C)	Properties and Uses
Tungsten .	W	183·85	19·3	3 410	Imparts very great hardness to steel and is the main constituent of "high-speed" steel. Its high m.pt. makes it useful for lamp filaments.
Uranium .	U	238·03	18·7	1 150	Used chiefly in the production of atomic energy.
Vanadium .	V	50·942	5·7	1 710	Added to steels as a "cleanser" (deoxidiser) and a hardener.
Zinc . .	Zn	65·37	7·1	419·5	Used widely for galvanising mild steel and also as a basis for some die-casting alloys. Brasses are copper–zinc alloys.
Zirconium .	Zr	91·22	6·4	1 800	Small amounts used in magnesium and high-temperature alloys. Also, un-alloyed, in atomic energy plant.

mass was taken as unity and the relative masses of the atoms of other elements calculated as multiples of this. Thus atomic weight became

$$\frac{\text{mass of an atom of the element}}{\text{mass of an atom of hydrogen}}.$$

Later it was found more convenient to adjust the atomic weight of oxygen (by far the most common element) to exactly 16·0000. On this basis the atomic weight of hydrogen became 1·008 instead of 1·0000. More recently chemists and physicists have agreed to relate atomic weights to that of a carbon isotope (C = 12·000) (see 1.90).

1.25. The most common metallic element in the Earth's crust is aluminium (Table 1.2), but in the metallic form it is not the cheapest.

TABLE 1.2.—*The Approximate Composition of the Earth's Crust (to the Extent of Mining Operations)*

Element	% by mass	Element	% by mass
Oxygen	49·1	Nickel	0·02
Silicon	26·0	Vanadium . . .	0·02
Aluminium . . .	7·4	Zinc	0·02
Iron	4·3	Copper . . .	0·01
Calcium . . .	3·2	Tin	0·008
Sodium	2·4	Boron	0·005
Potassium . . .	2·3	Cobalt	0·002
Magnesium . . .	2·3	Lead . . .	0·002
Hydrogen . . .	1·0	Molybdenum . .	0·001
Titanium . . .	0·61	Tungsten . . .	9×10^{-4}
Carbon . . .	0·35	Cadmium . . .	5×10^{-4}
Chlorine . . .	0·20	Beryllium . . .	4×10^{-4}
Phosphorus . . .	0·12	Uranium . . .	4×10^{-4}
Manganese . . .	0·10	Mercury . . .	1×10^{-4}
Sulphur . . .	0·10	Silver . . .	1×10^{-5}
Barium . . .	0·05	Gold . . .	5×10^{-6}
Nitrogen . . .	0·04	Platinum . . .	5×10^{-6}
Chromium . . .	0·03	Radium . . .	2×10^{-10}

This is because clay, of which aluminium is a constituent, is a very difficult substance to decompose chemically. Therefore our aluminium supply comes from the mineral bauxite (originally mined near the village of Les Baux, in France), which is a relatively scarce ore. It will be seen from the table that apart from iron most of the useful metallic elements account for only a very small proportion of the Earth's crust. Fortunately they occur in relatively concentrated deposits which makes their mining and extraction economically possible.

Chemical Reactions and Equations

1.30. Engineers will be familiar with the behaviour of simple chemical salts during electrolysis. The metal particles, or ions, being positively charged, are attracted to the negative electrode (or cathode), whilst the non-metal ions, being negatively charged, are attracted to the positive electrode (or anode). Metals are therefore said to be electropositive and non-metals electronegative.

In general, elements react or combine with each other when they possess opposite chemical natures. Thus the electropositive metals tend to combine with the electronegative non-metals, forming very stable chemical compounds. If one metal is more strongly electropositive than another the two may combine, forming what is called an intermetallic compound, though usually they only mix with each other, as they do in the majority of useful metallurgical alloys.

1.31. Atoms combine with each other in simple fixed proportions. For example, one atom of the gas chlorine (Cl) will combine with one atom of the gas hydrogen (H) to form one molecule of the gas hydrogen chloride. We can write down a **formula** for hydrogen chloride which expresses at a glance its molecular constitution, viz. HCl. Since one atom of chlorine will combine with one atom of hydrogen, its **valency** is said to be one, the term valency denoting the number of atoms of hydrogen which will combine with one atom of the element in question. Again, two atoms of hydrogen combine with one atom of oxygen to form one molecule of water (H_2O), so that the valency of oxygen is two. Similarly, four atoms of hydrogen will combine with one atom of carbon to form one molecule of methane (CH_4). Hence the valency of carbon in this instance is four. However, carbon, like several other elements, exhibits a variable valency, since it will also form the substances ethylene (C_2H_4) and acetylene (C_2H_2).

1.32. Compounds also react with each other in simple molecular proportions, and we can express such a reaction in the form of a chemical **equation** thus:

$$CaO + 2HCl = CaCl_2 + H_2O$$

The above equation tells us that one molecule of calcium oxide (CaO—usually known as "quicklime") will react with two molecules of hydrogen chloride (hydrochloric acid) to produce one molecule of calcium chloride ($CaCl_2$) and one molecule of water (H_2O). Though the total number of molecules may change due to a reaction, as is demonstrated above, the total number of atoms remains the same on either "side" of the equation. In other words, the equation must "balance" rather like a financial balance sheet.

1.33. By substituting the appropriate relative atomic masses (approximated values from Table 1.1) in the above equation we can obtain further useful information from it.

$$CaO \quad + \quad 2HCl \quad = \quad CaCl_2 \quad + \quad H_2O$$

$$\underbrace{40 + 16}_{56} \quad \underbrace{2(1 + 35{\cdot}5)}_{73} \quad \underbrace{40 + 2(35{\cdot}5)}_{111} \quad \underbrace{2(1) + 16}_{18}$$

Thus, 56 parts by weight of quicklime will react with 73 parts by weight of hydrogen chloride to produce 111 parts by weight of calcium chloride and 18 parts by weight of water. Naturally, instead of "parts by weight" we can express these reacting quantities in specific units of mass, such as kilograms or megagrams, as required. Incidentally, the value 56 is known as the "Molecular Weight" or "Relative Molecular Mass" of quicklime, being the sum of the atomic weights or relative atomic masses of the component atoms.

We will now deal with those chemical processes which concern us most when studying metallurgy.

Oxidation and Reduction

1.40. Oxidation is one of the most common of chemical processes. It refers, in its simplest terms, to the combination between oxygen and any other element—a phenomenon which is taking place all the time around us. In our daily lives we make constant use of oxidation. We inhale atmospheric oxygen and reject carbon dioxide (CO_2)—the oxygen we breathe combines with carbon from our animal tissues, releasing energy in the process. We then reject the waste carbon dioxide. Similarly, heat energy can be produced by burning carbonaceous materials, such as coal or petroleum. Just as without breathing oxygen animals cannot live, so without an adequate air supply fuel cannot burn. In these reactions carbon and oxygen have combined to form a gas, carbon dioxide (CO_2), and at the same time heat energy has been released—the "energy potential" of the carbon having fallen in the process.

1.41. Oxidation, however, is also a phenomenon which works to our disadvantage, particularly in so far as the metallurgist is concerned, since a large number of otherwise useful metals show a great affinity for oxygen and combine with it whenever they are able. We must therefore use expensive processes, such as painting, galvanising or plating, to protect many of our metals from the unwelcome attentions of oxygen.

1.42. It should be noted that, to the chemist, the term "oxidation" has a much wider meaning, and in fact refers to a chemical process in which the electronegative (or non-metallic) constituent of the molecule is increased. For example, ferrous chloride may be oxidised to ferric chloride—

$$2FeCl_2 + Cl_2 = 2FeCl_3$$

Ferrous   Ferric
chloride   chloride

The element oxygen is not involved in this reaction, yet we say that ferrous chloride has been "oxidised" by chlorine since the chlorine ion is electronegative and so the electronegative portion of the ferric chloride molecule is greater than that of the ferrous chloride molecule.

1.43. Whilst many metals exist in the Earth's crust in combination with oxygen as oxides, others are combined with sulphur as sulphides. The latter form the basis of many of the non-ferrous (that is, containing no iron) metal ores. The separation and removal of the oxygen or sulphur contained in the ore from the metal itself is often a difficult and expensive process. Most of the sulphide ores are first heated in air to convert them to oxides, e.g.—

$$2ZnS + 3O_2 = 2ZnO + 2SO_2$$

Zinc sulphide   Zinc oxide   Sulphur
("zinc blende")   dioxide (gas)

The oxide, whether occurring naturally or produced as indicated in the above equation, is then generally mixed with carbon in the form of coke or anthracite and heated in a furnace. In most cases some of the carbon is burned simultaneously in order to provide the necessary heat which will cause the reaction to proceed more quickly. Under these conditions carbon usually proves to have a greater affinity for oxygen than does the metal and so takes oxygen away from the metal, forming carbon dioxide and leaving the metal (often impure) behind, e.g.—

$$2ZnO + C = 2Zn + CO_2$$

This process of separating the atoms of oxygen from a substance is known as **reduction**. Reduction is thus the reverse of oxidation, and again, in the wider chemical sense, it refers to a reaction in which the proportion of the electronegative constituent of the molecule is decreased.

1.44. Some elements possess greater affinities for oxygen than have others. Their oxides are therefore more difficult to decompose. Aluminium and magnesium, for example, have greater affinities for oxygen than has carbon, so that it is impossible to reduce their oxides in the normal way using coke—electrolysis, a much more expensive process, must be used. Metals, such as aluminium, magnesium, zinc, iron and lead, which form stable, tenacious oxides are usually called **base** metals, whilst those metals which have little affinity for oxygen are called **noble** metals. Such metals include gold, silver and platinum, metals which will not scale or tarnish to any appreciable extent due to the action of atmospheric oxygen.

Acids, Bases and Salts

1.50. When an oxide of a non-metal combines with water it forms what we call an **acid**. Thus sulphur trioxide (SO_3) combines with water to form the well-known sulphuric acid (H_2SO_4), and so sulphur trioxide is said to be the **anhydride** of sulphuric acid. Though not all chemical acids are as corrosive as sulphuric, it is fairly well known that in cases of accident involving acids of this type it is necessary to neutralise the acid with some suitable antidote.

1.51. Substances which have this effect are called **bases.** These are metallic oxides (and hydroxides) which, when they react with an acid, produce water and a chemical compound which we call a **salt**. Typical examples of these acid–base reactions are—

$$H_2SO_4 \ + \ CaO \ = CaSO_4 \ + \ H_2O$$

| Sulphuric acid | Calcium oxide ("quicklime") | Calcium sulphate (a salt) | Water |

$$HCl \ + NaOH \ = \ NaCl \ + \ H_2O$$

| Hydrochloric acid | Sodium hydroxide ("caustic soda") | Sodium chloride ("common salt") |

We can generalise in respect of equations like this and say—

$$Acid + Base = Salt + Water$$

Similarly, the acid **anhydride** will often combine with a base, forming a salt, e.g.—

$$SO_3 + CaO = CaSO_4$$

This type of reaction occurs quite frequently during the smelting of metallic ores.

1.52. One of the most common elements in the Earth's crust is silicon, present in the form of its oxide, silica (SiO_2). Since silicon is a non-metal, its oxide is an acid anhydride, and though all the common forms of silica, such as sand, sandstone and quartz, do not seem to be of a very reactive nature at normal temperatures, they are sufficiently reactive when heated to high temperatures to combine with many of the metallic oxides (which are basic) and produce neutral salts called silicates. Silica occurs entangled with most metallic ores, and although some of it is rejected by mechanical means before the ore is charged to the furnace, some remains and could constitute a difficult problem in that its high melting point of 1 780° C would cause a sort of "indigestion" in the furnace. To overcome this a sufficient quantity of a basic flux is added in order to combine with the silica and produce a slag with a melting point low enough to allow it to run from the furnace. The cheapest metallic oxide, and the one in general use, is lime.

$$SiO_2 \ + \ 2CaO \ = \ 2CaO.SiO_2$$

$$\underset{\text{Acid}}{} \qquad \underset{\text{Base}}{} \qquad \underset{\text{Salt (calcium}}{}$$

Acid anhydride   Base   Salt (calcium silicate)

The formula for the slag, calcium silicate, is generally written $2CaO.SiO_2$ rather than Ca_2SiO_4, since lime and silica will combine in other proportions. When the formula is written in the former manner, the molecular proportions can be seen at a glance.

1.53. As we shall see in the next chapter, similar reactions can occur to our disadvantage between slags and furnace linings. Thus, since we do not wish to liquefy our furnace lining, we must make sure that it does **not** react with the charge or the slag covering it. In short, we must make sure that the furnace lining is of the **same** chemical nature as the slag, i.e. if the slag contains an excess of silica, and is therefore acid, we must line the furnace with a similar silica-rich refractory, such as silica brick or ganister; whilst if the slag contains an excess of lime or other basic material, we must line the furnace with a basic refractory, such as burnt dolomite (CaO.MgO) or burnt magnesite (MgO). If the chemical nature of the furnace lining is the same as that of the slag, then, clearly, no reaction is likely to take place between them. Since silica and ganister, on the one hand, and dolomite and magnesite, on the other, all have high melting points, they will also be able to resist the high temperatures encountered in many of the metallurgical smelting processes.

Solutions

1.60. It is well known that many substances will dissolve in water and that as the temperature rises so their solubility will increase. This is illustrated by the solution of the salt potassium chlorate in water (Fig. 1.1). If 25% by weight of potassium chlorate is added to water at room

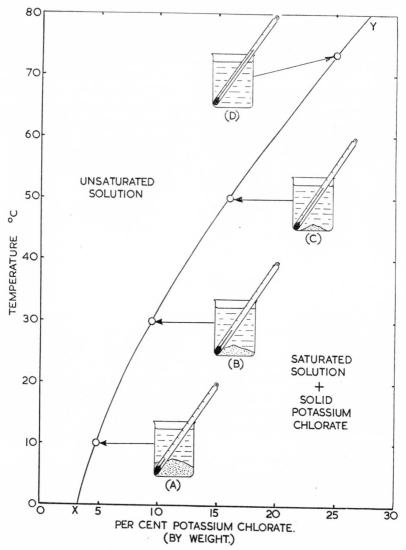

FIG. 1.1.—Solubility Curve for Potassium Chlorate in Water.

temperature (say 10° C), just under 5% will dissolve, and the remainder will settle to the bottom of the beaker, as shown in Fig. 1.1A.

If the beaker is heated, more and more of the potassium chlorate will go into solution, until at about 73° C the whole of the salt which we added will have dissolved and the beaker will contain a clear solution (Fig. 1.1D). As long as a minimum temperature of 73° C is maintained, all the potassium chlorate will remain in solution, but if the temperature is allowed to fall slowly the potassium chlorate will be precipitated from solution and will settle to the bottom of the beaker. At $10°$ C only the initial amount of potassium chlorate will remain in solution.

The solution must be cooled slowly if it is to remain in compositional "equilibrium" at any given temperature. If it is cooled too quickly some of the salt may not precipitate in accordance with the solubility curve, and the solution will become "supersaturated". That is, at any given temperature the solution will contain rather more potassium chlorate than is indicated by the appropriate point on the solubility curve XY. Usually if we allow the solution to stand for some time precipitation of the excess salt takes place and the contents of the beaker once more reach a state of compositional equilibrium.

In the above example the water is called the **solvent** and the potassium chlorate the **solute**.

1.61. Solution plays a big part in metallurgy. For example, just as potassium chlorate will dissolve in water, so will carbon, in the form of anthracite, dissolve in molten steel. In the case of metals, however, solution and precipitation can take place in the **solid** state, and therefore solution of one solid metal in another solid metal is known as "solid solution". Moreover, it is far easier for a metallic solid solution than for a liquid solution of a salt to become supersaturated and, in order that precipitation shall take place and lead to compositional equilibrium, it is necessary to cool the alloy much more slowly than is the case with the potassium chlorate solution mentioned above. By quenching a metallic structure from a high temperature it is possible to retain a supersaturated solid solution indefinitely.

Atomic Structure

1.70. Whilst the search for new fundamental particles continues and much still remains to be learned concerning the nature of various types of radiation, a reasonably satisfactory explanation of the chemical properties of the elements can be made if we assume that atoms are built up from the three fundamental particles mentioned earlier in this chapter (1.20), where it was stated that different types of atom consist of varying numbers of electrons, protons and neutrons.

1.71. The electron appears to be a very small particle with a mass of $9·11 \times 10^{-31}$ kg and carrying a charge of negative electricity of magnitude $1·602 \times 10^{-19}$ C. Electrons may be detached from atoms in a number of ways. When metals are heated or when substances undergo radioactive disintegration they emit electrons. Conversely, when a body contains an excess of electrons it is negatively charged, whilst a stream of electrons passing through a body is what we call an electric current. In fact the electron **is** negative electricity.

1.72. The proton is one of the particles contained in the nucleus of the atom. It is a particle some 1 840 times the mass of the electron and carries a charge equal to that of the electron, but positive. A stable atom always contains protons and electrons in equal numbers. It is therefore electrically neutral.

1.73. The other particle, contained in the nucleus of the atom, the neutron, is of a similar size and mass to the proton but carries no electrical charge. Its presence in the nucleus does not therefore affect the balance of electrical charges in the atom.

The properties of these three particles can be summarised thus—

TABLE 1.3

	Approximate mass (Carbon atom—12·00)	Electrical charge
Electron	1/1840	Negative ⎫
Proton	1	Positive ⎬ Equal
Neutron	1	Zero ⎭

1.74. Other particles, such as the meson, the positron (or "positive electron") and the neutrino, are now known to science, but they seem to have little direct effect on the **chemical** properties of matter and will not, therefore, be considered here.

1.75. Originally the atom was visualised as consisting of a relatively small nucleus containing both protons and neutrons, around which rotated a number of electrons in fixed orbits which varied in number with the complexity of the atom. The constitution of an atom was thus regarded as being similar to that of the solar system and containing about the same density of actual "matter". This idea has now been largely modified to the extent that the definite orbit is replaced by a mathematical function which represents the distribution of "electrons" in the space taken up by the atom. This distribution is referred to as the **orbital** of the electron. In fact many visualise the electron as being in the nature of a "cloud" of electricity rather than a single discrete particle, the orbital indicating the density of that cloud at any point within the atom. From the point of view of any diagram representing atomic structure it is more convenient to indicate the electron as being a definite particle travelling round the nucleus in a simple circular orbit. Such diagrams should be studied with the foregoing ideas on the actual nature of the electron in mind and should not be regarded as a "blueprint" of the internal structure of the atom.

1.76. Simplest of all atoms is the ordinary hydrogen atom. It consists of one proton with one electron in orbital around it. Since the positive charge of the proton is balanced by the equal but negative charge of the electron, the resultant atom will be electrically neutral. The mass of the electron being very small compared with that of the proton, the mass of the atom will be roughly that of the proton.

TABLE 1.4

Element	Number of protons	Electrons in each "shell"			
Hydrogen	1	1			
Helium	2	2			
Lithium	3	2	1		
Beryllium	4	2	2		
Boron	5	2	3		
Carbon	6	2	4		
Nitrogen	7	2	5		
Oxygen	8	2	6		
Fluorine	9	2	7		
Neon	10	2	8		
Sodium	11	2	8	1	
Magnesium	12	2	8	2	
Aluminium	13	2	8	3	
Silicon	14	2	8	4	
Phosphorus	15	2	8	5	
Sulphur	16	2	8	6	
Chlorine	17	2	8	7	
Argon	18	2	8	8	
Potassium	19	2	8	8	1
etc.					

1.77. An atom of ordinary helium comes next in order of both mass and complexity. Here the nucleus contains two protons which are associated with two electrons in the same "shell", i.e. similar orbitals surrounding the nucleus. The nucleus also contains two neutrons. However, in Table 1.4, which indicates the proton–electron make-up of some of the simpler atoms, neutrons have been omitted for reasons which will become apparent later (1.90). The number of protons in the nucleus, which is equal to the total number of electrons in successive shells, is called the Atomic Number of the element.

In Table 1.4 it will be noted that with the metal lithium a new electron shell is formed and that this "fills up" by the addition of a single electron with each successive element until, with the "noble"* gas neon, it contains a total of eight electrons. With the metal sodium another new shell then begins and similarly fills so that with the noble gas argon this

* In the chemical sense the term "noble" means that an element is not very reactive—thus, the "noble" metals, gold, platinum, etc., are not readily attacked by other reactive substances, such as corrosive acids.

third shell also contains eight electrons. The next shell then begins to form with the metal potassium.

1.78. In the case of the elements dealt with in Table 1.4, this periodicity in respect of the number of electrons in the outer shell is reflected in the chemical properties of the elements themselves. Thus the metals lithium, sodium and potassium each have a single electron in the outer shell and all are very similar chemically. They will all oxidise very rapidly and react readily with water, liberating hydrogen and forming soluble hydroxides. Each of these elements has a valency of one. Physically, also, they are very similar in that they are all light soft metals, more or less white in colour.

In a similar way the gases fluorine and chlorine, with seven electrons in the outer shell in each case, have like chemical properties. Both are coloured gases (at normal temperatures and pressures) with strongly non-metallic properties.

The noble gases helium, neon and argon occur in small quantities in the atmosphere. In fact it is only there where they are likely to exist under natural conditions, since these noble gases are similar in being non-reactive and, under ordinary circumstances, unable to combine with other elements. Chemical combination between elements is governed by the number of electrons present in the outer shell of each atom concerned. When the outer shell contains eight electrons it becomes, as it were, "saturated", so that such an atom will have no tendency to combine with others.

1.79. In elements with atomic numbers greater than that of potassium, rather complex shells containing more than eight electrons are present in the atom, but the general "periodicity" of properties relative to the number of electrons in the outer shell still persists throughout the ninety-two elements. This periodicity of properties was noticed by chemists quite early in the nineteenth century, and led to the advent of order in inorganic chemistry with the celebrated "Periodic Classification of the Elements" by the Russian chemist Demitri Mendeléef in 1864. Of more recent years this periodicity in the chemical properties of the elements has been explained in terms of theoretical atomic structure, as outlined very briefly above.

Chemical Combination and Valency

1.80. It was mentioned above that chemical combination between two atoms is governed by the number of electrons in the outer electron shell of each. Moreover, it was pointed out that those elements whose atoms had eight electrons in the outer shell (the noble gases neon and argon) had no inclination to combine with other elements and therefore

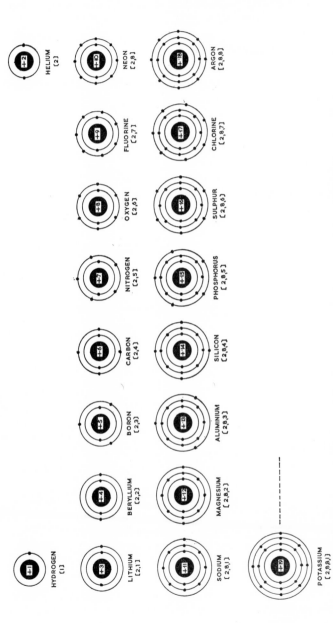

FIG. 1.2.—This Diagram Indicates the Electron–Proton Make-up of the Nineteen Simplest Atoms, But It Does Not Attempt to Illustrate the Manner in Which They Are Actually Distributed.

had no chemical affinity. It is therefore reasonable to suppose that the completion of the "octet" of electrons in the outer shell of an atom leads to a valency of zero. The noble gas helium, with a completed "duplet" of electrons in the single shell, behaves in a similar manner.

As far as the simpler atoms we have been discussing are concerned, the tendency is for them to attempt to attain this noble-gas structure of a stable octet (or duplet) of electrons in the outer shell. Their chemical

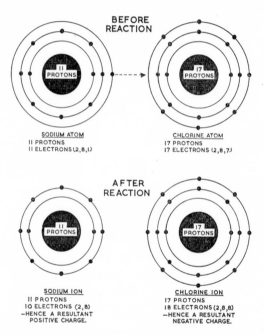

Fig. 1.3.—The Formation of the Electro-valent Bond in Sodium Chloride, by the Transfer of an Electron from the Sodium Atom to the Chlorine Atom.

properties are reflected in this tendency. With the more complex atoms the situation is not quite so simple, since these atoms possess larger outer shells which are generally sub-divided, to the extent that electrons may begin to fill a new outer "sub-shell" before the penultimate sub-shell has been completed. This would explain the existence of groups of metallic elements the properties of which are transitional between those of one well-defined group and those of the next. The broad principles of the electronic theory of valency mentioned here in connection with the simpler atoms will apply. On these general lines three main forms of combination exist.

1.81. Electro-valent Combination. In this type of combination a metallic element loses from the outer electron shell of its atom a number of electrons equal to its numerical valency. These lost electrons are transferred to the outer electron shells of the non-metallic atom (or atoms) with which the metal is combining. In this way a complete octet of electrons is left behind in the metallic particle and completed in the non-metallic particle.

Let us consider the combination which takes place between the metal sodium and the non-metal chlorine to form sodium chloride (common salt). The sodium atom has a single electron in its outer shell and this transfers to join the seven electrons in the outer shell of the chlorine

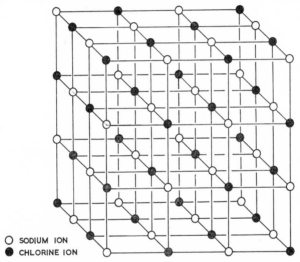

○ SODIUM ION
● CHLORINE ION

Fig. 1.4.—A Simple Cubic Crystal Lattice Such as Exists in Solid Sodium Chloride.
Each sodium ion is surrounded by six chlorine ions and vice versa. (The positions of only the cores of the ions are indicated.)

atom. When this occurs each resultant particle is left with a complete octet in the outer shell. (The sodium particle now has the same electron structure as the noble gas neon, and the chlorine particle has the same electron structure as the noble gas argon.) The balance of electrical charges which existed between protons and electrons in the original atoms is, however, upset. Since the sodium atom has lost a negatively charged particle (an electron), the remaining sodium particle must have a resultant positive charge, and since the chlorine atom has gained this electron, the resultant chlorine particle must carry a negative charge. These charged particles, derived from atoms in this manner, are referred to as "ions". In terms of symbols the sodium ion is written thus, Na^+, and the chlorine ion Cl^-.

1.82. The salt sodium chloride crystallises in a simple cubic form in which sodium ions and chlorine ions arrange themselves in the manner indicated in Fig. 1.4. Except for the force of attraction which exists between oppositely charged particles, no other "bond" exists between sodium ions and chlorine ions, and when a crystal of sodium chloride is dissolved in water separate sodium and chlorine ions are released and can move as separate particles in solution. Such a solution is known as an electrolyte because it will conduct electricity. If we place two electrodes into such a solution and connect them to a direct-current supply, the positively charged sodium ions will travel to the negative electrode and the negatively charged chlorine ions will travel to the positive electrode. The applied E.M.F. does not "split up" the sodium chloride, since it becomes ionised as soon as it dissolves in water.

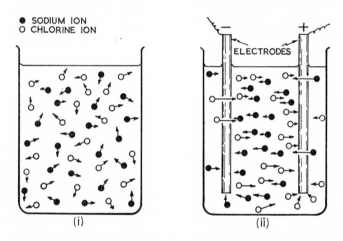

FIG. 1.5.—(i) A solution of sodium chloride in which the separate sodium ions and chlorine ions are moving independently within the solution. Note that ionisation of the salt has taken place on solution and does not depend upon the passage of an electric current. (ii) When E.M.F. is applied to the solution the charged ions are attracted to the appropriate electrode.

1.83. Thus, the unit in solid sodium chloride is the crystal, whilst in solution separate ions of sodium and chlorine exist. In reality there is no sodium chloride molecule and it is therefore incorrect to express the salt as "NaCl". Busy chemists are, however, in the habit of using symbols in this manner as a type of chemical shorthand. The author has in fact been guilty of this indiscretion earlier in this chapter when discussing formulae and equations in which electro-valent compounds are involved. For example, the equation representing the reaction between hydrochloric acid and caustic soda (1.51) would more correctly be written:

$$H^+ + Cl^- + Na^+ + OH^- = Na^+ + Cl^- + H_2O$$

| Hydrochloric Acid | Sodium hydroxide | Sodium chloride |

1.84. Co-valent Combination. In this type of combination there is no transfer of electrons from one atom to another. Instead a certain number of electrons are "shared" between two atoms. In a molecule of the gas methane, four hydrogen atoms are combined with one carbon atom. The carbon atom has four electrons in its outer shell, but these are joined by four more electrons, contributed singly by each of the four hydrogen atoms (Fig. 1.6). Thus the octet of the carbon atom is completed and at the same time, by sharing one of the carbon atom's electrons, each hydrogen atom is able to complete its "helium duplet". This sharing of electrons by two atoms binds them together, and a molecule is formed in which

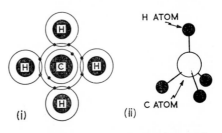

FIG. 1.6.—(i) Co-valent Bonding in a Molecule of Methane, CH_4. (ii) Spatial Arrangement of Atoms in the Methane Molecule.

atoms are held together by definite valency bonds. Each shared electron now passes from an orbital controlled by one nucleus into an orbital controlled by two nuclei and it is this control which constitutes the co-valent bond. Chemists express the structural formula for the methane molecule thus:

$$H-\overset{\displaystyle H}{\underset{\displaystyle H}{C}}-H$$

Each co-valent bond is indicated by the sign —. Co-valent compounds, since they do not ionise, will not conduct electricity and are therefore non-electrolytes. They include many of the organic* compounds, such as benzene, alcohol, turpentine, chloroform and members of the paraffin series.

As the molecule size of co-valent compounds increases, so the bond strength of the material increases, as indicated in complex compounds such as rubber and vegetable fibres. Sometimes simple molecules of

* Those compounds associated with animal and vegetable life and containing mainly the elements carbon, hydrogen and oxygen.

co-valent compounds can be made to unite with molecules of their own type, forming large chain-type molecules in which the bond strength is very high. This process is called "polymerisation". For example, the gas ethylene C_2H_4 can be made to polymerise forming polythene:

As the molecular chain increases in length, the strength also increases. Nylon (synthesised from benzene), polyvinylchloride (PVC) (synthesised from acetylene) and polyvinylacetate (the middle layer in "Triplex" safety glass) are all "super polymers", the strength of which depends upon a long chain of carbon atoms co-valently bonded.

Forces of attraction (1.86), acting between points where these "chain molecules" touch each other, hold the mass of them together. If such a substance is heated, the forces acting between the molecules are reduced and, under stress, the fibrous molecules will gradually slide over each other into new permanent positions. The substance is then said to be **thermoplastic**. Substances which (like water) are built up of simple molecules generally melt at a sharply defined single temperature since there is no entanglement, as exists with the ungainly molecules of the super polymers, and when the forces of attraction between simple molecules fall below a certain limit they can separate instantaneously. Super polymers, then, soften progressively as the temperature is increased rather than melt at a well-defined temperature.

Some polymers, when heated, undergo a chemical change which firmly anchors the chain molecules to each other by means of co-valent bonds. On cooling, the material is rigid and is said to be **thermosetting**. Bakelite is such a substance.

Rubber possesses elasticity (4.11) by virtue of the folded nature of its long-chain molecules. When stressed, one of these molecules will extend after the fashion of a spiral spring and, when the stress is removed, it will return to its original shape. In **raw** rubber the tensile stress will also cause the chain molecules to slide relative to each other, so that, when the stress is removed, some permanent deformation will remain in the material (Fig. 1.7). If the raw rubber is mixed with sulphur and heated it becomes "vulcanised". That is, sulphur causes the formation of co-valent links between the large rubber molecules which are thus held firmly together. Consequently, vulcanised rubber possesses elasticity due to the behaviour of its folded chain molecules, but it resists permanent deformation, since these molecules are no longer able to slide over each other into new positions.

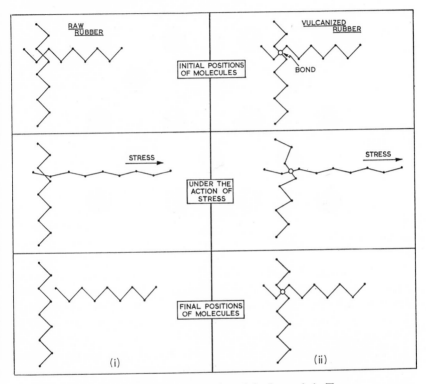

FIG. 1.7.—Rubber Molecules Are of the Long-chain Type.
Due to their "folded" form they become extended in tension, but return to their original shapes when the stress is removed. In raw rubber (i) a steady tensile force will cause separate molecules to glide slowly past each other into new positions, so that when the force is removed some plastic deformation remains, although the elastic deformation has disappeared. By "vulcanising" the raw rubber (ii) the chain molecules are bonded together so that no permanent plastic deformation can occur and only elastic deformation is possible.

1.85. The Metallic Bond. The absence of oppositely charged ions in the structure of a pure metal or alloy, together with the fact that there are not enough valency electrons to form an ordinary co-valent bond, leads to the sharing of valency electrons by more than two atoms. It is thought that valency electrons are no longer associated with any particular atom, but are contributed to form a negative electron "cloud" which is shared by the positive ions among which the electrons are free to move. The positive ions tend to repel each other and take up positions according to some geometrical pattern, but at the same time they are held together by their mutual attraction for the electron cloud (Fig. 1.8). Thus the metallic bond may be considered as a kind of

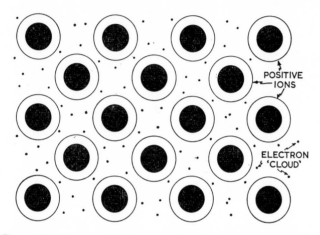

FIG. 1.8.—Diagrammatic Representation of the "Metallic Bond".

multiple co-valent bond involving a large number of atoms. Since the valency electrons are relatively free to travel in such an arrangement, this could explain the high electrical conductivity of metals, whereas co-valent compounds are non-conductors because all of the valency electrons are held firmly captive.

The opaque lustre of metals is due to the reflection of light by free electrons. A light wave striking the surface of a metal causes a free electron to vibrate and absorb all the energy of the wave, thus stopping transmission. The vibrating electron then re-emits the wave from the metal surface giving rise to what we term "reflection".

The very important property of most metals in being able to undergo considerable plastic deformation is also due to the existence of the metallic bond. Under the action of shearing forces, layers of positive ions can be made to slide over each other without drastically altering their relationship with the shared electron cloud.

1.86. Van der Waal's Forces are very small forces of attraction acting between atoms in cases where the formation of ionic or covalent bonds is not possible. Imagine two atoms of a noble gas such as argon. These will be unable to form a molecule in the normal way, since each has a complete octet of electrons in its outer shell. However, if the two atoms are in close proximity and their electron clouds happen to be concentrated so:

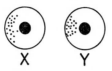

it is reasonable to suppose that the nucleus of atom X will be attracted by the electron cloud of atom Y, at the moment in time when the nucleus of X is unshielded by its own electron cloud. This situation will be continually changing as the electron clouds move, but a resultant weak force of attraction exists. Basically similar forces also act between atoms which are already bonded in neighbouring molecules, giving rise to weak Van der Waal's forces between long-chain molecules in polymers.

Isotopes

1.90. In the above discussion of the mechanism of chemical combination no mention was made of the part played by the neutron. This was because it has no apparent effect on the chemical properties of atoms. Chemical properties, as stated previously, are mainly a function of the electron structure of the atom. The principal role of the neutron is to affect the actual mass of an atom. Thus, the sodium atom with 12 neutrons in the nucleus, in addition to the 11 protons already mentioned, has a total nuclear mass of 23. The 11 electrons present are negligible in mass when compared with the massive protons and neutrons, so that the mass of the total atom is approximately 23 units, and it is from this value that the atomic weight is derived.

There are many instances, however, in which two atoms contain the same number of protons but unequal numbers of neutrons. Clearly, since they have equal numbers of protons, they will also have equal numbers of electrons and, chemically, such atoms will be identical. Differing numbers of neutrons, however, in respective atoms will cause these atoms to have unequal masses. An element possessing atoms which are chemically identical but which are of different mass is said to be **isotopic** and the different groups of atoms are known as **isotopes**.

1.91. Two such isotopes occur in the element chlorine. The chemical properties of these isotopes are identical because in each case an atom will contain 17 protons and 17 electrons. Only the relative weights of each atom will be different, since the nucleus of isotope II contains two more neutrons than that of isotope I (Fig. 1.9). Since there are rather more atoms of isotope I (usually written Cl^{35}) than of isotope II (written Cl^{37}), the atomic weight of chlorine "averages out" at 35·45.

The term "isotope" tends to be associated in modern technology with the release of nuclear energy by suitable elements. The properties of the isotopes of uranium in this connection are dealt with later in this book (18.83).

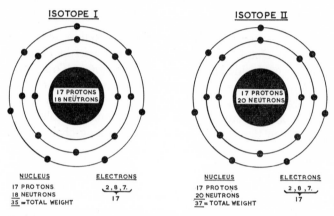

FIG. 1.9.—The Particle "Make-up" of the Two Isotopes of Chlorine.

EXERCISES

1. What is the difference between an *atom* and a *molecule*?
2. What is meant by the term *chemical compound*?
3. If the valency of aluminium is three, write down the formula for its oxide.
4. Why must aluminium be obtained from its ore by electrolysis instead of by the more usual method of reduction by coke?
5. A slag formed during a steel-making process will contain an excess of lime. With what refractories could the furnace be lined?
6. A solution of potassium chlorate at 80° C contains 15% by weight of the salt. At what temperature (approximately) will solid potassium chlorate begin to precipitate if the solution is cooled slowly? What is likely to happen if the solution is cooled rapidly?
7. Differentiate between "atomic number" and "atomic weight".
8. Without reference to any tables, draw a diagram representing the structure of an atom which contains fourteen protons.
9. Make a sketch representing the electron distribution associated with the co-valent bonds in a molecule of the gas ammonia, NH_3.
10. Define concisely the term "isotope".
11. Show how far modern atomic theory is successful in explaining not only many of the mechanical properties of a metal but also the fact that it is a conductor of electricity.

(West Bromwich Technical College, A1 year.)

CHAPTER II

THE PRODUCTION OF IRON AND STEEL

2.10. The history of the iron and steel trade in Britain makes a very romantic story, including, as it does, the achievements of such famous pioneers as Huntsman, Bessemer, Gilchrist and Thomas. We must not forget, however, that in the East—possibly in India—steel production of a sort was carried out many centuries earlier than in Britain; and that the iron trade was introduced into England by the Romans. In Roman times the iron trade grew up in Sussex, depending upon locally mined ore and charcoal made from the trees of the Ashdown Forest. There the industry remained until de-forestation of the land became so serious that, in order to preserve timber supplies for shipbuilding, Acts were passed by Elizabeth I and other monarchs, limiting the felling of timber for iron production. The introduction of coal for iron smelting early in the seventeenth century by Dud Dudley led to the transfer of the iron trade from the wooded districts to the coalfields of Scotland, the Midlands and South Wales. Towards the close of the great Victorian era Britain was producing the bulk of the world's steel; but diminishing supplies of ore in these islands, coupled with the development overseas of vast fields of high-grade ore, have altered the situation, and at present the United States is the world's leading producer, followed by the U.S.S.R., Japan and Western Germany, with Britain in fifth place and producing approximately 5% of the world's supply of steel.

2.11. Before primitive man learned to win metals from their chemical compounds in ores the only iron used was that occurring "free" or uncombined in nature. This was always meteoric iron, and it seems that meteorites were a source of supply of iron to the ancient Egyptians, since they called iron "the metal from Heaven". As late as 1894, when Commander Peary was exploring in Greenland, an Eskimo showed him the remains of a huge meteorite (of a mass of approximately 40 tonnes) which for probably a hundred years had provided Eskimo hunters with their weapons.

2.12. To-day, however, we are concerned only with iron as it occurs combined with other elements (mainly oxygen) in the Earth's crust, since such ores provide our present source of supply and it is no longer necessary to await the descent of a meteor.

The most important ores of iron are oxides. These are, as we have

27

seen in the previous chapter, compounds in which iron is combined
with oxygen; but in Britain we also make use of the less-rich carbonate
and hydroxide ores which occur in our East Coast areas and Northamp-
tonshire. In addition to the elements oxygen, carbon and hydrogen,
which are chemically combined with the iron, the ore also contains a
large amount of earthy waste material or "gangue", which is merely
entangled or mixed with the iron-bearing mineral. Since the amounts
of gangue vary, it follows that the iron content of ores will also vary,
and whilst in some parts of the world ores are being mined which contain nearly 70% iron, here in Britain we must do our best with ores containing as little as 20% iron. Such low-grade ores need to be enriched
with higher-grade ore from abroad in order to increase the output of the
blast furnace relative to the amount of coke used.

Types of Iron Ore

2.20. The most important varieties of iron ore are:

(a) *Magnetite* (Fe_3O_4), which contains $72 \cdot 4\%$ iron when pure but
usually rather less as mined. It is a magnetic mineral—a fact which
proves useful in locating deposits.

(b) *Hematite* (Fe_2O_3), which accounts for the bulk of ore mined in
the world and occurs in a variety of forms—red, brown or black. It
contains 40–65% iron.

(c) *Limonite* and other hydroxide types of ore, varying in composition from $2Fe_2O_3.H_2O$ to $Fe_2O_3.3H_2O$ and containing 20–55% iron.

(d) *Siderite* ($FeCO_3$) and other types of carbonate ore, usually of
rather low iron content.

2.21. The Ore-producing Regions of the World. In terms of iron
content the average British ores are inferior to those of Western Europe
as a whole. The Lorraine ores, for example, contain 32–33% iron,
whilst some of the north French deposits contain as much as 50% iron.
In Germany too the average is little better than in Britain. This disadvantage is to some extent offset by the cheap open-cast mining which
is possible in Britain.

2.22. The world's most important producing regions of iron ore are:

(a) The American Lake Superior deposits, which consist of high-
grade hematite and contain 30–65% iron.

(b) The large deposits in the U.S.S.R., consisting mainly of hematite in the Ukraine and magnetite in Siberia.

(c) The low-grade or "lean" sedimentary deposits of Britain's
East Coast, which are mainly limonite and siderite containing 20–
30% iron.

(d) The Lorraine deposits of "minette" (a type of hematite) in the north of France, containing 26–50% iron.

(e) The very high-grade deposits of magnetite in Sweden's "Land of the Midnight Sun". Kirunavara is a mountain consisting almost entirely of magnetite and containing 60–68% iron.

(f) Canadian deposits in Labrador, Quebec and Newfoundland. These are mainly of high-grade hematite of the Lake Superior type.

(g) Large reserves, mainly of hematite, occurring in the Amazon basin of Brazil. Exploitation of this very important source of ore is increasing rapidly. Much of it is exported to the United States as well as satisfying Brazil's own industrial undertakings in the São Paulo region.

(h) Indian deposits of high-grade ore which are more than enough to satisfy her present home demands and exports are being developed.

(i) Deposits in Venezuela of high-grade hematite (65% iron). Much of this ore is exported to the United States and Britain.

Iron ore is mined on a smaller scale in many other parts of the world.

2.23. In addition to home-produced ore, British blast furnaces depend upon ores from Sweden, Africa (Mauritania, Liberia, Algeria, Morocco, Sierra Leone and Tunisia), Canada and Venezuela, as well as upon smaller quantities from the U.S.S.R., Brazil, Norway, Spain and elsewhere. Since such ores are generally of very high grade, they make good to some extent the diminishing supplies of Cumberland hematite (which cannot last much longer even at the present reduced production rates) and also the complete extinction of ore production in Scotland, Shropshire, my native Staffordshire and other parts of Britain.

2.24. Iron content is not the only criterion of the value of an ore. The amount of phosphorus it contains is also very important; indeed, previous to the work of Thomas and Gilchrist, the only ores which could be used were those containing little or no phosphorus. At present the bulk of ore being mined, both here and in Western Europe, contains phosphorus, and this makes conversion of the resulting pig iron into steel a more costly operation, since phosphorus is difficult to remove.

Pre-smelting Treatments of Ores

2.30. After the initial mining and crushing operations, an ore may undergo preliminary treatments before it is charged to the blast furnace. These treatments may be necessary to increase the efficiency of the smelting process. They include:

2.31. Concentration, or the removal of as much as possible of the earthy waste or gangue, which would otherwise take up useful space in the blast furnace, thus reducing output. In the case of ores which have to be shipped over long distances, it is usually more economical to concentrate the ore at the mine rather than to pay transport charges on the crude ore, containing, as it does, a large amount of useless gangue. Magnetic ores, such as magnetite (and hematite, which has been "roasted" to convert it to magnetite), can be concentrated in a magnetic separator, whilst hematite ores are treated by washing to remove the lighter gangue.

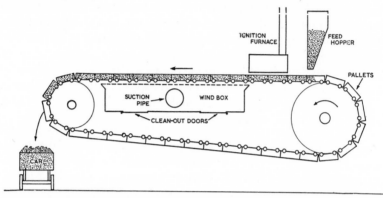

FIG. 2.1.—The Principle of the Dwight–Lloyd Sintering Machine.

2.32. Calcination is effected in a kiln of either the shaft or rotary type and is applied to those ores which contain a large amount of moisture or carbon dioxide. These substances may cause irregular working in the blast furnace and are best removed before the ore is smelted. Calcining also helps to remove sulphur by oxidising it to sulphur dioxide. Rotary kilns are fired at the lower end by either blast-furnace gas or oil, and the temperature is high enough to expel water and carbon dioxide from the ore, which is fed at the upper end so that it meets the rising hot gases. The calcined ore is ejected at the lower end of the kiln.

2.33. Agglomeration. The previous processes may result in the formation of a certain amount of dust. This dust is rich in iron, and must therefore be reclaimed. It cannot be charged to the blast furnace as it is, since it would tend to fill up the spaces between the lumps of the charge (thus impeding the upward flow of gases) or be blown out of the furnace. Hence the dust is generally dealt with by one of the following methods:

(*a*) *Sintering*. In this process the dust is mixed with a small amount of coke or breeze and fed to a Dwight–Lloyd machine (Fig. 2.1).

Ignition of the coke, assisted by air drawn through the pallets, generates sufficient heat to "frit" together the particles of ore, producing a porous and easily reducible sinter.

(b) *Pelletising*, in which the fine dust is mixed with a little starch and water and rolled in a drum so that small balls are formed. Coal dust is then added, and the balls rolled on to the grate of a sintering machine where they are first dried and then baked at 1 000° C to harden them.

(c) *Nodulising*, which is similar in some respects to pelletising except that a heated rotating kiln is employed to produce the agglomerate.

The Smelting Operation

2.40. Having been converted into a suitable condition, the ore is charged along with coke and limestone, to the blast furnace, where smelting takes place. Smelting is a twofold operation including:

(a) The chemical reduction of the iron oxide by the white-hot coke and the carbon monoxide gas arising therefrom.

(b) The liquefaction of the gangue by means of the flux (limestone) to form a fusible slag which will run from the furnace.

2.41. The blast furnace consists of a vertical steel shell lined with refractory materials and having a charging arrangement at the top and a means of running off the pig iron and slag at the bottom (Fig. 2.2). Air is blown in near the bottom of the furnace, and this forced draught increases the speed of combustion and maintains the necessary high temperature. The complete structure towers to 30 m, and often more.

2.42. During the production of pig iron large amounts of gas are evolved. This gas has a considerable calorific value due to the carbon monoxide which it contains, and much of it is burned in chequered-brickwork stoves (Cowper stoves) in order to pre-heat the air blast to the furnace. These stoves work in much the same way as the regenerators used with the open-hearth steel-making furnace dealt with later in this chapter (2.91). Pre-heating the air blast will, of course, mean that less fuel is required; moreover, less sulphur will enter the pig iron since coke always contains sulphur.

Whenever air is used in a combustion process the efficiency is inevitably lowered by the 79% by volume of nitrogen contained in air. This nitrogen enters the combustion zone cold and leaves it hot, thus carrying away heat, though undergoing no chemical change itself. Hence much development work is at present taking place involving the enrichment of the blast with oxygen. Previously the high cost of oxygen

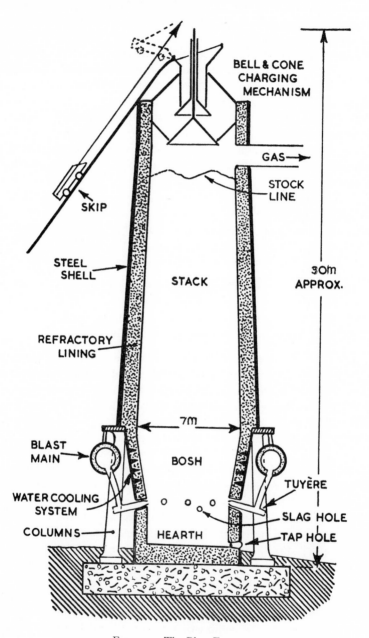

BELL & CONE
CHARGING
MECHANISM

SKIP

GAS →

STOCK
LINE

STEEL
SHELL

STACK

30m
APPROX.

REFRACTORY
LINING

7m

BLAST
MAIN

BOSH

TUYÈRE

WATER COOLING
SYSTEM

SLAG HOLE

COLUMNS

HEARTH

TAP HOLE

Fig. 2.2.—The Blast Furnace.

had precluded its use in this way, but the production of low-cost "tonnage oxygen" (99·5% pure) in recent years, coupled with the increased cost of metallurgical coke, has made it an attractive proposition. The increased scarcity—and consequently higher cost—of suitable coking coal has promoted experiments in injecting fuel oil, low-grade pulverised fuel and also natural gas at the tuyères. Both of these developments reduce production costs and increase output.

The general lay-out of the blast furnace and its ancillary equipment is shown in Fig. 2.3. Blast-furnace gas, which is produced in excess of the requirements of the Cowper stoves, is used in steam-raising for turbo-blowers and electric-power generators; as fuel for coke ovens and open-hearth steel furnaces, and various other uses in an integrated plant.

2.43. The air which is blown in at the bottom of the furnace causes partial combustion of the coke—

$$2C + O_2 = 2CO + \text{Intense heat}$$

As the carbon monoxide, which is a powerful reducing agent, rises through the charge, it chemically reduces the iron oxide—

$$Fe_2O_3 + 3CO = 2Fe + 3CO_2$$

This reaction takes place in the upper part of the furnace, where the temperature is too low to melt the iron formed. The iron therefore remains as a spongy mass until it moves down into the lower part of the furnace, where it melts and runs down over the white-hot coke, dissolving carbon, sulphur, manganese, phosphorus and silicon as it goes. Apart from carbon, which is absorbed from the coke, these elements are present due to the reduction of their compounds in the gangue.

2.44. At the same time as this reduction is taking place, the gangue, which is composed mainly of silica, combines with the lime (formed by decomposition of the limestone added with the charge) to produce a slag—

$$2CaO + SiO_2 = 2CaO.SiO_2$$
(Calcium silicate
slag)

This is an example of the type of acid–basic reaction mentioned in the previous chapter, in which the acid silica is neutralised by the basic lime in much the same way as stomach acids, which cause indigestion, are neutralised by magnesia. The important point in so far as the digestion of the blast furnace is concerned is that the calcium silicate slag has a much lower melting point than silica and can be run from the furnace quite easily. The quantity of limestone which must be added to flux the silica must be calculated, and this, together with the ore, is usually

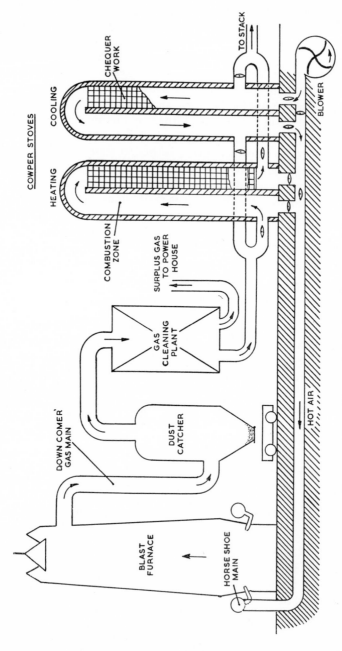

FIG. 2.3.—General Lay-out of the Blast Furnace and Ancillary Plant.

referred to as the "burden". The limestone has another function—the removal of sulphur; this, however, entails having an excess of lime in the slag and is often impracticable with low-grade ores, since they contain so much silica, which must be neutralised first. Hence in British practice the sulphur is often removed at a later stage by adding soda-ash (crude sodium carbonate) to the molten pig iron as it runs from the blast furnace into the ladle.

$$FeS \quad + \quad Na_2CO_3 = FeO + \quad Na_2S \quad + CO_2$$

(Iron sulphide in the pig iron)	(Soda-ash)	(Joins the slag)

2.45. The furnace is tapped at regular intervals, and the pig iron is run either into large ladles for transference, whilst still molten, to the steel-making plant or, alternatively, cast into pigs for subsequent use in the cupola or open-hearth furnace.

2.46. In the early years of the century the output of an average blast furnace was less than 100 tonnes per day, but at present outputs of 2 000 tonnes per day are general.* The approximate quantities in both charge and products associated with this output will be:

Charge (tonnes)		Products (tonnes)	
Ore (say, 50% iron) .	4 000	Pig iron . . .	2 000
Coke	1 800	Slag	1 600
Limestone . . .	800	Blast-furnace gas . .	10 800
Air	8 000	Dust	200
Total	14 600	Total	14 600

Blast-furnace slag is of comparatively low value, and in the early years of iron production was deposited as large, unsightly dumps in the vicinity of the furnace plant. Now much of it is used as railway ballast, or coated with tar for road-making. It is also used as a concrete aggregate and for the manufacture of slag wool, which is a valuable material for thermal and acoustical insulation. It should not, however, be confused with the "basic slag" used in agriculture as a source of phosphorus. This is a product of the basic open-hearth furnace used in steel-making.

Pig Iron

2.50. Pig iron is really a complex alloy. In addition to iron, it contains anything up to 10% of other elements, the chief of which are carbon, silicon, manganese, phosphorus and sulphur. These elements are all absorbed as the reduced iron melts and runs down through the white-hot coke and slag, and their relative quantities can, to some extent, be adjusted by different working conditions in the furnace. It is

* Japanese blast furnaces with daily outputs of 10 000 tonnes are now in operation.

[Courtesy of Messrs. Stewarts and Lloyds Ltd., Corby.

PLATE 2.1A.—General View of a Blast-furnace Plant.
The furnaces themselves are in the left-hand half of the picture.

[Courtesy of Messrs. Stewarts and Lloyds Ltd., Corby.

PLATE 2.1B.—Tapping the Slag from a Blast Furnace.

impossible to vary the amount of phosphorus in this way, and whatever phosphorus goes in at the top of the furnace comes out in the iron at the bottom.

2.51. The total amount of carbon which a pig iron contains is usually between 3 and 4%, and this carbon may be present either as the compound iron carbide Fe_3C (also known as cementite) or as uncombined carbon in the form of graphite. The form in which the carbon exists depends largely upon the relative amounts of other elements present. For example, a high silicon content and a low sulphur content favour the decomposition of the iron carbide into graphite, so that the resulting fracture of the iron will be grey, whilst if the silicon content is low and the sulphur content high, then the iron carbide will be stabilised and the fracture will be white, since iron carbide is a white compound. Intermediate compositions will give varying quantities of graphite and cementite and also modify the size of the graphite flakes, and hence the coarseness of the structure. These principles can be represented thus:

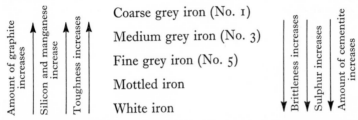

Coarse grey iron (No. 1)

Medium grey iron (No. 3)

Fine grey iron (No. 5)

Mottled iron

White iron

2.52. The rate of solidification will also affect the nature of the fracture. Rapid cooling tends to prevent the decomposition of the cementite from taking place, thus producing a white iron, whilst slow cooling allows decomposition to take place, thus producing a grey iron (always assuming that sufficient silicon is present to favour this). Hence pigs which have been chill-cast in a pig-casting machine cannot be graded by fracture; and chemical analysis is a more reliable criterion.

2.53. High-grade hematite ore will yield a pig iron low in sulphur and phosphorus. Such a pig iron may be employed in the manufacture of acid steel and high-duty iron castings. The low-grade phosphoric ores will produce a pig iron high in phosphorus—often termed a basic pig iron, since it is used mainly in the manufacture of steel by the basic process.

The Manufacture and Properties of Wrought Iron

2.60. Until the advent of the Bessemer process for the manufacture of steel in 1856, wrought iron was the most important ferrous material. In the middle of the nineteenth century some $2 \cdot 5 \times 10^6$ tonnes were produced annually, but to-day only a very small quantity is manufactured for special purposes.

At one time wrought iron was produced by a direct process from the iron ore, and at that time formed the basis for all iron and steel production. Since 1784, however, wrought iron has been manufactured from pig iron by a process originated by Cort, in which the impurities are successively removed by oxidation.

2.61. A suitable pig iron (see Table 2.1) is melted in the hearth of

TABLE 2.1.—*Typical Pig-iron Analysis*

	Combined carbon (%)	Graphitic carbon (%)	Silicon (%)	Manganese (%)	Sulphur (%)	Phosphorus (%)
Hematite Irons						
No. 1 . . .	0·5	3·5	2·5	0·55	0·02	0·04
No. 2 . . .	0·6	3·3	2·2	0·40	0·02	0·04
No. 3 . . .	0·8	3·1	2·0	0·45	0·04	0·04
No. 4 . . .	1·1	2·5	1·6	0·65	0·05	0·05
No. 5 . . .	1·3	2·1	0·8	0·50	0·11	0·05
Mottled . . .	1·8	1·4	0·6	0·50	0·15	0·05
White . . .	3·0	Trace	0·4	0·30	0·20	0·05
Basic Irons						
No. 1 . . .	0·1	3·2	2·7	1·9	0·03	1·1
No. 2 . . .	0·2	3·1	2·5	1·8	0·035	1·2
No. 3 . . .	0·2	3·0	2·3	1·5	0·04	1·2
No. 4 . . .	0·3	2·7	1·8	1·4	0·045	1·3
No. 5 . . .	0·4	2·6	1·5	1·4	0·06	1·3
Mottled . . .	1·7	1·7	1·4	1·3	0·07	1·3
White . . .	3·4	Trace	0·7	1·2	0·09	1·4
Acid Bessemer iron .	4·0–4·2		2·0–2·5	0·75–1·0	0.05 max.	0·05 max.
Basic Bessemer (Thomas) iron	3·0–3·6		0·5–0·8	0·5–0·8	0·04–0·07	1·5–2·0
Acid open-hearth iron .	3·0–4·0		1·2–2·7	0·5–3·0	0·05 max.	0·05 max.
Basic open-hearth iron .	2·5–3·5		1·0 max.	1·0–3·0	0·06 max.	0·8–1·5
Typical pig for the production of wrought iron	3·0–3·5		1·0–1·5	1·0–1·5	0·10 max.	1·0 max.

a puddling furnace (a furnace having an open hearth to which air can be admitted), and after the charge is melted, additions of mill-scale (iron oxide) are made to further the oxidation. Silicon and manganese oxidise first, since they have the greatest affinities for oxygen, and form an iron–manganese silicate slag on the surface of the molten iron. The carbon then begins to oxidise as the puddler stirs the mill-scale into the charge, and causes violent agitation as the bubbles of carbon monoxide burst through the slag with spurts of blue flame called "puddlers' candles". Finally, when all the silicon, manganese, carbon and phosphorus have been oxidised, almost pure iron remains in the furnace. The melting point of pure iron is 1 537° C, whereas that of the pig iron charged was only about 1 150° C (as we shall see later, alloying generally reduces the melting point of a metal). Since the furnace temperature is only between 1 300 and 1 400° C, it follows that as pure iron is formed it must solidify, or "come to nature".

2.62. When this happens the puddler collects the pasty iron together into balls of mass about 40 kg each and withdraws them from the furnace. They are transferred, together with much entangled slag, to the forging hammer, where much of the entrapped slag is expelled. After rolling, the crude bars are cut and piled together in a reheating furnace and again forged to make the slag distribution more even. The remaining slag is thus elongated into fibres along the direction of rolling. A representative analysis range of wrought iron would be:

Carbon	0·02–0·03%
Silicon. . . .	0·02–0·10%
Sulphur . . .	0·008–0·02%
Manganese . . .	Nil–0·02%
Phosphorus . . .	0·05–0·25%
Slag	0·05–1·50%
Iron	Balance

2.63. Despite the variable slag content (an unavoidable feature arising from the method of manufacture), wrought iron is a very dependable material, noted chiefly for its toughness and its resistance to shock. Moreover, since it can be forge welded with ease, it is an admirable material for the manufacture of anchor chains, railway couplings and lifting hooks.

2.64. As might be expected, attempts have been made from time to time to replace the puddling process by a more productive method. The increased demand for iron and steel during the middle of the nineteenth century had induced Henry Bessemer to experiment in the development of methods for producing wrought iron more quickly by blowing air into the hearth of the puddling furnace. Ultimately this led to the method of steel production for which Bessemer was later knighted. However, in more recent times the Aston–Byers process has been established in the United States. In this process molten pig iron is "blown" in a acid-lined converter to remove much of the silicon and carbon present. The refined iron is then poured into a vessel containing some molten slag at a temperature below the melting point of iron. This simulates conditions during the later stages in the hearth of the puddling furnace so that the metal "comes to nature". Balls of iron of mass 5–10 tonnes are formed and these are squeezed in a mechanical press to remove much of the slag.

The Manufacture of Steel

2.70. In the early days steel was manufactured by packing bars of wrought iron into cast-iron boxes with a certain amount of carbonaceous material and heating them in a furnace at about 1 000° C for a week. In these circumstances carbon was absorbed gradually into the surface of the iron, and when the bars were ultimately forged into whatever shape was required, the carbon became more or less evenly distributed throughout the structure. This process, called the Cementation Process, persisted until 1742, when Benjamin Huntsman, a watchmaker, decided that many of his watch springs were breaking because of a lack of homogeneity in the steel. He therefore melted the bars of "blister" steel (the product of the Cementation Process) in a crucible and, after casting and rolling, produced a high-grade steel, since most of the slag, present in the original bars of wrought iron and consequently in the blister steel, was lost during melting. High-quality steel is still made by a somewhat modified Huntsman process and continues to be known as "cast steel". In the broad sense all steel is now "cast" in so far as it is poured into ingot moulds and allowed to solidify prior to subsequent forming processes. The term "cast" used in this context is meant to indicate "crucible cast" as distinct from mass-produced steel made by the open-hearth, Bessemer or "oxygen" processes.

Until little more than a hundred years ago steel was an expensive material, produced in only small quantities for such articles as swords and watch springs, while structural work was carried out in cast iron or wrought iron. The first metal bridge in Europe was built over the River Severn (at Ironbridge) in 1779, and contains 400 tonnes of cast iron, whilst the erection of the Eiffel Tower involved the use of some 7 000 tonnes of wrought-iron girders. The ill-fated Tay Bridge consisted of 84 spans, each 60 m long, built entirely of wrought iron. It collapsed one stormy December night in 1879 when a train was passing over, with the loss of over a hundred lives. This disaster was not so much due to the low strength of wrought iron as to bad engineering design. Adequate allowances had not been made for the stresses of wind and waves.

2.71. The cementation and crucible methods were responsible for all the steel manufactured until the year 1856, when Henry Bessemer introduced the steel-making process which bears his name. This was the first process for the mass-production of steel. It was followed about ten years later by the Open-Hearth Process, and until 1952 these two processes were responsible for the manufacture of the bulk of the world's steel. To-day the newly developed "oxygen processes" are again changing the pattern of steel production both in Britain and elsewhere.

2.72. Steel-making processes are said to be either "acid" or "basic", these terms referring to the chemical natures of the slags and furnace linings. Until the researches of Thomas and Gilchrist made it possible to utilise phosphoric pig irons, only low-sulphur, low-phosphorus pigs could be used, and because these were usually rich in silicon, an **acid** slag, charged with silica, was produced. Since the slag was acid, it was necessary to line the furnace with a similar acid refractory (silica bricks) in order to prevent any reaction from taking place between the slag and the furnace lining.

2.73. The work of Thomas and Gilchrist enabled phosphoric pig irons to be used. In their processes an excess of lime is added to the slag in order to facilitate the removal of phosphorus. But lime is a basic substance which would quickly react with the ordinary silica-brick furnace lining, leading to its liquefaction in a similar way to that of the siliceous gangue when fluxed by the lime in the blast furnace. Hence in the basic steel-making processes it is necessary to line the furnace with a basic refractory, such as burnt dolomite (CaO.MgO) or burnt magnesite (MgO). These are expensive refractories, but their use enables the large quantities of British phosphoric ore to be turned into high-grade steel. Formerly basic steel was regarded as being inferior to acid steel, but of late years this traditional idea has been largely dispelled by the improved quality of basic steel which has resulted from technical improvements in the method of production.

The Bessemer Process

2.80. All steel-making processes rely on the oxidation of impurities contained in the pig iron or other raw material. In the Bessemer process this is effected by blowing air through the molten pig iron which constitutes the charge. The charge does not, as might be expected, become cool due to the passage of the cold air blast. On the contrary, it usually becomes much hotter, due to the heat produced by the oxidation of the impurities, which, in effect, are used as fuel to keep the charge molten. In the acid process silicon and carbon are the main impurities acting in this capacity, but for the basic process silicon must be fairly low, since it would tend to produce an acid slag attacking the basic lining of the converter. In the basic (or Thomas) process manganese, carbon and phosphorus provide the necessary heat of combustion to keep the charge molten during their removal to the slag. Representative analyses of pig irons used in both the acid Bessemer and Thomas processes are given in Table 2.1.

2.81. The Bessemer converter (Fig. 2.4) consists of a steel casing lined with the appropriate refractory. It can be rotated from a vertical to a horizontal position in order to facilitate charging and pouring. The converter is charged whilst its refractory lining is still hot from the previous charge (or if being started after relining, it is first gas heated).

When the molten pig iron has been added the blast is turned on and the converter slowly rotated into the vertical position so that air is discharged through the molten contents. Modern converters take charges of up to 60 tonnes.

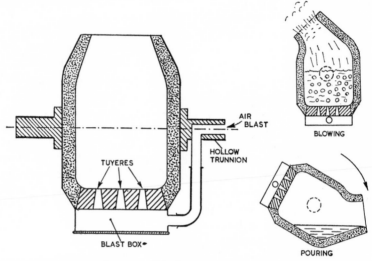

FIG. 2.4.—The Bessemer Converter.

2.82. Those impurities with the greater affinities for oxygen, namely silicon and manganese, are oxidised first. When the bulk of these elements have been converted to slag the carbon begins to oxidise, and this stage of the process is accompanied by a rumbling noise in the converter and the lengthening of the flame issuing from it. In the acid process the flame dies down approximately twenty minutes after the commencement of the "blow", and this indicates that the bulk of the impurities has been removed. The converter consequently contains "dead-mild" steel. In the basic (or Thomas) process, however, the blow is extended for a period of five minutes, known as the "after-blow", which is necessary for the removal of phosphorus (Fig. 2.5A). The phosphorus is oxidised to phosphorus pentoxide, P_2O_5, which, if not removed, would then combine with some of the iron oxide in the slag, forming iron phosphate. The presence of this compound, however, would be likely to allow some of the phosphorus to return to the steel; and to prevent this, lime is added to the slag, forming stable calcium phosphate instead of iron phosphate—

$$P_2O_5 + 3CaO = Ca_3(PO_4)_2$$
(Calcium phosphate—
joins slag)

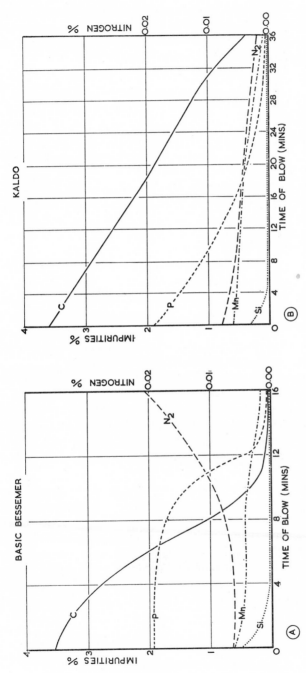

FIG. 2.5.—(A) The Progress of the Removal of Impurities During the Basic Bessemer (Thomas) Process. (B) The Removal of Impurities During the Kaldo Process. This is fairly typical of the other "Oxygen Processes".

2.83. At the end of the blow the converter is rotated into the horizontal position, the blast turned off and as much of the slag as possible is poured away. The steel is then poured into a ladle, ready for teeming into the ingot moulds, but before this happens it is necessary to remove the excess of iron oxide which is dissolved in the molten steel and which would, if allowed to remain, produce an unsound casting (4.21). This is done by "killing" the melt with ferromanganese (75–80% manganese; 6–7% carbon; balance iron).

$$FeO \; + \; Mn = Fe \quad + \quad MnO$$

(Soluble in steel)   (Insoluble in steel— joins slag)

Any excess manganese, together with the carbon, dissolves in the steel, so that, in addition to deoxidising, the ferromanganese acts as an agent to adjust the carbon content.

2.84. During the first half of the present century the Bessemer processes lost ground rapidly to the open-hearth processes. Initially this was due mainly to the greater degree of control on composition possible during the prolonged refining period in the open-hearth process. Moreover, in order to be suitable for Bessemer treatment, a pig iron must be of correct composition. It must contain sufficient impurities to burn and so provide heat to keep the charge molten during refining. The large volume of atmospheric nitrogen passing through the converter during the "blow" carries away useful heat and therefore limits the type of pig iron which can be used. Further, some of this nitrogen dissolves in the final product (Fig. 2.5A), giving rise to undesirable properties particularly in dead-mild steel destined for deep-drawing operations (11.35). The more recent demands by the motor-car and other industries for deep-drawing quality mild steel have given extra impetus to the development of the various "oxygen processes" now firmly established.

In this country the acid Bessemer process is now little used and then mainly for the production of steel castings in a modification known as the Tropenas process (1.35—Part II). The Thomas (basic Bessemer) process for a long time remained popular here because of the large amount of phosphoric pig iron available, but those plants which survived open-hearth competition are tending to be replaced by modern oxygen processes. In West Germany, Luxembourg and France, where suitable pig irons are available, the Thomas process continues to be used.

Oxygen Processes

2.85. Specific suggestions for the use of an oxygen-enriched blast were made as long ago as 1905 by a Belgian company, but until recently the high cost of oxygen production prevented any progress in that direction, and the industrial development of "oxygen steel-making" dates from 1952. In that year a process was finally established in the Austrian towns of Linz and Donawitz and, appropriately, became known as the L–D process.

2.86. The L–D Process. The L–D converter is a pear-shaped vessel not unlike a Bessemer converter, but closed at the bottom. It is charged with molten pig iron, steel scrap, lime and, sometimes, fluorspar (to increase slag fluidity), whilst in the vertical position. Oxygen is then introduced by means of a water-cooled "lance" which is lowered through the mouth of the converter to within 0·6 m of the surface of the molten charge. The molten iron reacts with the oxygen from the blast to form iron oxide and this combines with the lime to produce an oxidising slag. Slag–metal reactions then proceed and lead to the elimination of impurities from the charge, following broadly similar principles to those which obtain in the Thomas process except that the bulk of the phosphorus is oxidised earlier during the "blow"; but, more important, the nitrogen content falls progressively during the "blow" in the L–D process, whereas it increases considerably in the Thomas process (Fig. 2.5A).

Initially the L–D process was used largely for the treatment of basic pig irons fairly low in phosphorus (0·25%), but recent development has enabled ordinary basic Bessemer pig iron with up to 2·0% phosphorus to be treated by a modified method known as the L–D A–C process. In this process pulverised lime is injected through the oxygen lance and leads to a speedier and more effective phosphorus removal. This method is obviously of great use in treating British high-phosphorus pig iron, and a new plant of high output was recently installed at the Corby works to replace existing Thomas converters. At present L–D converters of up to 300 tonnes capacity are in operation. This size will undoubtedly be exceeded, since world output of steel by the L–D and associated processes is now more than 30×10^6 tonnes annually.

2.87. The Kaldo Process was developed in Sweden shortly after the introduction of the L–D process in Austria. It differs principally from the latter process in that the converter is rotated during the "blow" (Fig. 2.6). This causes intimate mixing of the slag and metal, so facilitating steel-making reactions. The charge consists of molten pig iron, together with lime and up to 40% scrap. The latter acts as a coolant to reduce the high temperature produced during the "blow", but at the same time improving the thermal efficiency of the process. Failing scrap, iron ore may be added for this purpose.

The chemical reactions (Fig. 2.5B) are fundamentally similar to those taking place in the L–D process. The high degree of control possible in the Kaldo process makes it suitable for the manufacture of all grades of steel. It can convert pig iron of Thomas quality into steel which is very low in phosphorus, sulphur and nitrogen.

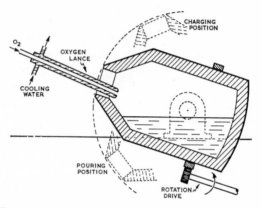

FIG. 2.6.—The Kaldo Converter in the Blowing Position.

2.88. The Rotor Process is of German origin and was developed at about the same time as the Kaldo process. It differs from other oxygen processes in that the oxygen is admitted in two stages (Fig. 2.7). This results in the complete combustion of carbon to carbon dioxide **before**

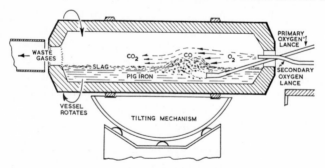

FIG. 2.7.—The Rotor Process.

it leaves the converter and so gives a much higher thermal efficiency than is the case when partial oxidation to carbon monoxide takes place, the latter then burning in the mouth of the converter and away from the metal bath. Slag–metal reactions follow the same general pattern of other oxygen processes.

The Rotor process was developed to use molten pig iron along with

PLATE 2.2A.—An "Oxygen" Converter During the "Blow".

WELLMAN 5 TON CHARGER

PLATE 2.2B.—Charging an Open-hearth Furnace.

high-purity Swedish ore as the coolant, though ore can be replaced by scrap for this purpose if suitable feeding arrangements are provided.

2.89. Spray Steel-making. At the moment an entirely new steel-making process is being developed at the Millom Hematite Ore and Iron Co., Ltd., Cumberland. It is a continuous method of steel-making originated by the British Iron and Steel Research Association. The principles of the process are indicated in Fig. 2.8. A stream of molten pig iron is allowed to fall into the reaction chamber where it meets "cross-fire" from a number of oxygen jets situated in a ring in the roof of the chamber. Lime and flux are blown into the stream as shown.

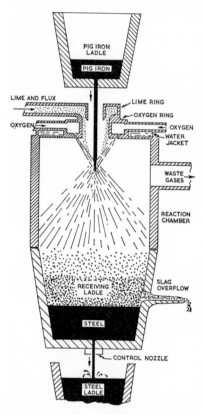

FIG. 2.8.—The Spray Steel-making Process.

The refining reactions, which are similar chemically to those in other oxygen processes, take place almost instantaneously so that continuous operation can be achieved. Combustion of carbon to carbon dioxide is almost complete. This leads to a high thermal efficiency so that up to 40% scrap can be used. Moreover, the equipment is relatively simple and of low cost, whilst wear on the refractory should be much less than in other oxygen processes because the reactions occur away from the walls of the reaction chamber. Since spray steel-making units could be sited adjacent to a blast furnace, a separate steel-making plant would become virtually unnecessary. Such a unit would lend itself well to continuous casting processes (2.341—Part II) and one can foresee that this process will become a very serious competitor of other newly established oxygen processes.

Thus, a number of advantages accrue from the use of these oxygen processes as compared with the original Thomas process. Raw materials of variable phosphorus content can be used and a wider range of steels produced. Since an oxygen blast is used, no nitrogen (a source of brittleness in steel) dissolves in the charge, and drawing-quality mild steel produced by these processes is judged to equal in quality that made in the basic open hearth. The high thermal efficiency, obtainable by the use of up to 40% scrap, offsets to a great extent the cost of tonnage oxygen used. Higher production rates are also possible than with the original Thomas process.

The Open-hearth Process

2.90. Like the Bessemer, this process can be either acid or basic depending upon the type of charge used and consequently the nature of the slag produced. Whereas the Bessemer and oxygen processes use no external source of heat, the open-hearth furnace is fired by producer gas and coke-oven gas (a by-product of blast-furnace coke manufacture) or by fuel oil. This independent source of heat allows a somewhat greater variation in the composition of the charge—particularly in terms of the proportion of steel scrap used—since the impurities are no longer needed to act as fuel. The acid open hearth is used mainly for the production of high-grade steel from low-phosphorus raw materials and for use as axles, wire ropes, springs, castings, piston rods and alloy steels; whilst the basic open hearth is generally used for the production of steel for a wide variety of work at less cost. It should be understood, however, that with modern practice the quality of basic open-hearth steel can be very high indeed.

2.91. In order to attain the necessary high temperature and to improve the fuel economy, a regenerative system is used to pre-heat the in-going air and gas. The furnace, together with its regenerative system, is illustrated diagrammatically in Fig. 2.9. The chequered-brickwork stoves on the right are being reheated by the hot waste gases as they leave the furnace. When they have attained a temperature between 900 and 1 200° C the change-over valves will be operated in order to reverse the flow of gases so that those stoves on the left, which up to now have been heating the in-going gases (and being cooled themselves in the process), will now receive the out-going gases and begin to heat up. In the meantime the pair of stoves on the right will come into action to pre-heat the in-going gas and air. The change-over will operate approximately every half-hour.

The furnace itself consists essentially of a hearth in the shape of a large, elongated basin able to hold 60 to 500 tonnes of steel according to size, though it is reported that furnaces of 10^3 tonnes capacity are operating in Russia. Along one side is a row of charging doors, whilst on the other side, which contains the tap hole, is the casting pit. At each end is a pair of ports (or inlets) for gas and air respectively.

2.92. The main requirements of the raw material for the acid open-hearth process are that it shall be low in sulphur and phosphorus, since neither of these elements will be removed during the process. The basic

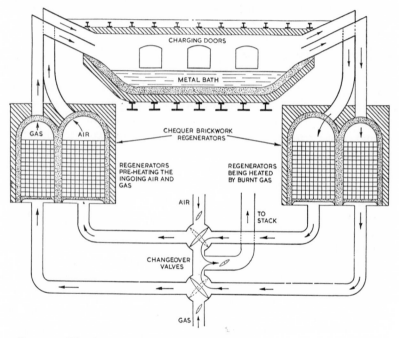

FIG. 2.9.—The Open-hearth Furnace, with a Schematic Representation of the Change-over Valve System.

open-hearth process, however, is the most adaptable of all steel-making processes, since apart from sulphur, the impurity content is not critical. Even a small amount of sulphur is allowable (up to about 0·05% overall), though its removal is not so definite as that of phosphorus. Sulphur has to be removed as manganese sulphide (MnS), and to facilitate this the manganese content must be fairly high. The basic open-hearth charge consists of phosphoric pig iron, burnt limestone and a high proportion of steel scrap.

When the charge, either acid or basic, is melted, oxide ore or mill-scale is added as an oxygen carrier, whilst in the basic process some lime is charged at this stage. From time to time test samples can be taken from the melt since the total time for the process is measured in hours rather than minutes, though with the introduction of modern "oxygen lancing", treatment times have been reduced by more than 50% so that a 450 tonne charge can now be refined in as little as four and a half hours. Impurities are progressively oxidised and join the slag which floats on the surface of the charge. In the basic process this slag becomes rich in calcium phosphate, which makes "basic slag" a valuable by-product as an agricultural fertiliser.

2.93. Since it can deal with phosphoric raw material of widely varying composition and turn it into high-quality steel, the basic open-hearth process produces by far the largest proportion of steel made in Britain at the moment. Apart from obvious considerations of ease of quality control, the utilisation of a high proportion of scrap by the open-hearth process made it more successful than the Bessemer, but the time cannot be far distant when the open-hearth too will be replaced in many instances by oxygen processes, which can now absorb in the region of 40% scrap by virtue of their high thermal efficiency.

2.94. The Ajax Process developed in Britain in 1957 is a highly modified basic open-hearth process designed to refine molten pig iron, though it can be used to deal with a mixed charge. It can best be regarded as an oxygen process, assisted when necessary by the use of fuel during the later stages of the refining process.

The Electric Furnace Process

2.95. Originally the electric-arc furnace was used for the manufacture of relatively small amounts of high-grade tool steels and alloy steels. In a modern integrated steel plant, however, it is widely used to dispose of process scrap and often to produce "open-hearth quality" steel from charges of 100% scrap. Though electricity is an expensive fuel, the utilisation of medium-grade scrap, which can be obtained cheaply, is possible in this process since conditions favour the removal of phosphorus **and sulphur**. The high cost of electricity used is thus largely offset.

2.96. The furnace used is of the arc type, and employs carbon rods striking an arc on to the charge (Fig. 2.10). The lining can be either acid or basic, though the latter is more generally employed, particularly when the purpose of the process is to refine the scrap charged, and basic slags must in consequence be used.

2.97. In the basic process the charge is melted under a basic **oxidising** slag consisting of lime and mill-scale. These conditions really duplicate those obtaining in the basic open-hearth process. Consequently the silicon, manganese and phosphorus present are absorbed into the slag, and when this process is found to be completed the slag is removed

PLATE 2.3A.—An Electric-arc Furnace, Pouring into the Ladle.

PLATE 2.3B.—Teeming the Finished Steel from the Ladle into the Ingot Moulds.
The molten metal is poured from bottom of the ladle and is being controlled by the
operator in the foreground.

completely, so that the only impurity present in the steel in appreciable amounts at the end of this stage is sulphur. Another slag is now formed on the surface of the melt, but this time a basic **reducing** slag composed

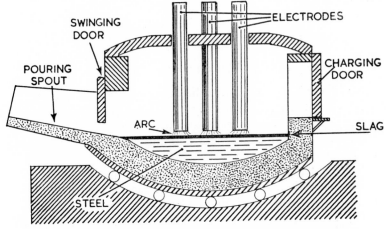

FIG. 2.10.—The Heroult Electric-arc Furnace.

of lime and anthracite. In all other steel-making processes conditions are oxidising during the refining stage, and it would be impossible, by virtue of the presence of atmospheric oxygen, to promote reducing conditions. In the electric furnace, however, conditions are under complete

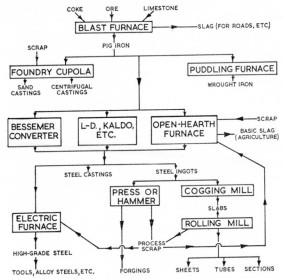

FIG. 2.11.—A Simplified "Flow Sheet" Indicating the Flow of Materials in the Main Channels of Ferrous Production.

control, and the production of a basic reducing slag is fairly simple. The main function of this slag is the effective complete removal of sulphur, and this process is the only steel-making process where this occurs.

$$FeS + C + CaO = Fe + CO + CaS$$

(Soluble in steel)   (Insoluble in steel— joins the slag)

2.98. Hence the main advantages of the electric process are:

(*a*) There is a definite and reliable removal of sulphur.

(*b*) Conditions are chemically clean, and there can be no contamination of the charge, as is possible even with gaseous fuels.

(*c*) The furnace atmosphere and the slag can be made either oxidising or reducing at will.

(*d*) The temperature can be controlled easily.

(*e*) The carbon content of the steel can be held constant, thus making its adjustment within fine limits possible.

(*f*) The addition of alloying elements can be made with precision.

RECENT EXAMINATION QUESTIONS

Unless otherwise stated, all questions in this book are taken from the final papers for the Higher National Certificates in Engineering of those Colleges to whom the questions are gratefully acknowledged. As far as possible, questions are arranged in the order in which material relevant in answering them is likely to be found in the text. Thus, questions are *not* graded either according to relative difficulty or source.

1. A typical British iron-blast-furnace burden may contain iron ores from world-wide sources such as Sweden and North Africa.

   (*a*) Give the reason why it is more economic to do this than to rely on iron ores that occur naturally in Britain.

   (*b*) Give in general outline the constituents of the burden of a modern iron blast furnace.

   (*c*) What treatments, if any, do the individual constituents receive before being charged into the furnace?

   (The Institution of Production Engineers Associate Membership Examination, Part II, Group A, Metallurgy.)

2. Outline the processes which may be applied to an iron ore to render it suitable for smelting in the blast furnace.

   (U.L.C.I., Paper No. C111–1–4.)*

* Union of Lancashire and Cheshire Institutes.

3. (a) Make a line diagram to illustrate the main features of a blast furnace for the production of iron. Describe briefly the method of operating the furnace.

(b) State in simple terms the chemical changes which occur in different zones within the furnace.

(City and Guilds of London Institute, Paper No. 293/6.)*

4. (a) What are the main functions of (i) the coke, (ii) the limestone, during the smelting of iron ore in a blast furnace?

(b) Explain why it is usual to pre-heat the air blast to a blast furnace.

(C. & G., Paper No. 293/6.)

5. (a) Describe, with the aid of sketches, the system used to pre-heat the air blast to a blast furnace.

(b) Explain the method of operation of this system.

(C. & G., Paper No. 293/6.)

6. (a) Give a typical analysis of a hematite pig iron.

(b) Outline a process used for the production of pig iron from an ore in which the metal occurs as an oxide.

(C. & G., Paper No. 293/6.)

7. Write a short account of the uses of limestone and lime in the manufacture of pig iron and steel.

(U.L.C.I., Paper No. B111–1–4.)

8. (a) Explain the terms *acid* and *basic* as applied to steel making.

(b) A pig iron has the following analysis: carbon, 3·3%; silicon, 0·65%; manganese, 0·72%; phosphorus, 1·9%; sulphur, 0·04%.

Suggest a suitable process by which it could be used to produce mild steel and give reasons for your choice.

(C. & G., Paper No. 293/6.)

9. Explain *one* sequence of operations for the production of steel from iron ore.

(Constantine College of Technology, Middlesbrough.)

10. Describe briefly the principal methods of steel-making. Give examples of suitable applications for the steels produced by THREE methods and explain why the methods are appropriate.

(Lanchester College of Technology, Coventry, H.N.D. Course.)

11. Describe the production of a heat of steel in a basic lined Bessemer furnace.

How do the properties of this steel differ from that made by the other methods available?

(Nottingham Regional College of Technology.)

12. Describe the production of steel by the basic open-hearth process.

(Constantine College of Technology, Middlesbrough.)

---

* City and Guilds of London Institute, Mechanical Engineering Technicians' Course.

13. (a) Why is most British steel produced by basic processes and in particular the basic open-hearth process?

(b) Compare and contrast the methods used to make steel by the
Bessemer, open-hearth and L–D (or oxygen) processes.

(C. & G. Paper No. 293/6.)

14. Write an essay on "Developments in the Steel Industry, due to the
introduction of Tonnage Oxygen".

(Lanchester College of Technology, Coventry.)

15. (a) Describe briefly the basic electric-arc process for the manufacture of
steel.

(b) Mention the principal advantages of such a process and state what
type of steel is generally made by this method.

(C. & G., Paper No. 293/6.)

CHAPTER III

THE PHYSICAL AND MECHANICAL PROPERTIES OF METALS AND ALLOYS

3.10. Metals constitute a majority of the chemical elements. They are generally characterised by their lustrous, opaque appearance. In fact, if we consider the more important physical properties we shall see that those of metal and non-metal are, in general, strongly contrasting.

(*a*) All metals (except mercury) are solid at normal temperatures. Non-metals include gases, a liquid (bromine) and solids.

(*b*) Metals generally are hard when compared with many of the non-metals, but at the same time they possess properties of malleability and ductility which are not possessed by non-metals.

(*c*) Metals generally have high densities.

(*d*) Metals are good conductors of heat and electricity, whilst non-metals are poor conductors.

(*e*) Metals have low specific heats compared with non-metals.

(*f*) Most metals reflect nearly all wavelengths of light equally well; for which reason they are white, or nearly white, in colour. Notable exceptions are gold and copper.

(*g*) Metals are comparatively difficult to penetrate with X-rays.

(*h*) Most metals are magnetic to some slight degree, but only in the metals iron, nickel and cobalt is magnetism strong enough to be of practical interest. The pronounced magnetism of this group is usually called "ferromagnetism".

Whilst many of these physical properties, such as conductivity, magnetism and melting point, dictate special uses to which metals will be put, it is chiefly upon the mechanical properties of strength, ductility and toughness that the usefulness of metals depends.

Fundamental Mechanical Properties

3.20. Before proceeding to a study of the standard methods of mechanical testing, let us be clear as to the nature of the fundamental mechanical properties of ductility, malleability and toughness. Ductility refers to the capacity of a substance to undergo deformation under tension without rupture, as in a wire- or tube-drawing operation. Malleability, on the other hand, is the capacity of a substance to withstand deformation under compression without rupture, as, for example, in forging or rolling. Whilst substances which are malleable are usually also ductile, this fact cannot be assumed, since a malleable substance may be weak in tension and liable to tear. Moreover, whilst malleability is usually increased by raising the temperature (for which reason metals and alloys are often hot-forged or hot-rolled), ductility is generally reduced by such a procedure, since the strength is also reduced.

3.21. Toughness refers to a metal's ability to withstand bending or the application of shear stresses without fracture. Hence copper, by this definition, is extremely tough, whilst cast iron is not. Toughness, therefore, should not be confused with either strength or hardness.

3.22. Since these fundamental mechanical properties of ductility, malleability and toughness cannot be expressed in simple numerical terms, it has become necessary to introduce certain mechanical tests which are related to these properties and which will allow of comparative numerical interpretation. Moreover, the engineer is generally more concerned with the **forces** which cause deformation in metals rather than with the deformation itself. Consequently tensile tests and hardness tests correlate the amounts of deformation produced with given forces in tension and compression respectively, whilst impact tests are an almost direct measurement of toughness. Such precise measurements of force–deformation values make it possible to draw up sets of specifications upon which the engineer can base his design.

The Tensile Test

3.30. It is not proposed here to describe in detail tensile testing equipment, since most engineering students will have carried out such tests; but it will be necessary to discuss the results of the test, and in particular the type of force–extension diagram obtained.

Fig. 3.1 shows a force–extension diagram of a kind with which the engineering student will be familiar. This is what might be called an ideal case, and is rarely encountered in the laboratory, except with wrought-iron or mild-steel specimens. With other alloys the different stages of elastic and plastic deformation are not so clearly marked and often merge one into the other. The initial straight-line part in Fig. 3.1 indicates that the extension produced is directly proportional to the force applied. It thus obeys Hooke's Law, which states that, for an elastic body, the strain produced is proportional to the stress applied. Therefore the values $\dfrac{\text{Stress}}{\text{Strain}}$ = a constant, usually known as Young's Modulus of Elasticity (E). Consider a test piece of original length L and cross-sectional area A, stretched elastically by an amount l under a force P acting along the axis of the specimen, then

$$E = \frac{\text{Longitudinal stress}}{\text{Longitudinal strain}} = \frac{P/A}{l/L} = \frac{PL}{Al}$$

If at any point on the part of the curve under consideration the force is removed the test piece will return to its original length, the extension so far being entirely elastic.

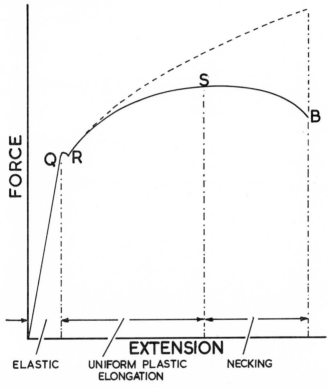

FIG. 3.1.—Type of Force–Extension Diagram Obtained for a Material such as Mild Steel in the Soft Condition.

3.31. As the force is increased a point Q is ultimately reached where the curve begins to deviate from the straight line, and if the force is now removed the specimen will not return to its original length but will be found to have acquired a small permanent plastic extension or "permanent set". Q is therefore called the "Limit of Proportionality", and if the force is increased beyond this point a stage is reached where a sudden extension takes place with no increase in force. This is known as

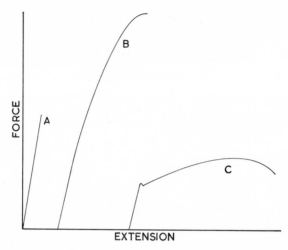

FIG. 3.2.—The Effects of Heat-treatment on the Force–Extension Diagram of a Carbon Steel.

(A) is in the quenched condition; (B) is quenched and tempered; and (C) represents the annealed condition.

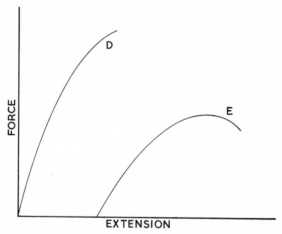

FIG. 3.3.—Typical Force–Extension Diagrams for a Non-ferrous Alloy, Showing the Absence of a Well-defined Yield Point.

(D) represents the cold-worked condition, and (E) the fully annealed condition.

the "Yield Point" R. From R onwards the specimen stretches plastically, at first uniformly along the whole gauge length, and finally, locally at a "neck", until fracture occurs. The diagram indicates that as the specimen approaches its breaking point the applied force necessary to produce extension falls. This, of course, is because the diameter of the test piece is decreasing and the **actual** stress per unit area of cross-section would be similar to that indicated by the broken line.

3.32. As has been stated, the type of diagram dealt with is really a special case. Most alloys, particularly if they have been heat-treated or cold-worked, show neither a definite elastic limit nor a yield point and

[Courtesy of Messrs. W. & T. Avery Ltd., Birmingham.

PLATE 3.1.—The Avery Servo-controlled Tensile Testing Machine, with an applied force of up to 600 kN.

The straining unit, which is shown on the left, embodies a double-acting hydraulic cylinder and ram. The force on the test piece is measured by load cells and is indicated on the upper linear scale in the centre of the control desk (which is shown on the right). The extension is measured either by an extensometer or by a built-in transducer and is indicated on the lower scale.

In addition to the indicating equipment, the control desk carries the associated electronic equipment. The right-hand panel contains the controls for a built-in programming unit, whilst that on the left carries operating buttons for selecting the programming mode.

give, on test, diagrams of the types shown in Figs. 3.2 and 3.3. The yield point, however, is possibly of greater importance to the engineer than the tensile strength itself, so it becomes necessary to specify a stress which corresponds to a definite amount of permanent extension as a substitute for the yield point. This is commonly called the "Proof Stress", and is derived as shown in Fig. 3.4. A line BC is drawn parallel to the line of proportionality, from a predetermined point B. The stress corresponding to C will be the proof stress—in the case illustrated it will be known as the "0·1% proof stress", since AB has been made equal to 0·1% of the gauge length. The material will fulfil the specification therefore if, after the proof force is applied for fifteen seconds and removed, a permanent set of not more than 0·1% of the gauge length has been produced. Proof lengths are commonly 0·1 and 0·2% of the gauge length, depending upon the type of alloy. The time limit of 15 seconds is

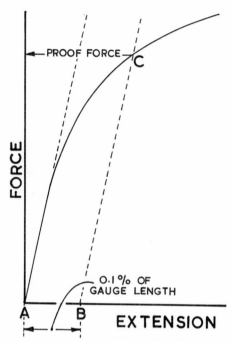

FIG. 3.4.—Method Used to Obtain the 0·1% Proof Stress.

specified in order to allow sufficient time for extension to be complete under the proof force.

3·33. In an industrial test the two ends of the broken test piece will be fitted together (Fig. 3.5) so that the total extension can be measured and

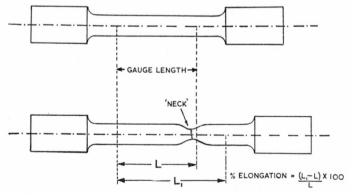

FIG. 3.5.—Method Used to Obtain the Percentage Elongation.

the percentage elongation calculated on the gauge length used. It follows that if the percentage elongation is to be a comparative figure, the dimensions of the test piece must be standardised so that a specified relationship between gauge length and cross-sectional area exists. Test pieces which are geometrically similar and fulfil these conditions are known as *proportional* test pieces. They are generally circular in cross section. BSI lays down that, for proportional test pieces:

$$\text{Gauge length} = 5 \cdot 65 \sqrt{\text{Cross-sectional area}}$$

This formula has been accepted by international agreement, and SI units are used. For test pieces of circular cross-section it gives a value:

$$\text{Gauge length} \simeq 5 \times \text{diameter}$$

Thus a test piece 200 mm² in cross-sectional area will have a diameter of 15·96 mm and a gauge length of 80 mm. Old tensile testing machines are calibrated in "tons force". Since 10 kN = 1·003 61 tonf, dual-value scales are not necessary, since, within the accuracy that most tensile machines are usually read, 1 tonf is equivalent to 10 kN.

3.34. The smallest diameter of the local neck is measured, and from it the percentage reduction in area calculated, so that from our complete set of observations we can derive the following standard results:

(*a*) $\text{Yield stress} = \dfrac{\text{Yield force}}{\text{Original cross-sectional area}}$

(As indicated already, it may be necessary to substitute proof stress for this value.)

(*b*) $\text{Tensile strength} = \dfrac{\text{Maximum force}}{\text{Original cross-sectional area}}$

(*c*) $\text{Percentage elongation} = \dfrac{\text{Extension of gauge length} \times 100}{\text{Original gauge length}}$

(*d*) Percentage reduction in area =
$$\dfrac{(\text{Original area of section} - \text{Final area}) \times 100}{\text{Original area of cross-section}}$$

Hardness Tests

3.40. Early attempts to measure the hardness of metals involved the use of loaded diamond points to produce a scratch on the surface of the test piece, as in the original Turner Sclerometer, but whilst such methods possibly bear a truer relationship to the fundamental conception of hardness, they have been almost entirely abandoned in favour of methods in which the size of an indentation, produced under static pressure, is measured.

3.41. The Brinell Test is possibly the best known of hardness tests, and was devised by a Swede, Dr. Johan August Brinell. In this test a hardened steel ball is forced into the surface of the test piece under a suitable standard load. The diameter of the impression is then measured and the Brinell Hardness Number H found from:

$$H = \frac{\text{Load } P}{\text{Surface area of impression}}$$

The surface area of the impression is $\pi\dfrac{D}{2}(D - \sqrt{D^2 - d^2})$, where D is the diameter of the ball and d the diameter of the impression, so that we can derive H—

$$H = \frac{P}{\pi\dfrac{D}{2}(D - \sqrt{D^2 - d^2})}$$

To obviate tedious calculations, however, H is generally found by reference to tables drawn up for different loads and ball diameters.

3.42. A certain amount of common sense and forethought are necessary in making a Brinell test. If, for example, in testing a soft metal, we use a load which is too great, relative to the diameter of the ball, we shall get an impression similar to that indicated in Fig. 3.6A. Here the ball has sunk to its

[Courtesy of Messrs. W. & T. Avery Ltd., Birmingham.
PLATE 3.2.—The Avery Visual Hardness Testing Machine, for both Brinell and Vickers tests.

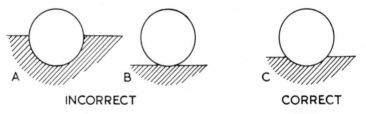

INCORRECT   CORRECT

FIG. 3.6.—The Influence of Depth of Impression on the Accuracy of a Brinell Determination.

full diameter, and the result is obviously meaningless. The impression shown in Fig. 3.6B, on the other hand, would be obtained if the load were too light relative to the ball diameter, and here the result would be likely to be very uncertain. For different materials, then, the ratio P/D^2 has been standardised in order to obtain accurate and comparative results. P is measured in kgf, and D in mm.

Material	P/D^2
Steel	30
Copper alloys	10
Aluminium alloys and pure copper . . .	5
Lead and tin alloys	1

Thus in testing a piece of steel we can use either a 10-mm ball in conjunction with a 3 000 kgf load; a 5 mm ball with a 750 kgf load; or a 1 mm ball with a 30 kgf load. In the interests of accuracy it is always advisable to use the largest ball diameter that is possible. The limiting factors will be the width and thickness of the test piece, and the small ball would be used for thin specimens, since by using the large ball we would probably be, in effect, measuring the hardness of the table supporting our test piece. The thickness of the specimen should be more than seven times the depth of the impression for hard alloys and more than fifteen times the depth of the impression for soft alloys.

3.43. The Vickers Pyramid Hardness Test eliminates the necessity of deciding the correct ratio of P/D^2 by using a diamond square-based pyramid which will give geometrically similar impressions under different loads. Moreover, the diamond is more reliable for hard materials which have a hardness index of more than 500, since it does not deform under pressure to the same extent as a steel ball. Using the diamond point, however, does not eliminate the necessity of ensuring that the thickness of the specimen is sufficient, relative to the depth of the impression.

In this test the diagonal length of the square impression is measured by means of a microscope which has a variable slit built into the eyepiece. The width of the slit is adjusted so that its edges coincide with the corners of the impression and the relative diagonal length of the impression then obtained from a small instrument attached to the slit which works on the principle of a revolution counter. The ocular reading thus obtained is converted to Vickers Pyramid Hardness Number by reference to tables. The size of the impression is related to hardness in the same way as is the Brinell number.

3.44. The Rockwell Test was devised in America, and is particularly suitable for rapid routine testing of finished material, since it indicates the final result direct on a dial. There are three scales on the dial—Scale B, which is used in conjunction with a $\frac{1}{16}$ in diameter ball and a 100 kgf load; Scale C, which is used with a diamond cone and a 150 kgf load; and Scale A, which is used in conjunction with the diamond cone and a 60 kgf load.

3.45. The Shore Scleroscope (Greek—"skleros", meaning "hard") tests the metal very near to its surface. The instrument embodies a small diamond-pointed hammer weighing $\frac{1}{12}$ oz which is allowed to fall from the standard height of 10 in inside a graduated glass tube. The height of rebound is taken as the hardness index. Since the Shore Scleroscope is a small, portable instrument, it is very useful for the determination of hardness of large rolls, castings and gears, and other large components which could not easily be placed on the testing tables of any of the more orthodox testing machines.

Table 3.1 give representative hardness numbers, together with other mechanical properties, for some of the better-known metals and alloys.

[Courtesy of Messrs. W. & T. Avery Ltd., Birmingham.

PLATE 3.3.—The Avery Universal Impact Testing Machine.
In addition to standard and sub-standard Izod tests, this machine can be used for Charpy and Impact tension tests.

Impact Tests

3.50. Impact tests indicate the toughness of a material and its capacity for resisting shock. Brittleness, resulting from incorrect heat treatment or other causes, may not be revealed by the tensile test but will usually be evident in an impact test.

3.51. The Izod Impact Test. In this test a standard notched specimen is held in a vice and a heavy pendulum, mounted on ball bearings, is allowed to strike the specimen after swinging from a fixed height. The striking energy of 167 J (120 ft lbf) is partially absorbed in breaking the specimen and, as the pendulum swings past, it carries a pointer to its highest point of swing, thus indicating the amount of energy used in fracturing the test piece.

3.52. The Charpy Test, developed originally on the Continent but now gaining favour in Britain, employs a test piece mounted as a simply-supported beam instead of in the cantilever form used in the Izod test (Fig. 3.7). The striking energy is 300 J (220 ft lbf).

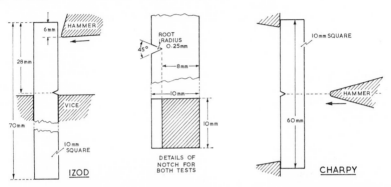

FIG. 3.7.—Dimensions of Standard Test Pieces for both Izod and Charpy Tests. In the Izod test piece, notches 28 mm apart may be cut in three different faces so that a more representative value is obtained.

3.53. To set up stress concentrations which ensure that fracture does occur, test pieces are notched. It is essential that notches always be standard, for which reason a standard gauge is used to test the dimensional accuracy of the notch. Fig. 3.7 shows standard notched test pieces for both the Izod and Charpy impact tests.

3.54. The results obtained from impact tests are not always easy to interpret, and some metals which are ductile under steady loads behave as brittle materials in an impact test. As mentioned above, however, the impact test gives a good indication of how reliable the material is likely to be under conditions of shock.

Creep

3.60. Mechanical tests, such as the tensile test and the Izod impact test, render information as to the behaviour of a material when it undergoes short-time tests. When loaded over long periods, however, a metal may exhibit gradual extension and ultimately fail by this process at a **stress value well below the tensile strength.** This phenomenon of continuous gradual extension under a steady force is known as "creep". The effects of creep must be considered seriously, particularly in the design of gas and steam turbines; of steam and chemical plant; and of furnace equipment operating at elevated temperatures, where creep takes place more easily. Some of the softer metals, however, exhibit measurable creep at room temperatures; for example, lead sheeting, which may have been protecting church roofs for centuries, is sometimes found to be slightly thicker near the eaves than at the ridge. The lead has "crept" under its own weight, and the effect over centuries reaches measurable proportions.

TABLE 3.1.—*Typical Mechanical Properties of Some Metals and Alloys*

Metal or alloy	Condition	0·1% proof stress		Tensile strength		Percentage elongation on 50 mm (2 in)	Brinell hardness number	Fatigue limit	
		N/mm²	tonf/in²	N/mm²	tonf/in²			N/mm²	tonf/in²
Lead . .	Soft sheet	—	—	17·7	1·15	64	4	2·78	0·18
Aluminium .	Wrought and annealed	—	—	58·7	3·8	60	15	30·9	2·0
Duralumin .	Extruded and heat-treated	278	18	432	28	15	115	170	11
Magnesium with 10% aluminium	Cast and heat-treated	116	7·5	247	16	2	80	61·8	4·0
Copper . .	Wrought and annealed	46·3	3·0	216	14	60	42	66·4	4·3
70–30 brass (Cartridge Metal) . .	⎰Annealed sheet	84·9	5·5	324	21	67·5	62	114	7·4
	⎱Deep drawn	371	24	463	30	19·5	132	151	9.8
Phosphor-bronze (95% Cu; 5% Sn)	Hard rolled	649	42	711	46	5·5	188	185	12
Wrought iron .	Hot-rolled bar	201	13	309	20	30·0	100	185	12
Mild steel .	Hot-rolled sheet	0·5% Proof stress 232	15	309	20	28·0	100	185	12
Nickel–chromium–molybdenum steel (2·5% Ni; 0·75% Cr; 0·5% Mo)	Hardened and tempered	1310	85	1540	100	10·0	444	432	28
Stainless steel	Softened	185	12	463	30	30·0	170	263	17
Cast iron (grey)	Cast	—	—	309	20	0·0	250	137	8·9

3.61. The Limiting Creep Stress of a material at any given temperature is that stress below which no **measurable** creep takes place, but here we are limited by the sensitivity of the recording equipment, and it is possible that creep is taking place at even lower stresses. In practice, from a series of creep tests at a fixed temperature, a limiting creep stress is estimated which would lead to a certain arbitrary small rate of creep, and a factor of safety introduced for subsequent use in design.

3.62. The type of relationship which exists between the rate of creep and the applied static stress is shown in Fig. 3.8. At 380 N/mm² creep is relatively rapid and the material will ultimately fail, although this stress is well below the tensile strength. At 150 N/mm² the creep curve soon levels out, due apparently to work-hardening.

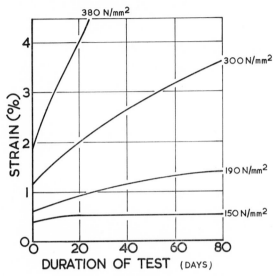

FIG. 3.8.—The Rate of Creep, at Various Stresses, of a 0·45% C Steel at 450° C.

3.63. In creep testing, the specimen, in form rather like a tensile test piece, is enclosed in a thermostatically controlled electric tube furnace which can be maintained with accuracy over long periods at any given temperature up to 900° C or more. A simple lever system is often used to load the specimen, and some form of extensometer employed to measure the resultant extension at suitable time intervals. The extensometer is sometimes of the optical type, using two mirrors, scales and telescopes.

3.64. Ordinary creep tests involve rather long periods of time extending into months, and are sometimes replaced by slow tensile methods of testing in order to obtain results more quickly. Such tests do not measure the true limiting creep stess, but serve only to indicate approximately the behaviour of a metal in creep. For example, the "time-yield" value is approximately equivalent to a rate of creep of one part in a million per hour. When subjected to the "time-yield" stress, the specimen must not show an extension greater than 0·50% of the gauge length during the first twenty-four hours, and in the next forty-eight hours must not show any further extension within a sensitivity of measurement of 0·002 5 mm on a 50 mm gauge length. For many purposes it is sufficient to take a half to two-thirds of the "time-yield" value as a safe working stress under continuous load, but for other classes of work longer-time creep tests are necessary to determine the stress which will produce a given amount of elongation in a given time at a particular temperature.

Fatigue

3·70. Whereas creep is a phenomenon to be reckoned with in members under the action of a dead load, fatigue refers to failure of a member under the action of repeated cycles of stress. The stress necessary to cause failure in this way is much less than that required to break the specimen by means of a **single** steady pull. Although the name of the German engineer Wöhler is usually associated with early attempts at fatigue testing, it was Sir William Fairbairn who in 1861 carried out the initial research in this direction. He found that raising and lowering a 3-ton load on to a wrought-iron girder some 3 million times would cause the girder to break, yet a static load of 12 tons was necessary to cause failure. From further work he concluded that there was some load below 3 tons which could be raised and lowered an infinite number of times without causing failure. Some ten years later Wöhler did further work in this direction and developed the fatigue-testing machine which bears his name. The effects of fatigue in respect of vibrational stresses have been illustrated only too vividly in recent years in the loss of the early Comet air-liners.

3·71. The fatigue limit of a metal is the maximum range of stress which the test piece will endure without failure after an infinite number of reversals about a mean stress of zero. Let us assume that the maximum load applied in one direction is W. If we now reverse the direction of this force, the load on the member will first fall to zero and then increase to W in the opposite direction, so that the average load has been zero and the maximum range of load $2W$. Fatigue tests are usually carried out in the Wöhler type of machine, the principle of which is shown in Fig. 3.9A. Here the specimen is in the form of a cantilever carrying the load W acting at the end via a ball-race. Since the load continues to act in the same direction, the direction of stress in the test piece will be reversed each time the test piece turns through 180°. Under these conditions the mean stress will be zero and the stress range that derived from the load $2W$. To find the fatigue limit a number of specimens are tested in this way, each at a different stress range, until failure occurs, or alternatively, until about 20 million reversals have been endured. From the results a stress–reversal curve is plotted as in

Fig. 3.9B. The curve becomes horizontal at a stress which is known as the fatigue limit or endurance limit.

3.72. A fatigue fracture has a characteristic type of surface and consists of two parts (Fig. 3.9C). One is quite smooth and burnished, and shows ripple-like marks radiating outwards from the centre of crack formation, whilst the other is coarse and crystalline, indicating the final

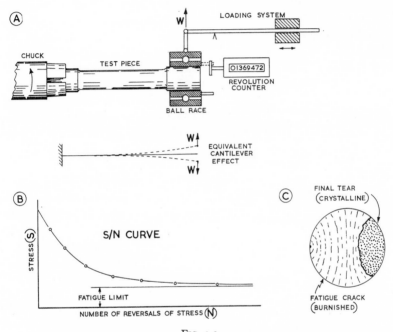

FIG. 3.9

(A) The Principle of a Simple Fatigue-testing Machine.
(B) The S/N Curve as a Method of Assessing the Fatigue Limit.
(C) The Appearance of a Shaft which has Failed Due to Fatigue.

fracture of the remainder of the cross-sectional area which could no longer withstand the load. Fatigue cracks usually start from some point of stress concentration, such as a keyway, sharp fillet, microstructural defect or even a bad tool mark. It must be realised, however, that fatigue cracks are not necessarily the result of faulty material. Sometimes considerations of design will limit the working section of some member subjected to alternating stresses. A knowledge of the behaviour of the material in fatigue will allow an assessment of the useful life of the component or structure to be made so that it can be withdrawn from service—or "junked" as our American friends put it—after an appropriate working period.

Non-destructive Tests

3.80. The mechanical tests already mentioned are obviously applicable only to test pieces produced from a batch of material which it is proposed to use for the manufacture of components. Such tests will naturally be of little use to indicate the properties of castings, forgings, welded joints, etc., each of which, by the very nature of their manufacturing processes, vary in quality of manufacture and which cannot be destructively tested. Provided the expenditure is justified, individual tests for flaws and unsoundness can be applied to such components. The most important methods of testing are by the use of magnetic dusts, and by means of X-rays, γ-rays and other specialised methods.

3.81. The Magnetic-dust Method consists in laying the steel specimen across the arms of a magnetising machine and then sprinkling it with a special magnetic powder. The excess powder is blown away, and any cracks or defects are then revealed by a bunch of powder sticking to the area on each side of the crack. Since the crack lies across the magnetic field, each side of it will have become a magnetic pole which collects the dust.

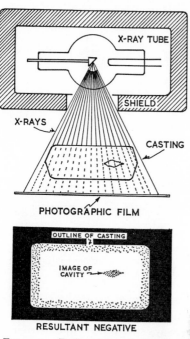

Fig. 3.10.—Radiography of a Casting Using X-rays.

3.82. X-rays are used widely in metallurgical research in order to investigate the nature of crystal structures in metals and alloys. Their use is not confined to the research laboratory, however, and many firms use X-rays in much the same way as they are used in medical radiography, that is, for the detection of cavities, flaws and other discontinuities in castings, welded joints and the like.

X-rays used in metallurgical radiography are "harder" than those used in medicine. That is, they are of shorter wavelength and better able to penetrate metals. At the same time their properties make them more dangerous to human body tissue, and plant producing radiation of this type needs to be carefully shielded in order to prevent the escape of stray radiations which would cause damage.

Like light, X-rays travel in straight lines, but, whilst metals are opaque to light, they are moderately "transparent" to X-rays, particularly those of short wavelength. Fig. 3.10 illustrates the principle of X-radiography. A casting is interposed between a shielded source of X-rays and a photographic plate. Some of the radiation will be absorbed by the metal so that the density of the photographic image will vary with the thickness of metal through which the rays have passed.

3.83. X-rays are absorbed logarithmically—

$$I = I_o e^{-\mu d}$$

where I_o and I are the incident and emergent intensities respectively d the thickness and μ the linear coefficient of absorption of radiation. μ is lower for radiations of shorter wavelength.

A cavity in the casting will result in those X-rays which pass through the cavity being less effectively absorbed than those rays which travel through the entire thickness of metal. Consequently the cavity will show as a dark patch on the resultant photographic negative in the same way that a greater intensity of light produces a darker area on an ordinary photographic negative.

A fluorescent screen may be substituted for the photographic plate so that the resultant radiograph may be viewed instantaneously. This type of fluoroscopy is obviously much cheaper and quicker, but is less sensitive than photography and its use is usually limited to the less-dense metals and alloys.

3.84. γ-**rays** can also be used in the radiography of metals. Since they are of shorter wavelength than X-rays, they are able to penetrate more effectively a greater thickness of metal. Hence they are particularly useful in the radiography of steel, which is more effective than the light alloys in absorbing radiation.

3.85. The emission of γ-rays occurs during nuclear fission. A number of elements also undergo natural radioactive disintegration. During such spontaneous disintegration different forms of radiation are emitted from the radioactive source. The phenomenon generally referred to as "α-rays" is in fact the effect produced by a stream of moving particles— α-particles. An α-particle is, in effect, of the same constitution as the nucleus of a helium atom. Similarly, β-rays consist of a stream of fast-moving β-particles (in actual fact these are electrons). Only γ-rays are in fact true electromagnetic radiation, and since this radiation is of short wavelength it is able to penetrate considerable thicknesses of metal.

All three forms of radiation are emitted at some stage in the natural radioactive disintegration of uranium, U^{238}. Obviously the loss of an α-particle from the nucleus of a U^{238} atom will cause a change in both its atomic mass number and atomic number and consequently in its properties. In this general way U^{238} changes in a series of steps to radium, and finally to a stable, non-radioactive isotope of lead (Pb^{206}). Each

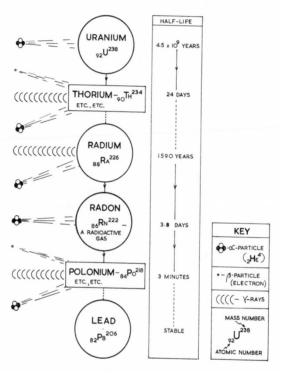

FIG. 3.11.—Stages in the Radioactive Disintegration of Uranium, $_{92}U^{238}$.

stage of the radioactive decay is accompanied by the emission of one or more of the types of radiation described above. A somewhat simplified representation of this process is indicated in Fig. 3.11.

3.86. Some stages in the process of radioactive disintegration take place more quickly than others, and each separate decay process is governed by the expression—

$$-\frac{dn}{dt} = \lambda n$$

where n is the number of atoms of the species present and λ is a constant for that "species". Integration of this expression and evaluation of the time necessary for half of the atoms present, initially, to decompose gives what is called the "half-life period" of the species.

Thus—

$$t_{\frac{1}{2}} = \frac{0.6932}{\lambda}$$

Each radioactive species is characterised by its half-life period. Thus it takes 4.5×10^9 years for half of the U^{238} atoms present in a mass of uranium to change to the next uranium isotope in the series, whilst it takes only 1 590 years for half of the radium atoms present to change to the radioactive gas radon.

3.87. Initially, radium and radon were used as a source of radiation, but artificially activated isotopes of other substances are now widely used as a source of γ-rays. These activated isotopes are prepared by "irradiating" the element in an atomic pile. A nucleus struck by a

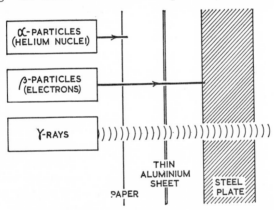

FIG. 3.12.—Diagram Showing the Relative Penetration Obtained by Various Forms of Radiation Derived from Radioactive Decay.

neutron absorbs it and then possesses an excess of energy. Subsequent radioactivity then releases γ-rays. One of the most useful activated isotopes is cobalt Co^{60}, since it has a convenient half-life of 5.3 years, and it is in this respect preferable to the tantalum isotope Ta^{182} which was formerly used and which has a half-life of only 112 days. Many other active isotopes can be manufactured, including the notorious "strontium 90" which is a dangerous "fall-out product" of atomic explosions.

3.88. Manipulation of the isotope as a source of γ-radiation in metallurgical radiography is in many respects more simple than is the case with X-rays, though security arrangements are extremely important in view of the facts that γ-radiation is "harder" than X-radiation and that

76 ENGINEERING METALLURGY

it takes place continuously from the source without any outside stimulation. γ-rays can be used to radiograph considerable thicknesses of steel, the technique used being similar to that used with X-rays.

Recent Examination Questions

1. Define the following terms: (a) toughness; (b) hardness; (c) tensile strength; (d) ductility.

   Describe briefly the tests you would employ to measure these properties. (Constantine College of Technology, Middlesbrough.)

2. Discuss one standard material test. State the physical properties it is designed to test, and show how these properties are related to both design and manufacture.

   (Institution of Production Engineers Associate Membership Examination, Part II, Group B, Production Metallurgy.)

3. Draw a typical stress–strain diagram for a tensile test on mild steel. Discuss the information which may be obtained from such a test and its practical importance. (University of Glasgow, B.Sc. (Eng.).)

4. During a tensile test on a non-ferrous alloy the following figures were obtained for force and corresponding extensions:

Extension (in.)	0·002	0·004	0·006	0·008	0·010	0·015	0·020	0·025
Force (tonf)	1·17	2·34	3·51	4·68	5·84	7·91	9·12	9·87

Extension (in.)	0·030	0·035	0·040	0·045	0·050	0·055	0·056	
Force (tonf)	10·40	10·73	10·86	10·70	10·29	9·54	9·29	(Break)

Original diameter of specimen—0·564 in. Gauge length—2 in.

Draw the load–extension diagram (on the squared paper provided) and determine:

(a) the 0·1% proof stress;
(b) the tensile strength;
(c) the percentage elongation, for the material.

(C. & G., Paper No. 293/6.)

5. Sketch and discuss the types of force–extension diagram obtained during the tensile testing of the following materials:

(a) 0·4% C steel in the normalised condition;
(b) 0·4% C steel in the hardened condition;
(c) 0·4% C steel in the hardened and tempered condition;
(d) a non-ferrous alloy in the annealed condition;
(e) a non-ferrous alloy in the work-hardened condition.

What information can be derived from these diagrams?
Define "proof stress" and discuss its significance.

(West Bromwich Technical College, A1 year.)

6. Explain how modern methods of testing may be considered to represent a true evaluation of surface hardness.

Discuss the Brinell hardness test and describe the precautions which must be taken in its operation.

(West Bromwich Technical College, A1 year.)

7. (a) Why are hardness tests conducted and what useful information can be obtained from them?

(b) What are the basic differences between a Shore Scleroscope and a Rockwell hardness test? (C. & G., Paper No. 293/6.)

8. Suggest a method of hardness determination for each of the following components. Give reasons to justify your choice.

(a) A microspecimen.
(b) A small cast-iron casting.
(c) A roll *in situ.*
(d) Mass-produced small finished components.

(Constantine College of Technology, H.N.D. Course.)

9. Describe with the aid of sketches, the apparatus, specimen and procedure to be followed for one type of impact test.

Explain briefly the properties of the material which are indicated by the test. (C. & G., Paper No. 293/31.)

10. Give an account of the various factors which influence the creep behaviour of metals. (Nottingham and District Technical College.)

11. Define "limiting creep stress" and show how it can be measured.

Discuss the importance of both the formal limiting creep stress and what are usually called "time-yield" values in engineering design.

(West Bromwich Technical College, A1 year.)

12. Define the term "fatigue limit" as applied to a fatigue test on a mild-steel specimen. In what way may this property be affected by the environment of the specimen? (Derby and District College of Technology.)

13. Write a short account of the importance of a consideration of fatigue in engineering design. (West Bromwich Technical College, A1 year.)

14. Write an essay on the fatigue of metals.

(Brunel College,* Dip. Tech.)

15. Describe TWO methods which may be used for the detection of surface defects which could lead to fatigue failure in metal components.

What are the advantages and limitations of the two methods described?

(The Polytechnic, London W.1., Diploma Course.)

16. Describe suitable non-destructive methods which may be used to detect flaws on, or close to, the surface of a metal component.

(C. & G., Paper No. 293/6.)

* Now Brunel University, awarding the degree of B.Sc.(Eng.).

17. Write a short account of the methods commonly used for the detection of internal cavities in aluminium castings.

(C. & G., Paper No. 293/6.)

18. (*a*) What is meant by non-destructive testing of cast components?

(*b*) List the methods of non-destructive testing which are commercially available.

(*c*) Describe one method in detail. (C. & G., Paper No. 293/7.)

19. (*a*) Describe in detail how X-rays may be employed as a means of flaw detection.

(*b*) Describe briefly *one* method of locating the exact position of a flaw in a casting using this process. (C. & G., Paper No. 293/8.)

20. (*a*) Describe fully the radiographical methods available for non-destructive examination of engineering components and explain in detail one method employed for the exact location of the flaw.

(*b*) Compare briefly the advantages and disadvantages of X-ray and γ-ray methods of radiological examination.

(Institution of Production Engineers Associate Membership Examination, Part II, Group B, Production Metallurgy.)

21. Outline the principle underlying radiographic methods of non-destructive testing. Discuss the use of these methods in the engineering industry.

(S.A.N.C.A.D.)*

22. What inspection techniques would be applied to find the following defects in cast products:

(i) internal cavities in a large steel casting;
(ii) cracks in grey iron castings;
(iii) fine cracks in aluminium alloy castings;
(iv) internal porosity in aluminium alloy castings;
(v) pressure tightness in an automobile cylinder head?

Describe briefly the methods chosen and give reasons for your choice.

(Institution of Production Engineers Associate Membership Examination, Part II, Group B, Production Metallurgy.)

* Scottish Association for National Certificates & Diplomas.

CHAPTER IV

THE CRYSTALLINE STRUCTURE OF METALS

4.10. All elements can exist as either solids, liquids or gases depending upon the conditions of temperature and pressure which prevail. Thus, at atmospheric pressure, the metal zinc melts at 419° C and boils at 907° C. In the gaseous state particles are in a state of constant motion and the pressure exerted by the gas is due to the impact of these particles with the walls of the container. As the temperature of the gas is increased, the velocity of the particles is increased and so the pressure exerted by the gas increases, assuming that the container does not allow the gas to expand. If, however, the gas is allowed to expand the distance apart of the particles increases and so the potential energy increases.

Engineers will understand the term "potential energy" as being that energy possessed by a body by virtue of its ability to do work. Similarly, matter possesses potential energy by virtue of its state. As the distance apart of particles increases, so the potential energy increases. In a gas such as oxygen the "particles" referred to here are, in fact, molecules, each of which consists of two atoms held together by a co-valent bond, but in a metallic gas these "particles" consist of single atoms, each with its own compliment of electrons. In the gaseous state the metallic bond does not exist.

On condensation to a liquid the atoms come into contact with each other to form bonds (Fig. 4.1), but there is still no orderly arrangement of the atoms, though a large amount of potential energy is given up in the form of latent heat. When solidification takes place there is a further discharge of latent heat, and the potential energy falls even lower as the atoms take up orderly positions in some geometrical pattern which constitutes a crystal structure. The rigidity and cohesion of the structure is then due to the operation of the metallic bond as suggested in 1.85.

4.11. Substances can be classified as either "amorphous" or "crystalline". In the amorphous state the elementary particles are mixed together in a disorderly manner, their positions bearing no fixed relationship to those of their neighbours. The crystalline structure, however, consists of atoms, or, more properly, ions, arranged according to some regular geometrical pattern. This pattern varies, as we shall see, from one substance to another. All metals are crystalline in nature. If a metal, or other crystalline solid, is stressed below its elastic limit, any distortion produced is temporary and, when the stress is removed, the solid will return to its original shape. Thus, removal of stress leads to removal of strain and we say that the substance is elastic.

The amorphous structure is typical of all liquids in that the atoms or molecules of which they are composed can be moved easily with respect

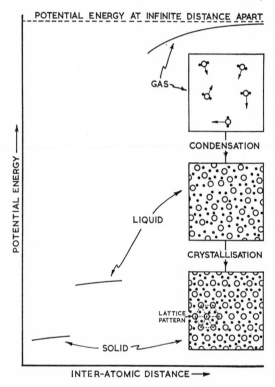

FIG. 4.1.—Relative Potential Energy and Atomic Arrangements in the Three States of Matter.

In the gaseous and liquid states these arrangements are disorderly, but in the solid state the ions conform to some geometrical pattern. (Note that in this diagram ions are indicated thus: O, whilst *valency* electrons are indicated so: •, i.e. the metal is assumed to be bivalent.)

to each other, since they do not conform to any fixed pattern. In the case of liquids of simple chemical formulae in which the molecules are small, the forces of attraction between these molecules are not sufficient to prevent the liquid from flowing under its own weight; that is, it possesses high "mobility". Many substances, generally regarded as being solids, are amorphous in nature and rely on the existence of "long-chain" molecules, in which all atoms are co-valently bonded, to give them strength as in certain super polymers (1.84). In these substances the large thread-like molecules, often containing many thousands of atoms each, are not able to slide over each other as is the case with relatively simple molecules in a liquid. The forces of attraction* between these large molecules are much greater and, since there will be considerable mutual entanglement by virtue of their fibrous nature, the resultant amorphous mass will lack mobility and will be extremely viscous, or "plastic". An amorphous structure, therefore, does not generally possess elasticity but only plasticity. A notable exception to this statement is provided by rubber, in which the "folded" nature of the long-chain molecules gives rise to elasticity (1.84).

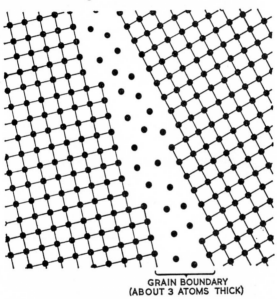

GRAIN BOUNDARY
(ABOUT 3 ATOMS THICK)

FIG. 4.2.—Diagrammatic Representation of a Grain Boundary.
The atoms (ions) here are farther apart than those in the crystals themselves.

A piece of metal consists of a mass of separate crystals irregular in shape but interlocking with each other rather like a three-dimensional jig-saw puzzle. Within each crystal the atoms are regularly spaced with respect to one another. The state of affairs at the crystal boundaries has long been a subject for conjecture, but it is now widely held that in these regions there exists a film of metal, some three atoms thick, in which the atoms do not conform to any pattern (Fig. 4.2). This crystal boundary film is in fact of an amorphous nature. The metallic bond acts within and across this crystal boundary. Consequently, the crystal boundary is not necessarily an area of weakness except at high tempera-

* Van der Waal's Forces (1.86).

tures when inter-atomic distances increase, and so bond strength decreases. Thus at high temperatures metals are more likely to fail by fracture following the crystal boundaries, whilst at low temperatures failure by transcrystalline fracture is common.

4.12. When a pure liquid solidifies into a crystalline solid, it does so at a fixed temperature called the freezing point. During the crystallisation process the atoms assume positions according to some geometrical pattern (Fig. 4.1), and whilst this is taking place, heat (the latent heat of solidification) is given out without any fall in temperature taking place. A typical cooling curve for a pure metal is shown in Fig. 4.3.

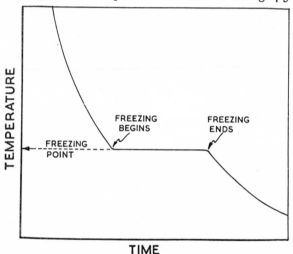

FIG. 4.3.—Typical Cooling Curve of a Pure Metal.

4.13. Atoms are very small entities indeed, and it has been calculated that approximately 8 470 000 000 000 000 000 atoms are contained in 1 mm³ of copper. Thus an individual copper atom measures approximately $1·92 \times 10^{-10}$ m in diameter, putting it far beyond the range of an ordinary optical microscope with its maximum magnification of only 2 000. However, modern high-resolution electron microscopes, capable of magnifications of a million or so, can show planes of atoms in metals; whilst field-ion microscopy producing magnifications of several millions reveals individual atoms in the structures of some metals.

We have said that the atoms in a solid metal are arranged according to some geometrical pattern. How, then, was this fact ascertained and what form do these patterns take?

Little work was possible in this direction until in 1911 Max von Laue employed X-rays in an initial study of the structures of crystals. Since

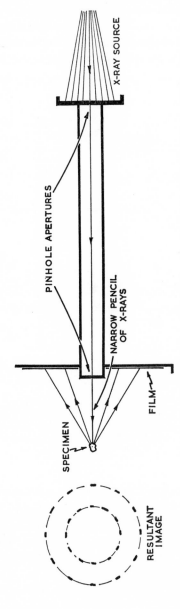

FIG. 4.4.—One of the Methods Used to Determine the Lattice Structure of a Metal by X-ray Diffraction. This is termed the "back-reflection" method.

then, X-rays have found an increasing application in the study of crystal structures, including those of metals. When a beam of monochromatic* X-rays is directed as a narrow "pencil" at a specimen of the metal in question, diffraction takes place at certain of the crystallographic planes. The resultant "image" is recorded on a photographic film as a series of

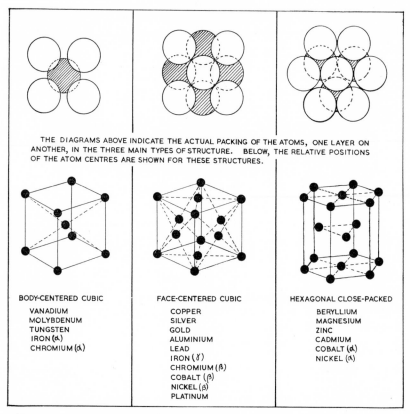

THE DIAGRAMS ABOVE INDICATE THE ACTUAL PACKING OF THE ATOMS, ONE LAYER ON ANOTHER, IN THE THREE MAIN TYPES OF STRUCTURE. BELOW, THE RELATIVE POSITIONS OF THE ATOM CENTRES ARE SHOWN FOR THESE STRUCTURES.

BODY-CENTERED CUBIC	FACE-CENTERED CUBIC	HEXAGONAL CLOSE-PACKED
VANADIUM	COPPER	BERYLLIUM
MOLYBDENUM	SILVER	MAGNESIUM
TUNGSTEN	GOLD	ZINC
IRON (α)	ALUMINIUM	CADMIUM
CHROMIUM (α)	LEAD	COBALT (α)
	IRON (γ)	NICKEL (α)
	CHROMIUM (β)	
	COBALT (β)	
	NICKEL (β)	
	PLATINUM	

FIG. 4.5.—The Three Principal Types of Structure in Which Metallic Elements Crystallise.

spots, and an interpretation of the patterns produced leads to a reconstruction of the original crystal structure of the metal. One such method, that of "back reflection", is shown in Fig. 4.4. Other methods are in use, but only this brief mention is possible here.

4.14. There are several types of pattern or "space lattice" in which metallic atoms can arrange themselves on solidification, but the three most common are shown in Fig. 4.5. Of these the hexagonal close-

* As in the case of light, the term "monochromatic" signifies radiation of a single wavelength.

packed represents the closest packing which is possible with atoms. It is the sort of arrangement obtained when one set of snooker balls is allowed to fall in position on top of a set already packed in the triangle. (This is illustrated at the top right-hand corner of Fig. 4.5.) The face-centred cubic arrangement is also a close packing of the atoms, but body-centred cubic is relatively "open"; and when, as sometimes happens, a metal changes its crystalline form as the temperature is raised or lowered, there is a noticeable change in volume of the body of metal. An element which can exist in more than one crystalline form in this way is said to be *allotropic*. Thus pure iron can exist in three separate crystal-line forms, which are designated by letters of the Greek alphabet: "alpha" (α), "gamma" (γ) and "delta" (δ). Alpha-iron, which is body-centred cubic and exists at normal temperatures, changes to gamma-iron, which is face-centred cubic, when heated to 910° C. At 1 400° C the face-centred cubic structure reverts to body-centred cubic delta-iron. (The essential difference between alpha-iron and delta-iron, therefore, is only in the temperature range over which each exists.) These allotropic changes are accompanied by changes in volume—contraction and expansion respectively, as shown in Fig. 4.6A.

The contraction which takes place as the body-centred cubic struc-ture changes to face-centred cubic can be demonstrated with the simple apparatus shown in Fig. 4.6B. A wire is held taut under a steady load, and an electric current, sufficient to heat it above the alpha–gamma change point, is passed through it. As the change point is reached, the instantaneous contraction of the wire is indicated by a sharp "kick" of the pointer to the left. As the wire cools again, when the current is switched off, there is a kick to the right, accompanied by a brightening of the red glow emitted by the wire. This brightening is particularly noticeable when the experiment is made in a darkened room. This indicates that the gamma-to-alpha change is accompanied by a libera-tion of heat energy, known as "recalescence". In the actual experiment a steel wire is used for the sake of convenience, since it is more easily obtainable than a pure iron wire. (The alpha-to-gamma change will be exhibited in a similar way but at a lower temperature than if a pure iron wire were used.) It is this allotropic change in iron which makes possible the hardening of carbon steels by quenching. Thus, if iron did not chance to be an allotropic element, can one imagine that Man would have reached his present state of technological development? A world without steel is a prospect difficult to visualise.

4.15. The space lattices indicated in Fig. 4.5 represent the simplest units which can exist in the three main types mentioned. In actual fact a metallic crystal is built up of a continuous series of these units, each

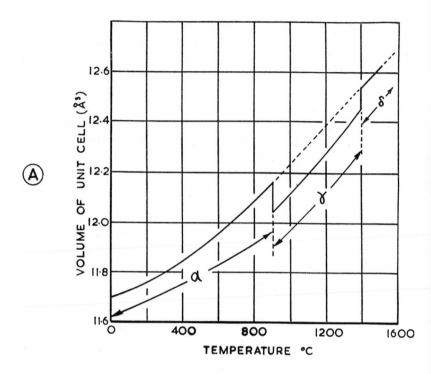

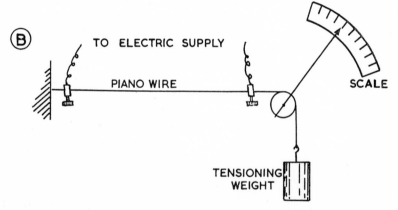

FIG. 4.6.—The Effect of Allotropic Transformations on the Expansion of Pure Iron.

(A) The close packing of the γ phase causes a sudden decrease in volume of the unit cell at 910° C and a corresponding increase at 1 400° C when the structure changes to δ. (B) The transformation at 910° C can be demonstrated by the apparatus shown.

face of a unit being shared by an adjacent unit. Any atom will belong to several different crystallographic planes cutting in different directions through a crystal. In order to be able to specify these planes some system of reference is required. The system used is that of "Miller indices".

These are the smallest whole numbers proportional to the reciprocals of the intercepts which the plane under consideration makes with the three crystal axes (X, Y and Z). Consider a simple cubic structure (Fig. 4.7). Each face intersects only one axis so that the intercepts are

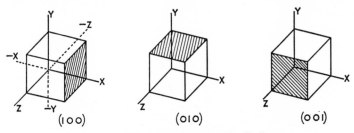

FIG. 4.7.—The Derivation of Miller Indices.

$(1,\infty,\infty)$, $(\infty,1,\infty)$ and $(\infty,\infty,1)$ respectively. The reciprocals of these numbers are $(1,0,0)$, $(0,1,0)$ and $(0,0,1)$, and these are the Miller indices of the three planes coinciding with the three faces of the cube under consideration. For quick reference they are usually written (100), (010) and (001). The three opposite faces in each case would have negative signs, e.g. (Ī00). In Fig. 4.8A the plane indicated is represented by Miller indices (110) and that in Fig. 4.8B by (111).

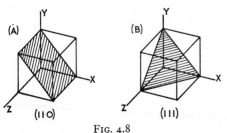

FIG. 4.8

4.16. When a pure metal solidifies, each crystal begins to form independently from a nucleus or "centre of crystallisation". The nucleus will be a simple unit of the appropriate crystal lattice, and from this the crystal will grow. The crystal develops by the addition of atoms according to the lattice pattern it will follow, and rapidly begins to assume visible proportions in what is called a "dendrite". This is a sort of crystal skeleton, rather like a backbone from which the arms begin to grow in other directions, depending upon the lattice pattern. From these secondary arms, tertiary arms begin to sprout, somewhat similar to the branches and twigs of a fir-tree. In the metallic dendrite, however, these branches and twigs conform to a rigid geometrical pattern. A metallic crystal grows in this way because heat is dissipated more quickly from a point, so that it will be there that the temperature falls most quickly leading to the formation of a rather elongated skeleton (Fig. 4.9).

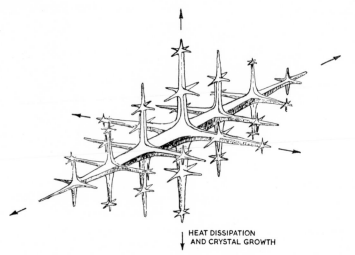

HEAT DISSIPATION
AND CRYSTAL GROWTH

FIG. 4.9.—The Early Stages in the Growth of a Metallic Dendrite.

The dendrite arms continue to grow and thicken at the same time, until ultimately the space between them will become filled with solid. Meanwhile the outer arms begin to make contact with those of neighbouring dendrites which have been developing quite independently at the same time. All these neighbouring crystals will be orientated differently due to their independent formation; that is, their lattices will meet at odd angles. When contact has taken place between the outer arms of neighbouring crystals further growth outwards is impossible, and solidification will be complete when the remaining liquid is used up in thickening the existing dendrite arms. Hence the independent formation of each crystal leads to the irregular overall shape of crystals. The dendritic growth of crystals is illustrated in Fig. 4.10. In these diagrams, however, the major axes of the crystals are all shown in the same horizontal plane, i.e. the plane of the paper, whereas in practice this would not necessarily be the case. It has been shown so in the illustration for the sake of clarity.

4.17. If the metal we have been considering is pure we shall see no evidence whatever of dendritic growth once solidification is complete, since all atoms are identical. Dissolved impurities, however, will often tend to remain in the molten portion of the metal as long as possible, so that they are present in that part of the metal which ultimately solidifies in the spaces between the dendrite arms. Since their presence will often cause a slight alteration in the colour of the parent metal, the dendritic

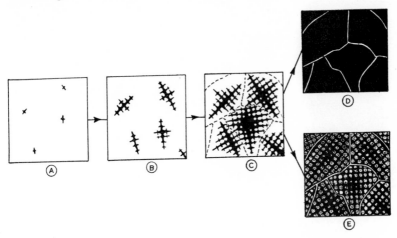

FIG. 4.10.—The Dendritic Growth of Metallic Crystals from the Liquid State.

A solid pure metal (D) gives no hint of its dendritic origin since all atoms are identical, but an impure metal (E) carries the impurities between the dendritic arms, thus revealing the initial skeleton.

structure will be revealed on microscopical examination. The areas containing impurity will appear as patches between the dendrite arms (Fig. 4.10E). Inter-dendritic porosity may also reveal the original pattern of the dendrites to some extent. If the metal is cooled too rapidly during solidification, molten metal is often unable to "feed" effectively into the spaces which form between the dendrites due to the shrinkage which accompanies freezing. These spaces then remain as cavities following the outline of the solid dendrite. Such shrinkage cavities can usually be distinguished from blow-holes formed by dissolved gas. The former are of distinctive shape and occur at the crystal boundaries, whilst the latter are quite often irregular in form and occur at any point in the crystal structure (Fig. 4.11).

4.18. The rate at which a molten metal is cooling when it reaches its freezing point affects the size of the crystals which form. A slow fall in temperature, which leads to a small degree of undercooling at the onset of solidification, promotes the formation of relatively few nuclei, so that

PLATE 4.1A.—Dendrites on the Surface of an Ingot of Antimony.

As solidification proceeded, shrinkage caused the remaining molten metal to be drawn away, leaving these partly formed crystals exposed. × 3.

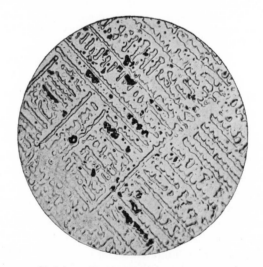

PLATE 4.1B.—Shrinkage Cavities (Black Areas) in Cast Tin Bronze.

These follow roughly the shape of the original dendrites and occur in that part of the alloy to solidify last. × 200. Etched in ammonia–hydrogen peroxide.

the resultant crystals will be large (they are easily seen without the aid of a microscope). Rapid cooling, on the other hand, leads to a high degree of undercooling being attained, and the onset of crystallisation results in the formation of a large "shower" of nuclei. This can only mean that the final crystals, being large in number, are small in size. In the language of the foundry, "chilling causes fine-grain castings". (Throughout this book the term "grain" and "crystal" are used synonymously.)

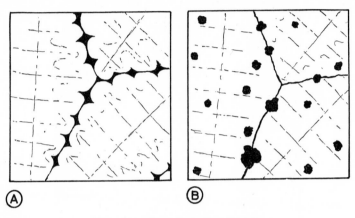

FIG. 4.11.—Porosity in Cast Metals.

Shrinkage cavities (A) tend to follow the shape of the dendrite arms and occur at the crystal boundaries, whilst gas porosity (B) is usually of irregular shape and occurs at almost any point in the structure.

Thus the crystal size of a pressure die-casting will be very small compared with that of a sand-casting. Whilst the latter cools relatively slowly, due to the insulating properties of the sand mould, the former solidifies very quickly, due to the contact of the molten metal with the metal mould. Similarly, thin sections, whether in sand- or die-casting, will lead to a relatively quicker rate of cooling, and consequently smaller crystals.

In a large ingot the crystal size may vary considerably from the outside surface to the centre (Fig. 4.12). This is due to the variation which exists in the temperature gradient as the ingot solidifies and heat is transferred from the metal to the mould. When metal first makes contact with the mould the latter is cold, and this has a chilling effect which results in the formation of small crystals at the surface of the ingot. As the mould warms up, its chilling effect is reduced, so that the formation of nuclei will be retarded as solidification proceeds. Thus crystals towards the centre of the ingot will be larger. In an intermediate position the rate of cooling is favourable to the formation of elongated columnar crystals, so that we are frequently able to distinguish three separate zones in the crystal structure of an ingot, as shown in Fig. 4.12.

A number of defects can occur in cast structures. The more important are dealt with below.

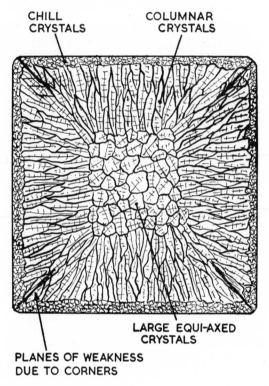

FIG. 4.12.—The Crystal Structure in a Section of a Large Ingot.

Blow-holes

4.20. These are caused by furnace gases which have dissolved in the metal during melting, or by chemical reactions which have taken place in the melt. Gas which has dissolved freely in the molten metal will be much less soluble in the solid metal. Therefore, as the metal solidifies, gas will be forced out of solution. Since dendrites have already formed, the bubbles of expelled gas become trapped by the dendrite arms and are prevented from rising to the surface. Most aluminium alloys and some of the copper alloys are susceptible to "gassing" of this type, caused mainly by hydrogen dissolved from the furnace atmosphere. The difficulty can be overcome only by making sure that there is no dissolved gas in the melt prior to casting.

4.21. Porosity may arise in steel which has not been completely de-oxidised before teeming. Although carbon present will tend to reduce most of the oxide there according to the following equation:

$$FeO + C = Fe + CO$$

a small amount of FeO can exist in equilibrium with the iron and carbon in solution. As the ingot begins to solidify, it is almost pure iron of which the initial dendrites are composed. This causes an increase in the concentration of carbon and the oxide, FeO, in the remaining molten metal, thus upsetting chemical equilibrium so that the above reaction will commence again. The bubbles of carbon monoxide formed are trapped by the growing dendrites, producing blow-holes. The formation of blow-holes of this type is prevented by adequate "killing" of the steel before it is cast—that is, by adding a sufficiency of a deoxidising agent such as ferromanganese. This removes residual FeO and prevents the FeO–C reaction from occurring during subsequent solidification. In some cases the FeO–C reaction is utilised as in the production of "rimmed" ingots (2.21—Part II).

4.22. Surface blow-holes may be caused in ingots by the decomposition of oily mould dressing, particularly when this collects in the fissures of badly cracked mould surfaces. The gas formed forces its way into the partially solid surface of the metal ingot, producing extensive porosity.

Shrinkage

4.30. When a metal solidifies its volume decreases, and if the mould is of a design such that isolated pockets of liquid remain when the outside surface of the casting is solid, shrinkage cavities will form. Hence the mould must be so designed that there is always a "head" of molten metal which solidifies last and can therefore "feed" into the main body of the casting as it solidifies and shrinks. Shrinkage is also responsible for the effect known as "piping" in cast ingots. Consider the ingot mould (Fig. 4.13A) filled instantaneously with molten steel. That metal which is adjacent to the mould surface solidifies almost immediately, and as it does so it shrinks. This causes the level of the remaining metal to fall slightly, and as further solidification takes place the process is repeated, the level of the remaining liquid falling still further. This sequence of events continues to be repeated until the metal is completely solid and a conical cavity or "pipe" remains in the top portion of the ingot. With an ingot shaped as shown it is likely that a secondary pipe would be formed due to the shrinkage of trapped molten metal when it solidifies. It is usually necessary to shape large ingots in the way shown in Fig. 4.13A, that is, small end upwards, so that the mould can be lifted from the solidified ingot. Therefore various methods of minimising the pipe must be used, one of the most important of which is to pour the metal into the mould fairly slowly, so that solidification almost keeps pace with pouring. In this way molten metal feeds into the pipe formed by the solidification and consequent shrinkage of the metal. Smaller ingots can be cast into moulds which taper in the opposite direction to that shown in Fig. 4.13A, i.e. large end upwards (Fig. 4.13B), since these can be trunnion-mounted to make ejection of the ingot possible.

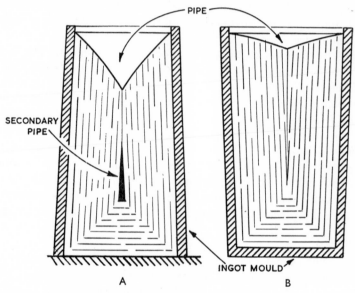

FIG. 4.13.—The Influence of the Shape of the Mould on the Extent of Piping in a Steel Ingot.

Segregation of Impurities

4.40. There is a tendency for dissolved impurities to remain in that portion of the metal which solidifies last. The actual mechanism of this type of solidification will be dealt with later (8.33), and it will be sufficient here to consider its results.

4.41. The dendrites which form first are of almost pure metal, and this will mean that the impurities become progressively more concentrated in the liquid which remains. Hence the metal which freezes last at the crystal boundaries contains the bulk of the impurities which were dissolved in the original molten metal. This local effect is known as *minor* segregation (Fig. 4.14A).

4.42. As the columnar crystals begin to grow inwards, they will push in front of them some of the impurities which were dissolved in the molten metal from which they themselves solidified. In this way there is a tendency for much of the impurities in the original melt to become concentrated in the central pipe. If a vertical section of an ingot is polished and etched, these impurities show as V-shaped markings in the area of the pipe (Fig. 4.14B). The effect is called *major* segregation.

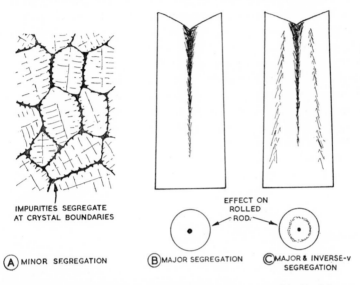

IMPURITIES SEGREGATE
AT CRYSTAL BOUNDARIES

EFFECT ON
ROLLED
ROD.

(A) MINOR SEGREGATION   (B) MAJOR SEGREGATION   (C) MAJOR & INVERSE-V
SEGREGATION

FIG. 4.14.—Types of Segregation Which May Be Encountered in Steel Ingots.

4.43. With very large ingots the temperature gradient may become very slight towards the end of the solidification process, and it is common for the band of metal which has become highly charged with impurities, just in front of the advancing columnar crystals, to solidify last. As has been mentioned already (2.61), the presence of impurities reduces the freezing point of a metal considerably, so that this band of impure metal, just in advance of the growing columnar crystals, has a much lower freezing point than the relatively pure molten metal at the centre. Since the temperature gradient is slight, this metal at the centre may begin to solidify in the form of large equi-axed crystals, so that the impure molten metal is trapped in an intermediate position. This impure metal therefore solidifies last, causing inverted V-shaped markings to appear in the etched section of such an ingot. It is known as "*inverse-vee*" segregation (Fig. 4.14C).

4.44. Of these three types of segregation, minor segregation is probably the most deleterious in its effect, since it will cause overall brittleness of the castings and, depending upon the nature of the impurity, make an ingot hot- or cold-short, that is, liable to crumble during hot- or cold-working processes.

4.50. From the foregoing remarks it will be evident that a casting, suffering as it may from so many different types of defect, is one of the least reliable and least predictable of metallurgical structures. In some cases we can detect the presence of blow-holes and other cavities by the use of X-rays, but other defects may manifest themselves only during subsequent service. Such difficulties are largely overcome when we apply some mechanical working process during which such defects, if serious, will become apparent by the splitting or crumbling of the material undergoing treatment. At the same time a mechanical working process will give a product of greater uniformity in so far as structure and mechanical properties are concerned. Thus, all other things being equal, a forging is likely to be more reliable in service than a casting. Sometimes, however, such factors as intricate shape and cost of production dictate the choice of a casting. We must then ensure that it is of the best possible quality.

Recent Examination Questions

1. Compare and contrast the following:
   (a) co-valent and metallic bonding;
   (b) atom packing in face-centred cubic and hexagonal close-packed structures.   (University of Strathclyde, B.Sc.(Eng.).)
2. Make sketches of the THREE most important types of spatial arrangement encountered in the lattice structures of metals. Mention metals in each group and explain fully how the mechanical properties of ductility and malleability can be affected by the type of lattice structure.
   (U.L.C.I., Paper No. C111–1–4.)
3. Produce clear sketches to show the arrangements of atoms in the following types of crystals: (a) body-centred cubic; (b) face-centred cubic; (c) close-packed hexagonal.
   Name one metal which crystallises in each of the above crystal forms.
   State the number of atoms per unit cell of each of the two cubic forms, showing your reasoning.
   Of the three crystal forms listed above, which possesses the closest atomic packing?   (The Polytechnic, London W.1., Diploma Course.)
4. Write an essay on "The Crystalline Nature of Metals".
   (Brunel College, Dip. Tech.)
5. What is meant by "allotropy"?
   Discuss the term with particular reference to iron, showing what connections allotropy has with the engineering uses of the metal.
   How could you demonstrate one of the manifestations of allotropy in iron?   (West Bromwich Technical College, A1 year.)

6. (a) What do you understand by the term dendritic crystallisation?

(b) Show how certain mechanical properties in castings can be explained by reference to this type of crystal structure.

(Institution of Production Engineers Associate Membership Examination, Part II, Group A.)

7. Describe the solidification of a pure metal. If a pure metal is cast into (a) a chill mould, and (b) a sand mould, explain giving reasons how the resultant microstructures would differ.

(Constantine College of Technology, Middlesbrough, H.N.D. Course.)

8. A liquid pure metal is cast from a high temperature into a chill mould. Describe the manner of solidification you would expect to occur in such a case. What defects may occur in the final ingot and how could these be overcome? (Brunel College, Dip. Tech.)

9. Show how the rate of cooling affects the macrostructure of an ingot. Describe three principal defects which may arise in ingots and indicate how they may be minimised.

(Derby and District College of Technology.)

10. State four serious defects which may occur in metal ingots.

How do these defects arise and what steps are taken to minimise or eradicate them?

(Institution of Production Engineers Associate Membership Examination, Part II, Group B, Production Metallurgy.)

11. Explain how the following defects may occur in a steel ingot: (a) piping. (b) segregation; and (c) blow-holes.

What steps may be taken to minimise these defects?

(C. & G., Paper No. 293/6.)

12. Explain, with the aid of sketches, how a "pipe" forms in a steel ingot during solidification.

Outline TWO methods which may be used to minimise this pipe.

(U.L.C.I., Paper No. B111–1–4.)

13. Discuss the formation and distribution of gas porosity in (i) cast steels, and (ii) cast aluminium alloys.

Show in each case how this form of defect may be minimised.

(West Bromwich Technical College, A1 year.)

14. Write an essay on "Gases in Metals, their solubility, sources, effects and removal". (Constantine College of Technology, H.N.D. Course.)

CHAPTER V

THE EFFECTS OF MECHANICAL DEFORMATION AND RECRYSTALLISATION

5.10. When studying the mechanical properties of metals in Chapter III, we learned that deformation can occur either by elastic movement or by plastic flow. In elastic deformation a limited distortion of the crystal lattice is produced, but the atoms do not move permanently from their ordered positions, and as soon as the stress is removed the distortion

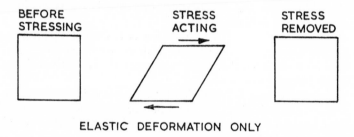

ELASTIC DEFORMATION ONLY

ELASTIC AND PLASTIC DEFORMATION

Fig. 5.1.—Diagrams Illustrating the Difference, in Action and Effect, of Deformation by Elastic and Plastic Means.

disappears. When metal is stressed beyond the elastic limit plastic deformation takes place and there must, clearly, be some movement of the atoms into new positions, since considerable permanent distortion can be produced. Our present task, then, is to consider ways in which this extensive rearrangement of atoms within the lattice structure can take place to give rise to this permanent deformation.

5.11. Plastic deformation proceeds in metals by a process known as "slip", that is, by one layer or plane of atoms gliding over another. Imagine a pile of pennies as representing a *single* metallic crystal. If we apply any force which has a horizontal component, i.e. any force not acting vertically, the pile of pennies will be sheared as one slides over another, provided that the horizontal component is sufficient to overcome friction between the pennies.

It is possible, under controlled conditions, to grow single crystals of some metals, and under application of adequate stress such crystals behave in a manner very similar to that of the pile of pennies. An "off-set" on one side of the crystal is balanced by similar offset on the other side (Fig. 5.2), and both off-sets lie on a single continuous plane called a "slip plane". Whilst in the case of the pile of pennies the force necessary to cause slip was that required to overcome friction between them, in a single metallic crystal the force necessary to cause slip is related to that required to overcome the resistance afforded by the metallic bond.

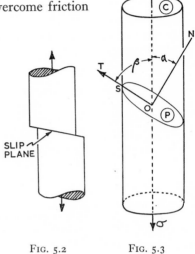

FIG. 5.2   FIG. 5.3

For a single crystal the shear component, T, of a tensile stress, σ, resolved along the direction of slip, OS, on a slip plane, P, can be calculated. ON is normal to P.

$$\text{Component of } \sigma \text{ along } OS = \sigma \cdot \cos \beta$$

$$\text{Area of projection on } C \text{ of unit area of } P = \cos \alpha$$

$$T = \sigma \cos \alpha \cdot \cos \beta$$

Experiments show that different single crystals of the same metal slip at different angles and different tensile stresses, but if the stresses are resolved along the slip plane, all crystals of the same metal slip at the same critical value of resolved shear stress.

5.12. The results of slip in a polycrystalline mass of metal may be observed by microscopical examination. The direction of the slip planes is indicated in such a piece of metal after deformation by the presence of slip bands which form on the surface of the metal. If a piece of soft iron is polished and etched and then squeezed in a vice so that the polished surface is not scratched, these slip bands can be seen on the surface.

Their formation is indicated in Fig. 5.4. Such slip bands are all parallel in a single crystal but differ in orientation from one crystal to another. It has been shown by electron microscopy that a single visible slip band consists of a group of roughly 10 steps on the surface, each about 40 atoms thick and approximately 400 atoms high. If the deformation has been excessive, the presence of slip bands is apparent even when a specimen is polished and etched *after* deformation. Heavily stressed parts of the crystal dissolve more quickly during etching, revealing these so-called "strain bands".

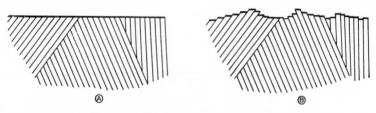

FIG. 5.4.—The Formation of Slip Bands.

(A) Indicates the surface of the specimen before straining, and (B) the surface after straining. The relative slipping along the crystallographic planes is apparent as ridges (visible under the microscope) on the surface of the metal.

5.13. All metals of similar crystal structure slip on the same crystallographic planes and in the same crystallographic directions. Slip occurs when the shear stress resolved along these planes reaches a certain value—the critical resolved shear stress. This is a property of the material and does not depend upon the structure. The process of slip is facilitated by the presence of the metallic bond, since there is no need to break direct bonds between individual atoms as there is in co-valent or electro-valent structures, nor is there the problem of repulsion of ions of like charge as there is for an electro-valently bonded crystal. When slip occurs in co-valent or electro-valent structures it does so with much greater difficulty than in metals. Some types of crystal are more amenable to deformation by slip than others. For example, in the face-centred cubic type of structure there are a number of different planes along which slip could conceivably take place, whereas with the hexagonal close-packed structure there is only one possibility, namely, slip along the basal plane of the hexagon. Thus, metals with a face-centred cubic structure, such as copper, aluminium and gold, are far more malleable and ductile than metals with a hexagonal close-packed lattice like zinc.

5.14. It will be evident to engineers that slip cannot be quite so simple a process as is outlined above. Whilst the gliding of one part of the crystal over the other is in agreement with microscopical observations, it does not attempt to explain the work-hardening which takes place during mechanical deformation, and which leads to a stage where further application of stress results in fracture. Nor does it explain the fact that the actual stress required to produce slip is far less than that which is theoretically necessary. (It is possible to calculate this value in terms of the strength of the metallic bond.) A perfect material in plastic deformation would presumably deform without limit at a constant yield stress as indicated in Fig. 5.5A, but in practice a stress–strain relationship of the type indicated in Fig. 5.5B exists.

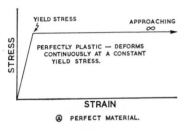
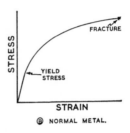

FIG. 5.5

5.15. Earlier theories which sought to explain slip by the simultaneous gliding of a complete block of atoms over another have now been discarded and the modern conception is that slip occurs step by step by the movement of so-called "dislocations" within the crystal.

If the reader has ever tried his hand at paper-hanging he will know that wrinkles have a habit of appearing in the paper when it is laid on the wall. Attempts to smooth out these wrinkles by pulling on the edge of the paper would undoubtedly prove fruitless, since the tensile force necessary to cause the *whole sheet of paper to slide* would be so great as to tear it. Instead, gentle coaxing of the wrinkles individually with the aid of a brush or cloth leads to their successful elimination, causing them to glide and "pass out of the system". The movement of dislocations on a slip plane in a metallic crystal probably follows a similar pattern.

Dislocations are faults or distorted regions (Fig. 5.6) in otherwise perfect crystals and the step-by-step movement of such faults explains why the force necessary to produce slip is of the order of 1 000 times less than the theoretical, assuming simultaneous slip over a whole plane.

The fundamental nature of a dislocation is illustrated in Fig. 5.7 (in which for the sake of clarity only the centres of atoms are indicated). Assume that a shearing stress, σ, has been applied to the crystal causing the top half, *APSE*, of the face, *ADHE*, to move inwards on a slip plane, $PQRS$, by an amount PP_1 (one atomic step). There has been no corresponding slip, however, at face $BCGF$. Consequently the top half of the crystal contains an extra half-plane as compared with the bottom

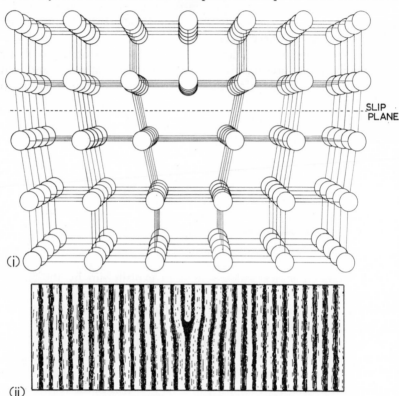

FIG. 5.6

(i) A "Ball-and-wire" Model of an Edge Dislocation.
(ii) Crystallographic Planes Containing an Edge Dislocation, as they Appear in Aluminium at a Magnification of Several Millions. (A sketch of the structure as revealed by a high-resolution electron microscope.)

part of length P_1Q. In Fig. 5.7 this is the half-plane $WXYZ$ and the line XY is known as an *edge dislocation*. It separates the slipped part $PXYS$ of the slip plane from the unslipped part $XQRY$. During slip this "front" XY moves to the right through the crystal.

The movement of such an edge dislocation under the action of a shearing force is illustrated in Fig. 5.8. Here an edge dislocation exists already (i), and the application of the shearing stress (ii) causes the dislocation to glide along the slip plane in the manner suggested. In this case it has been assumed that the dislocation glides out of the crystal completely, producing a slip step of one atom width at the edge of the

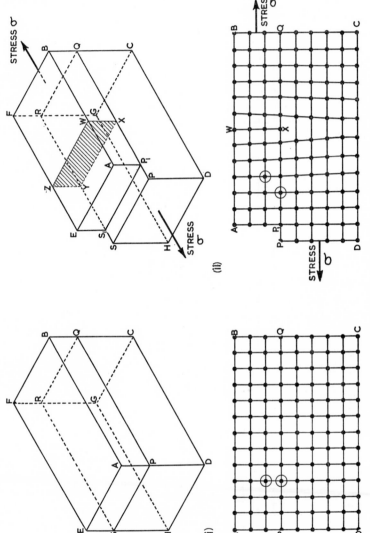

FIG. 5.7.—The Formation of an Edge Dislocation by the Application of Stress.

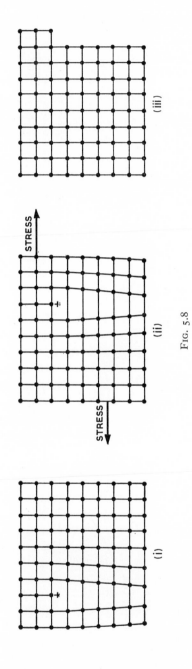

STRESS

STRESS

(i)

(ii)

(iii)

Fig. 5.8

crystal (iii). The dislocation can be moved through the crystal with relative ease, since only one plane is moving at a time and then only through a small distance.

Slip can also take place by the movement of "screw" dislocations. These differ from edge dislocations in that the direction of movement of the dislocation is *normal* to the direction of formation of the slip step. The mechanism of the process is indicated in Fig. 5.9. Here a shear stress, σ, had displaced P to P_1 and Q to Q_1 on one face of the crystal so that P_1Q_1YX has slipped relative to $PQYX$. In this case the screw dislocation XY separates the slipped part $PQYX$ from the unslipped part $XYRS$ of the slip plane $PQRS$. It will be noted that XY is moving in a direction which is normal to the direction in which slip is being produced.

It is also possible for slip to take place by the combination of a screw dislocation with an edge dislocation. A curved dislocation is thus evolved which moves across the slip plane (Fig. 5.10).

5.16. Until comparatively recently much of the foregoing commentary concerning dislocations was of a speculative nature. Whilst many of the properties of a metal could be best explained by postulating the existence of dislocations, no one had actually seen a dislocation in a metallic structure since dimensions approaching atomic size are involved. True, dislocations had been observed in structures of crystalline complex compounds and it was reasonable to suppose that similar dislocations occurred in the crystalline structures of metals. During the last few years the rapid development of electron microscope techniques has made it possible to observe crystallographic planes within metallic structures. With the aid of high-resolution electron microscopy, photographs of edge dislocations in metals such as aluminium have been produced (Fig. 5.6(ii)).

Since the stress required to produce dislocations is great, it is assumed that they are not generally initiated by stress application but that the majority of them are formed during the original solidification process. During any subsequent cold-working process, dislocations have a method of reproducing themselves from what are called Frank–Read sources, so that in the cold-worked metal the number of dislocations has greatly increased.

The relationship between the force necessary to initiate dislocations and to move those which already exist is notably demonstrated by the tensile properties of metallic "whiskers". These are hair-like single crystals grown under controlled conditions and generally having a single dislocation running along the central axis. If a tensile stress is applied along this axis the dislocation is unable to slip. As no other dislocations

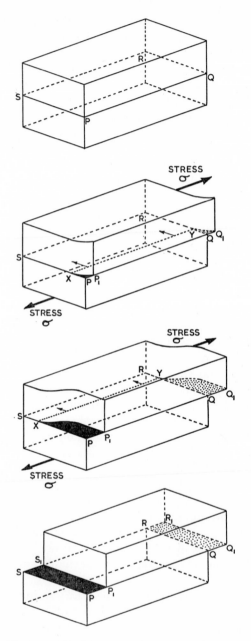

FIG. 5.9.—The Movement of a Screw Dislocation.

are available, the crystal cannot yield until a dislocation is initiated at E (Fig. 5.11). The stress then falls to that necessary to move the dislocation (Y), and dislocations then reproduce rapidly so that plastic flow proceeds.

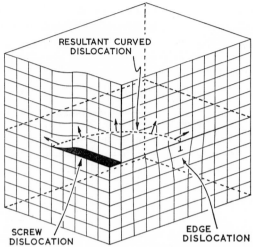

FIG. 5.10.—The Combination of a Screw Dislocation with an Edge Dislocation.

5.17. A brief mention of experimental evidence supporting the operation of slip in a polycrystalline mass of metal was made earlier in this chapter (5.12), but so far we have been dealing mainly with the

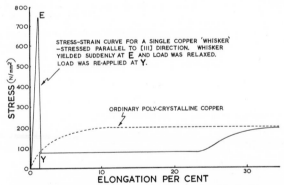

FIG. 5.11.—Differences in Tensile Properties Between a Single Copper "Whisker" and Ordinary Polycrystalline Copper.

methods by which slip can take place in individual crystals. If slip occurred completely over a whole plane, as was assumed above, the dislocation would pass right out of the crystal and there would be no reason for any change in mechanical properties. Yet, as we know, metals, which have solidified under ordinary conditions and which consist of a mass of individual crystals, become harder and stronger as the amount of cold work to which they are subjected increases, until a point is reached where they ultimately fracture.

Consequently, it is assumed that many dislocations remain in every crystal and that increase in hardness results from their mutual interference and the building up of a transcrystalline "traffic jam". Increase in hardness and strength is due to the greater difficulty in moving new dislocations against the jammed ones, whilst a "pile up" of jammed dislocations may propagate a fracture (Fig. 5.12). Dislocations will be

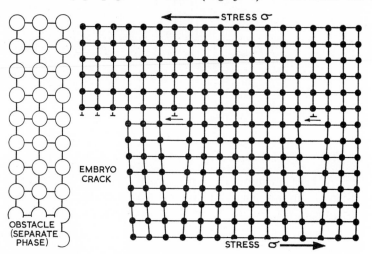

Fig. 5.12.—A Pile-up of Dislocations at Some Obstacle Leading to the Formation of a Crack Which Will Be Propagated if the Stress, σ, is increased.

unable to escape at crystal boundaries by forming steps because of the adherence of crystals in terms of the amorphous film (4.11). Moreover, the amorphous film will act as an effective barrier preventing dislocations from passing from one crystal to another, even supposing that the lattice structures of two neighbouring crystals were suitably aligned to make this possible. Since individual crystals develop at random in a polycrystalline mass, this will rarely be so.

Individual crystals in a polycrystalline metal deform by the same mechanisms as do single crystals, but since they are orientated at random, shear stress on slip planes will attain the critical value in different crystals at different loads (acting on the metal as a whole). Thus there is considerable difference between the deformation of a single crystal and that of a polycrystalline mass. The yield stress of the latter is generally higher and glide occurs almost simultaneously on several

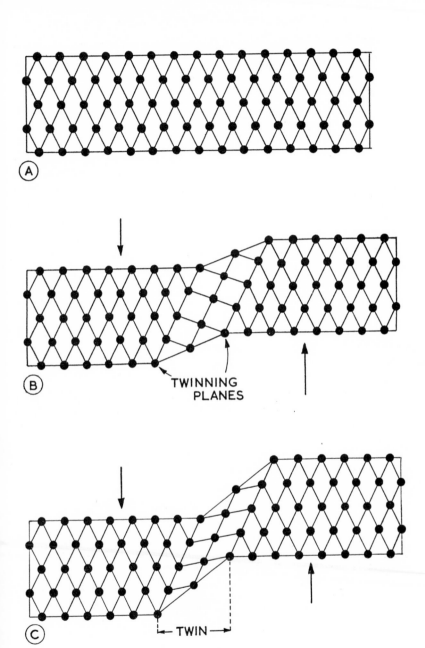

TWINNING
PLANES

TWIN

Fig. 5.13.—The Formation of "Mechanical Twins".

different systems of slip planes. Dislocations cannot escape from individual crystals and so they jam. A "pile up" on one slip plane will prevent dislocations in intersecting planes from moving, producing a situation not uncommon in the case of motor-traffic conditions at most crossroads near the South Coast on almost any Sunday in the summer.

For these reasons, strain-hardening proceeds much more rapidly in a polycrystalline metal than it does in a single crystal. A decrease in crystal size leads to strain-hardening taking place more quickly so that, though the material is stronger, it is less ductile. Strain-hardening, however, is supremely important. If a metal did not strain-harden but continued to slip, as would any perfectly plastic material (Fig. 5.5A), failure would be inevitable, immediately it were loaded above its yield point.

5.18. In addition to deformation by slip, some metals, notably zinc, tin and iron, deform by a process known as "twinning". The mechanism of this process is illustrated in Fig. 5.13. In deformation by slip all atoms in one block move the same distance, but in deformation by twinning, atoms in each successive plane within a block will move different distances, with the effect of altering the direction of the lattice so that each half of the crystal becomes a mirror image of the other half. It is thought that twinning also proceeds by the movement of dislocations.

The twins thus formed are called "mechanical twins" to distinguish them from the "annealing twins" which become apparent in copper alloys during an annealing operation which follows cold work. The mechanical twins formed in iron by a sudden shock are called "Neumann bands". Twin formation in a bar of tin can actually be heard as the bar is bent, and is referred to as "The Cry of Tin".

Energy of Mechanical Deformation

5.19. As deformation proceeds, the metal becomes harder and stronger, and, whether by slip or by twinning, a stage is reached when no more deformation can be produced. An increase in the applied force will then lead only to fracture. In this state, when tensile strength and hardness have reached a maximum and elongation a minimum, the material is said to be work-hardened. Thus we can summarise the effects of mechanical deformation briefly as follows—if sufficient stress is applied to a metal, slip (or twinning) will take place in individual crystals. As deformation proceeds, the capacity for slip decreases so that the force necessary to produce it must increase. A point is reached, coinciding with the maximum resistance to slip (the maximum strength and hardness), where no more deformation is possible and fracture will take place. The material must then be annealed if further cold work is to be carried out on it.

During a cold-working process approximately 90% of the mechanical energy employed is turned into heat, whilst the remaining 10% is stored inside the material as mechanical potential energy. The bulk of this stored energy (9% of that originally employed) is that associated with the number of dislocations present. These have energy because they distort the lattice and cause atoms to occupy positions of higher-than-minimum energy. The remaining stored energy (1% of the energy originally employed) exists as locked-up residual stresses arising from elastic strains internally balanced.

The increased-energy state of a cold-worked metal makes it chemically more active and consequently less resistant to corrosion. This is particularly true of micro-stresses acting at grain boundaries and leading to increased intercrystalline corrosion by causing the grain boundaries to corrode more quickly than the rest of the material. This stored potential energy is also the main driving force of recrystallisation.

Annealing and Recrystallisation

5.20. A cold-worked metal is in a state of considerable mechanical stress, resulting from elastic strains internally balanced. These elastic strains are largely due to unhomogeneous deformation having taken place during cold-working. If the metal is heated to a sufficiently high temperature these strains will be removed; at the same time the tensile strength and hardness of the metal will fall to approximately their original value and the capacity for cold-work return. This form of heat-treatment is known as annealing, and is made use of when the metal is required for use in a soft but tough state or, alternatively, when it is to undergo further cold deformation. Annealing takes place in three stages as follows:

5.30. Stage I—The Relief of Stress. This occurs at relatively low temperatures at which atoms, none the less, are able to move to positions nearer to equilibrium in the crystal lattice. Such small movements can reduce internal mechanical stress without, however, producing any visible alteration in the distorted shape of the cold-worked crystals. Moreover, hardness and tensile strength will remain at the high value produced by cold-work, and may even increase as shown in the curve for cold-worked 70–30 brass (Fig. 5.14). It is found that a controlled low-temperature anneal at, say, 200° C applied to hard-drawn 70–30 brass tube will effectively reduce its tendency to "season-crack" (16.33) without reducing strength or hardness.

5.40. Stage II—Recrystallisation. As mentioned above, a low-temperature anneal to relieve internal stress may sometimes be used,

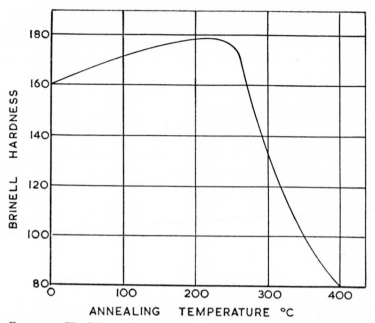

FIG. 5.14.—The Relationship between Hardness and Annealing Temperature (Cartridge Brass).

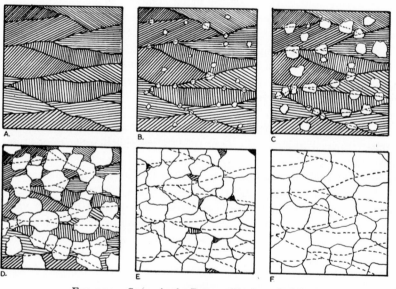

FIG. 5.15.—Stages in the Recrystallisation of a Metal.
(A) represents the metal in its cold-rolled state. At (B) recrystallisation has commenced with the formation of new crystal nuclei. These grow at the expense of the old crystals until at (F) recrystallisation is complete.

but generally annealing involves a definite and observable alteration in the crystal structure of the metal. If the annealing temperature is increased a point is reached when new crystals begin to grow from nuclei produced in the deformed metal. These nuclei are formed at points of

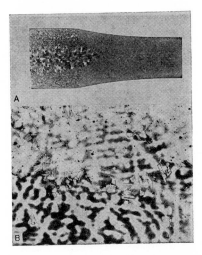

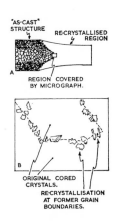

PLATE 5.1.—A Cast 70–30 Brass Slab (38 mm thick) Was Being Cold-rolled When the Mill Was Stopped. The Partly Rolled Slab Was then Annealed at 600° C.

The photomicrograph (A) reveals two regions: (1) the coarse-grained region which suffered no cold-work and consequently did not recrystallise on annealing (hence it shows the original coarse "as-cast" structure); (2) the heavily cold-worked part of the slab which has recrystallised on annealing so that the crystals are much too small to be visible in the photomicrograph.

The photomicrograph (× 100) (B) is taken from a region which suffered very little cold work. On annealing, recrystallisation has just begun at points on the original crystal boundaries where, presumably, a pile-up of dislocations had just commenced. Small twinned crystals have been formed at these points, whilst the remainder of the structure is still the original "as-cast".

high energy, such as crystal boundaries and dislocation entanglements. The crystals so formed are at first small, but grow gradually until they absorb the entire distorted structure produced originally by cold-work (Fig. 5.15). The new crystals are equi-axed in form, that is, they do not show any directional elongation, as did the distorted cold-worked crystals which they replace. They are, in fact, of equal axes.

This phenomenon is known as recrystallisation, and it is the principal method employed, in conjunction with cold-work, of course, to produce a fine-grained structure in non-ferrous metals and alloys. Only in rare cases—notably in steels and aluminium bronze, where certain structural changes take place in the solid state—is it possible to refine the grain size by heat-treatment only.

5.41. The minimum temperature at which recrystallisation will take place is called the recrystallisation temperature. This temperature is lowest for pure metals, and is generally raised by the presence of other elements. Thus, pure copper recrystallises at 200° C, whilst the addition of 0·5% arsenic will raise the recrystallisation temperature to well above 500° C; a useful feature when copper is to be used at high temperatures and must still retain its mechanical strength, as, for example, in railway-locomotive boiler tubes and fire-boxes. Similarly, whilst cold-worked commercial-grade aluminium recrystallises in the region of 150° C, that of "six nines" purity (99·999 9% pure) appears to recrystallise below room temperature and, consequently, does not cold work.

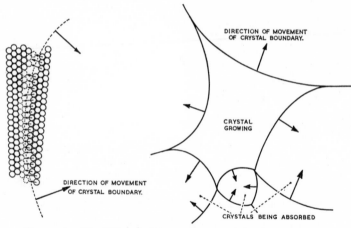

FIG. 5.16.—The "Cannibalising" of Small Crystals by Larger Ones.

Other metals, for example, lead and tin, recrystallise below room temperature, so that it is virtually impossible to cold-work them, since they recrystallise even whilst mechanical work is taking place. Thus, they can never be work-hardened at normal temperatures of operation. This phenomenon is essentially the basis of hot-working, as we shall see later.

5.42. The recrystallisation temperature is dependent largely on the degree of cold-work which the material has previously undergone, and severe cold-work will generally result in a lower recrystallisation temperature. It is therefore not possible accurately to specify a recrystallisation temperature for any metal. For most metals, however, the recrystallisation temperature is between one-third and one-half of the melting-point temperature, measured on the *absolute* scale of temperature. Thus the mobilities of all metallic atoms are approximately equal at the same fraction of their melting points (K).

5.50. Stage III—Grain Growth. If the annealing temperature is above the recrystallisation temperature of the metal, the newly formed crystals will continue to grow by absorbing each other cannibal-fashion, until the structure is relatively coarse-grained, as shown in Fig. 5.15. Since the crystal boundaries have higher energies than the interiors of the crystals, a polycrystalline mass will reduce its energy if some of the grain boundaries disappear. Consequently, at temperatures above that of recrystallisation large crystals grow by absorbing small ones. As

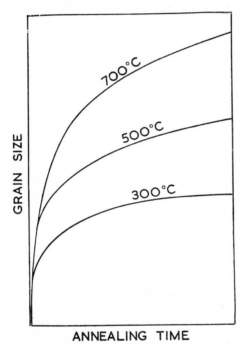

FIG. 5.17.—The Relationship between Grain Size and Annealing Time.

indicated in Fig. 5.16, a crystal boundary tends to move towards its centre of curvature in order to shorten its length. To facilitate this, atoms move across the boundary to positions of greater stability where they will be surrounded by more neighbours in the concave crystal face of the growing crystal. Thus the large grow larger and the small grow smaller. The same type of mechanism (in this case surface tension along a liquid film) causes small bubbles to be absorbed by the larger ones in the froth in a glass of beer. The extent of grain growth is dependent to a large degree on the following factors:

(*a*) The annealing temperature—as the temperature increases, grain size increases (Fig. 5.17).

(*b*) The duration of annealing—grains grow rapidly at first and then more slowly (Fig. 5.17).

(*c*) The degree of previous cold-work. In general, heavy deformation will lead to the formation of a large number of regions of high energy within the crystals. These will give rise to the production of many nuclei on recrystallisation and consequently the grain size will be small. Conversely, light deformation will give rise to few nuclei and the resulting grain size will be large (Fig. 5.18).

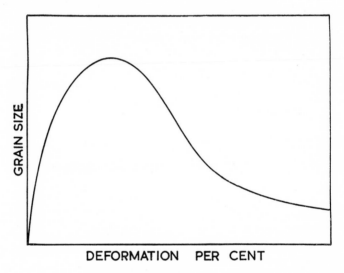

FIG. 5.18.—The Relationship between Grain Size and the Degree of Original Deformation.

(*d*) The influence of insoluble additions to the metal or alloy. An example is the use of thoria (thorium oxide) in tungsten lamp filaments. Here the films of thoria prevent excessive brittleness, which would otherwise result from grain growth in lamp filaments maintained at high temperatures over long periods. Similarly, the presence of vanadium carbide particles impedes the grain growth of austenite during the heat-treatment of high-speed and other steels.

(*e*) The addition of certain alloying elements, of which the most widely used is nickel. This element limits grain growth during annealing and other heat-treatment processes in steels and some non-ferrous alloys.

5.51. Of these factors the one requiring the most accurate control, in normal conditions, is the annealing temperature. Some alloys, particularly the brasses, are exceptionally sensitive to variations in the annealing temperature and an error of 100 °C, on the high side, may increase crystal size five-fold for a given annealing time. This coarse grain will, in turn, lead to loss in ductility and the formation of a rough, rumpled surface (known as "orange peel") during a subsequent forming operation. When coarse grain has been thus produced in material of finished size, it cannot be refined, as is possible with steel, by heat-treatment. The only remedy would be to remelt the brass and go through the complete stages of rolling and annealing, until once more the finished size was reached.

RECENT EXAMINATION QUESTIONS

1. Describe the solidification of a pure metal showing how the lattice structure and grain boundaries are formed. Show how elastic and plastic deformation depend on the metallic structure.
   (University of Strathclyde, B.Sc.(Eng.).)

2. What visual evidence is there to support the view that permanent deformation in metals takes place by a process of "slip"?
   Outline the current theory which seeks to explain the essential nature of slip. (West Bromwich Technical College, A1 year.)

3. (a) Discuss the effect that various microstructures have upon the cold-working capacity of materials.
   (b) Sketch a simple mechanism to illustrate slip in a single crystal and from this derive the theory behind work-hardening.
   (Aston Technical College, H.N.D. Course.)

4. Explain how the dislocation theory accounts for the plastic deformation of metals under comparatively low stresses.
   (University of Strathclyde, B.Sc.(Eng.).)

5. Outline the theory which seeks to describe the simple linear propagation of a dislocation in a pure metal.
   What effect does the presence of impurity atoms have on the propagation of a dislocation and hence upon the overall properties of a metal?
   (U.L.C.I., Paper No. C111–1–4.)

6. (a) Write a brief account of the dislocation theory of plastic deformation.
   (b) How is the theory used to explain the mechanism of strain-hardening? (Derby and District College of Technology.)

7. (*a*) Give a concise account of the physical changes that occur in a pure metal when it is subject to plastic deformation.

(*b*) Show how these changes can be used to produce certain desirable properties in metals, illustrating your answer with special reference to the deep drawing of commercially pure copper.

(Institution of Production Engineers Associate Membership Examination, Part II, Group A.)

8. (*a*) Compare the effects of hot-working and cold-working on the microstructures, the mechanical properties and the physical properties of metal.

(*b*) Explain clearly the meaning of the terms "strain-hardening" and "recrystallisation".

(Institution of Production Engineers Associate Membership Examination, Part II, Group B, Production Metallurgy.)

9. Discuss the effects of (*a*) cold-working, (*b*) subsequent heat-treatment, on the structure and mechanical properties of a pure metal.

Explain the terms *stress relief, recrystallisation* and *grain growth*.

(U.L.C.I., Paper No. C111–1–4.)

10. Describe the changes which occur during the annealing of a cold-worked metal.

Explain why the rate of working is important during a hot-working process. (Rutherford College of Technology, Newcastle upon Tyne.)

11. What is the effect of the following operations on the microstructure and mechanical properties of a ductile single-phase alloy initially in the cast and annealed condition?

(*a*) Increasing amounts of cold-work.

(*b*) Annealing at successively higher temperatures after cold work.

Discuss briefly the mechanism of strain-hardening and state the factors that influence recrystallisation.

(Constantine College of Technology, Middlesbrough.)

12. Discuss the relationship between the degree of cold-work and the recrystallisation temperature for a solid solution alloy. How is this relationship affected as a result of hot working the alloy?

(Derby and District College of Technology, H.N.D. Course.)

13. A heavily cold-worked bar of copper is placed with one end in a furnace at 900° C while the other end is maintained at room temperature. Assuming there is a uniform temperature gradient along the whole length, describe the structures you would expect to find along the bar.

(Brunel College, Dip. Tech.)

14. A nail is driven through a piece of sheet tin which has previously been cold-rolled and annealed under conditions which resulted in a small crystal size being produced. After puncture the specimen is annealed just above its recrystallisation temperature.

Sketch and describe the type of crystal structure which would now be found in the region of the nail hole and account for the results you describe.   (U.L.C.I., Paper No. 237–2–8.)

CHAPTER VI

THE INDUSTRIAL SHAPING OF METALS*

6.10. Metals and alloys may be shaped into something approaching the final required form by one of the following operations:

- (*a*) casting into either a sand or a metal mould;
- (*b*) casting as an ingot followed by a hot-working process;
- (*c*) casting as an ingot followed by a cold-working process;
- (*d*) sintering from a powdered metal.

There are, it is true, other processes, such as electro-deposition and the condensation of metal vapours, but these operations are usually confined to the surface treatment of metallic components rather than to their actual shaping. Electro-deposition is sometimes used to build up parts which have become badly worn. Any one of the shaping methods mentioned above may be followed by some form of metal-cutting process, referred to generally as "machining". The study of this wide field of operations is the province of the production engineer, but a brief treatment of the metallurgical aspects of machinability is given in Chapter XX.

Sand-casting and Die-casting

6.20. Large numbers of components are produced by casting the metal or alloy into a shaped sand mould (Fig. 6.1), but if numbers warrant it and the process is technically suitable, a far superior product is obtained by casting into a metal mould.

In permanent-mould casting (formerly known as *gravity* die-casting) the metal is allowed to run into the metal mould under gravity, whilst in die-casting (previously called *pressure* die-casting) the charge is forced in under considerable pressure as illustrated in Fig. 6.2. The processes are confined mainly to zinc- or aluminium-base alloys, and the product is metallurgically superior to a sand-casting in that the internal structure is more uniform and the grain much finer because of the rapid cooling rates which obtain due to the metal mould. Moreover, output rates are much higher when using a permanent mould than they are

---

* Although the processes mentioned in this chapter are dealt with in detail in Part II, an introduction to the more important methods of metal shaping was considered to be desirable here.

when using sand moulds. Greater dimensional accuracy and a better surface finish are also obtained by die-casting. However, some alloys which can be sand-cast cannot be die-cast because of their high shrinkage

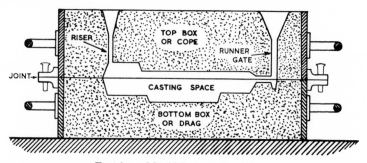

FIG. 6.1.—Mould for a Sand-casting.

A wooden pattern is used, the top box being raised so that the pattern can be removed.

coefficients during solidification. Such alloys would inevitably crack, due to their high contraction during solidification.

It is, of course, more economical to use a sand-casting process where

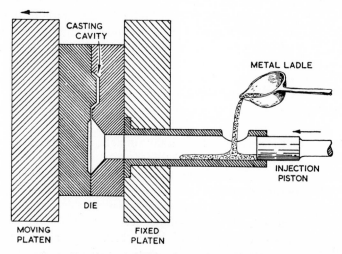

FIG. 6.2.—A Pressure Die-casting Machine.

The molten metal is forced into the die by means of the piston. When the casting is solid the die is opened and the casting moves away with the moving platen, from which it is ejected by "pins" passing through the platen.

the number of castings involved is small; whilst many intricate shapes could only be produced by sand-casting because of the necessity of destroying the core in order to remove it.

6.21. Pipes and similar shapes can be produced by "centrifugal" casting. In this process a permanent metallic cylindrical mould, without any cores, is spun at high speed and liquid metal poured into it. Centrifugal force flings the metal to the surface of the mould, thus producing a hollow cylinder of uniform wall thickness. The product has a uniformly fine-grained outer surface, and is considered superior to a similar shape which has been sand-cast. Cast-iron pipes for use as water and gas mains are made by the centrifugal method, whilst cast-iron piston rings are cut from centrifugally cast hollow cylinders.

6.22. A sand-casting is, in many respects, one of the less reliable of metallurgical structures because of the many variable influences which come into play, not only during the melting of the metal but also during its casting and subsequent solidification in the mould. Blowholes, shrinkage cavities and oxidation defects are all likely to be present in a sand-casting due to poor casting technique. Moreover, the inherent characteristics of a casting, particularly in respect of coarse grain and segregation, are such that its mechanical properties are inferior to those of a wrought structure.

If, therefore, mechanical strength and toughness are prime factors, a casting should properly be used only if the component is of such intricate shape that it cannot be produced by any other means, or if the alloy used is not amenable to any working process. In commerce the cost of manufacture must always be considered, however, so that where only a few components are required, the choice of operation will frequently fall upon sand-casting, nothwithstanding its shortcomings.

Hot-working Processes

6.30. Any increase in the temperature of a metal leads to an increase in atomic spacings so that the bond strength will decrease slightly. Moreover, a dislocation does not produce as much distortion and can move more easily through the crystal. Consequently the yield strength falls as the temperature rises.

However, hot-working operations invariably take place above the recrystallisation temperature of a metal or alloy. The importance of this will be at once apparent. Deformation and recrystallisation will be taking place simultaneously, so that a considerable speeding-up of the process is possible, with no tedious inter-stage anneals as are necessary in cold-deformation processes.

Moreover, with most alloys, malleability and plasticity are considerably increased at high temperatures, so that far less power is needed to produce deformation. With some alloys it is essential to use a hot-working process, since they are hard or brittle when cold, owing to the presence of a hard micro-constituent which is absorbed at the hot-working temperature.

The main hot-working processes are dealt with below.

6.31. Hot-rolling. The *rolling mill* was adapted by Henry Cort

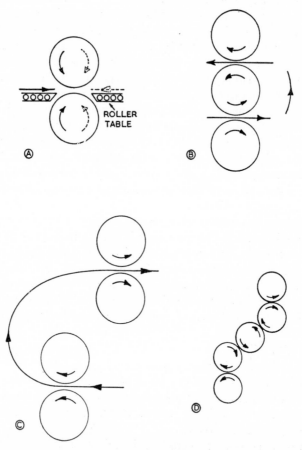

FIG. 6.3.—Roll Arrangements in Various Types of Rolling Mill.
(A) The "two-high" reversing mill.
(B) A "three-high" mill, in which reversing is not necessary.
(C) The "double duo" type of mill.
(D) Arrangement of gears connecting the rolls in the "double duo" mill.

in 1783 for the manufacture of wrought-iron bar. It was subsequently used for the "finishing" of Bessemer steel in 1856 and adapted as a "three-high" mill in 1857. Ramsbottom introduced the forerunner of the modern reversing mill at Crewe in 1866. Such progress in the development of rolling-mill technique was necessary in order to keep pace with the increasing amount of steel being made by new mass-production processes in the second half of the nineteenth century.

Hot-rolling is universally applied to the "breaking-down" of large steel ingots to sections, strip, sheet and rod of various sizes. In fact, the only conditions under which cold-work is applied to steel are when the section is too small to retain its heat, or when a superior finish is required in the product.

A steel-rolling shop consists of a powerful "two-high" reversing mill, to break down the white-hot ingots, followed by trains of rolls which will be either plain or grooved according to the type of product being manufactured. Hot-rolling is similarly applied to most of the non-ferrous alloys in the initial breaking-down stages, but the finishing operations are more likely to involve cold-work.

6.32. Forging. The simplest and most ancient working process known to metallurgical industry is essentially that which was employed by Tubal Cain (Genesis IV, 22), and which is still employed by a skilled blacksmith. Although power-driven hammers are now used, skill is still necessary on the part of the smith, since he works with comparatively simple shaping tools known as "swages" (Fig. 6.4).

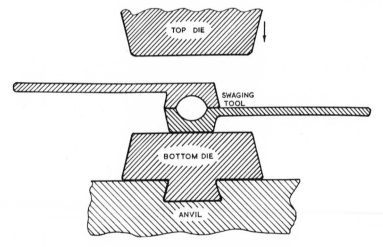

FIG. 6.4.—The Forging Process.

Wrought iron is always associated with the blacksmith, but many other ferrous and non-ferrous alloys can be shaped by both hand and mechanical forging processes. During forging, the coarse "as-cast" structure is broken down and replaced by one which is of relatively fine grain. At the same time impurities are redistributed in a more or less fibrous form. Therefore it is more satisfactory, all other things being equal, to forge a component than to cast it to shape.

6.33. Drop-forging. If a large number of identical forged components are required, then it is convenient to make them by a drop-

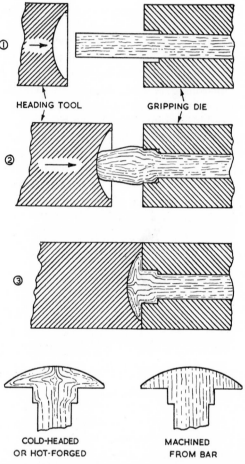

Fig. 6.5.—The Formation of a Bolt Head by a Hot-forging Process.

forging process. In this process a shaped die is used, one half being attached to the hammer and the other half to the anvil. With more complicated shapes a series of dies may be used.

The hammer, working between two vertical guides, is mechanically lifted some distance above the anvil and allowed to fall under its own weight on to the metal to be forged. This consists of a hot bar of metal held on the anvil by tongs. As the hammer falls it forges the metal between the two halves of the die. A modification of drop-forging employs either mechanical or steam power to force the hammer downwards, thus increasing the power of the blow.

6.34. Heading. "Heading" or "up-set" forging is employed extensively for the manufacture of bolts, rivets and other components, where an increase in diameter is necessary without loss of strength and shock-resistance. The stock bar is heated for a portion of its length and then forged, as shown in Fig. 6.5, in a machine which also parts off the component. It will be obvious that a bolt head forged in this way will be much stronger than one which has been machined from a bar, since forging does not cut into the fibrous structure of the material and introduce planes of weakness into the finished bolt head.

6.35. Hot-pressing. Hot-pressing is a development of the drop-forging process, but is generally applied in the manufacture of more simple shapes. The hammer of drop-forging is replaced by a hydraulically driven ram, so that, instead of receiving a rapid succession of hammer blows, the metal is gradually squeezed by the static pressure of the ram. This downwards thrust is sometimes as great as 100 MN.

The main advantage of hot-pressing over drop-forging is that working is no longer confined to the surface layers, as it is with drop-forging, but is transmitted uniformly to the interior of the metal being shaped. This is particularly important when forging very large components, such as marine propeller shafts, which would otherwise suffer from having a non-uniform internal structure.

6.36. Extrusion. The extrusion process is now used for shaping a variety of ferrous and non-ferrous metals and alloys. Its most important feature is that we are able to force the metal through a die, and, in a single process from the cast billet, to obtain quite complicated sections of tolerably accurate dimensions. The metal billet is heated to the required extrusion temperature (700–800° C for brasses; 350–500° C for aluminium alloys) and placed in the container of the extrusion press (Fig. 6.6). The ram is then driven hydraulically with sufficient pressure to force the metal through a hard alloy-steel die. The solid metal section issues from the die in a manner similar to the flow of toothpaste from its tube.

Using this process, a wide variety of sections can be produced, including round rod, hexagonal brass rod (for parting off as nuts), brass curtain rail, small-diameter rod (for drawing still further to wire), tubes and many stress-bearing sections in aluminium alloys (mainly for aircraft construction).

The most serious defect from which extruded sections suffer is called the "extrusion defect" or "back-end defect", because of its occurrence in the last part of the section extruded. It is caused by some of the surface scale of the original billet being drawn into the core of the section by the turbulent flow of metal in the container of the extrusion press. A number of methods are effective in reducing it, including the "indirect" process of extrusion, where, in effect, the die is pushed into

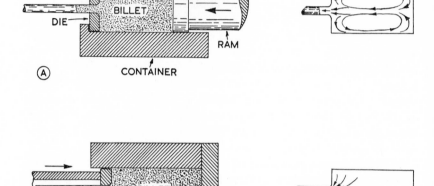

FIG. 6.6.—Extrusion by (A) the "Direct" and (B) the "Indirect" Process. The "flow" of metal in the container is indicated in each case.

the billet, instead of, as in the ordinary direct process, the billet being forced through the die. In the latter process there will be relative movement between the billet and the walls of the container, setting up a turbulent flow of metal due to frictional forces, and so drawing surface scale into the main stream of extruded metal. Since relative movement between billet and container is reduced to a minimum in the indirect process, frictional forces are much smaller. Consequently, not only is less turbulence produced in the surface layers of the metal being extruded but also the power requirements are less than in the direct process where much more energy is lost in overcoming friction.

Cold-working Processes

6.40. Cold-working from the ingot to the finished product, with, of course, the necessary intermediate annealing stages, is applied only in the case of a few alloys. These include both alloys which are very malleable in the cold and, on the other hand, those which become brittle when heated.

Cold-working is more often applied in the finishing stages of production. Then its functions are:

(a) to enable accurate dimensions to be attained in the finished product;

(b) to obtain a clean, smooth finish;

(c) by adjusting the amount of cold-work in the final operation after annealing, to obtain the required degree of hardness, or "temper", in alloys which cannot be hardened by heat-treatment.

Raising the temperature of an alloy generally increases its malleability, but reduces its ductility because of the attendant reduction in yield strength. Thus in hot-working processes we are always *pushing* the alloy into shape, whilst in cold-working operations we frequently make use of the high ductility of some alloys when cold, by *pulling* them into their required shapes. Therefore, processes involving the pulling or "drawing" of metal through a die are always cold-working operations.

There are many cold-working processes, but the principal ones used in metallurgical industries are discussed below.

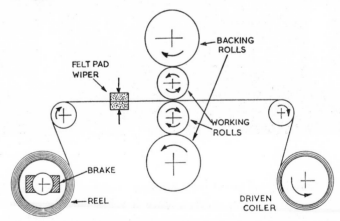

Fig. 6.7.—The Type of Rolling Mill Used in the Production of Thin Strip and Foil.

It is a "four-high" mill, the large-diameter backing rolls supporting the working rolls.

6.41. Cold-rolling. Cold-rolling is applied during the finishing stages of production of both strip and section and also in the production of very thin materials such as foil. The type of mill used in the production of the latter is shown in Fig. 6.7. In most other cases cold-rolling mills are similar in design to those used for hot-rolling.

The production of mirror-finished metal foil is carried out in rolls enclosed in an "air-conditioned" cubicle, and the rolls themselves are polished frequently with clean cotton wool. Only by working in perfectly clean surroundings, with highly polished rolls, can really high-grade foil be obtained.

In addition to quality of finish, the objects of cold-rolling are accuracy of dimensions and the adjustment of the correct temper in the material.

[*Courtesy of Messrs. W. H. A. Robertson & Co. Ltd., Bedford.*

PLATE 6.1.—A Four-high Non-reversing Mill for the Production of Stainless-steel Sheets.

6.42. Drawing of Solid and Hollow Sections. Drawing is exclusively a cold-working process, since it relies entirely on the high ductility of the material being drawn. Both solid sections and tubes are produced by drawing through dies, and all wire is manufactured by this process. In the manufacture of tubes the bore may be maintained by the use of a mandrel, as shown in Fig. 6.8. Rods and tubes are drawn at a long draw-bench on which a power-driven "dog" pulls the material through the die. Wire, and such material as can be coiled, is pulled through the die by winding it on to a rotating drum or "block". In each case the material is lubricated with oil or soap before it enters the die.

Drawing dies (Fig. 6.9) are made, for different purposes: from high-carbon steel; from tungsten and molybdenum steels; from tungsten carbide (a tungsten carbide "pellet" enclosed in a steel case); and, for very fine gauge copper wire, from diamond.

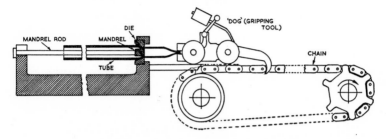

FIG. 6.8.—A Draw-bench for the Cold-drawing of Tubes.

6.43. Cold-pressing and Deep-drawing. These processes are so closely allied to each other that it is very difficult to define each separately. The operations range from making a suitable pressing in one stage to cupping followed by a number of re-drawings, as shown in Fig. 6.10. In each case the components are produced from sheet stock, and range from mild-steel motor-car bodies to cartridge cases (70–30 brass), bullet envelopes (cupro-nickel) and aluminium milk churns.

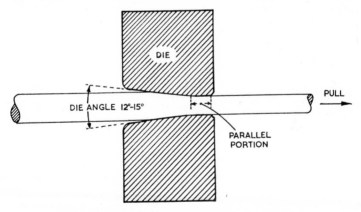

FIG. 6.9.—The Design of a Typical Wire-drawing Die.

Deep-drawing demands very high ductility in the sheet stock, and only a limited range of alloys are, therefore, available for the process. The best known are 70–30 brass, cupro-nickel, pure copper, pure aluminium and some of its alloys, and some of the high-nickel alloys. Mild steel can be pressed and, to a limited extent, deep-drawn. It is used in the manufacture of innumerable motor-car and cycle parts by these methods.

Typical stages in a deep-drawing process are shown in Fig. 6.10, but it should be noted that wall-thinning may or may not take place in such a process. If wall-thinning is necessary, then one of the more ductile alloys must be used. Though Fig. 6.10 shows the processes of shearing and primary cupping as being effected in different machines, usually a combination tool is used, so that both processes take place in one machine.

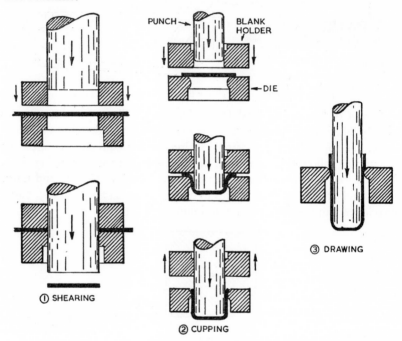

FIG. 6.10.—Stages in the Cupping and Deep-drawing of a Component.

Cold-pressing is very widely used, and alloys which are not quite sufficiently ductile for deep-drawing are generally suitable for shaping by simple press-work.

6.44. Coining. Coining, or embossing, is really a cold-forging process and, as its name implies, it is used for the production of coinage, medallions and the like. It has also been used on an experimental scale for producing engineering components to exact dimensions. A malleable alloy is essential for this type of process if excessive wear on the dies is to be avoided.

6.45. Spinning. Spinning is a relatively simple process in which a circular blank of metal is fixed to the spinning chuck of a lathe. As the blank rotates, it is forced up into shape by means of hand-operated tools of blunt steel or hardwood, supported on a hand-rest. The function of the hand-tool is to press the metal into contact with a former of the desired shape. This former is also fixed to the rotating chuck.

The operation is in some respects similar to that of a potter's wheel in which the lump of clay is replaced by a flat disc of metal. Large reflectors, aluminium teapots and hot-water bottles, and other domestic hollow-ware are frequently produced by spinning.

Sintering from a Powder

6.50. This method of producing metallic structures has become increasingly important in recent years, and is particularly useful when there is a big difference in the melting points of the metals to be alloyed, or when a metal has an extremely high melting point. As an example, products containing a high proportion of tungsten (m.pt 3 410° C) are usually formed by sintering, since this metal is difficult and expensive to melt on a commercial scale.

6.51. The metal to be sintered is obtained as a fine powder, either by grinding; by volatilisation and condensation; or by reduction of its powdered oxide. Any necessary mixing is carried out, and the mixed powdered metals are then placed in a hardened steel die and compressed. The pressures used vary with the metals to be sintered, but are usually between 70 and 700 N/mm^2. The brittle compressed mass is then heated in a small electric furnace to a temperature which will cause sintering to take place and produce a mechanically satisfactory product. Sintering will often occur below the melting point of either metal, though frequently it is above the melting point of one of them.

6.52. Tungsten is moulded in this way, and the resulting sintered rod is drawn down to wire for the manufacture of electric-lamp filaments. The demand by engineers for tools which would be superior to high-speed steels for cutting and machining metals led to the development of sintered tungsten carbide products. Tools of this type are made by mixing tungsten carbide powder with up to 13% of powdered cobalt; the mixture is pressed into blocks and heated in a hydrogen atmosphere to a temperature of 700–1 000° C. After this treatment the blocks are still soft enough to be cut and ground to the required shape. They are then heated in hydrogen to a temperature of 1 400–1 500° C, when sintering takes place, producing a material harder than high-speed steel. The function of cobalt is to provide a tough, shock-resistant, bonding material between the tungsten carbide particles. In addition to their use in cutting tools, cemented carbides of this type are employed in dies, other wear-resistant parts, percussion drills and armour-piercing projectiles. When making tools for machining steels, particularly with fine cuts or at high speeds, up to 20% titanium carbide may be incorporated.

6.53. Sintering is also used to produce an even distribution of some insoluble constituent in a metallic structure. In so-called oil-less bearings powdered graphite is compressed with powdered copper and tin to make a bronze bearing which is, as a result, impregnated with graphite. This is then sintered at about 800° C and subsequently quenched in lubricating oil. Such treatment results in a structure resembling a metallic sponge, which, when saturated with lubricating oil, produces a self-oiling bearing. In some cases the amount of oil held in the bearing is sufficient to last the lifetime of the machine. These self-lubricating bearings are used in the automobile industry, but find particular application in many domestic articles, such as vacuum cleaners, electric clocks and washing-machines; in all of which, long service with a minimum of maintenance is necessary.

Recent Examination Questions

1. Discuss the factors involved when making a choice between the available methods of casting a solid article.   (The Polytechnic, London W.1.)
2. What are the advantages and limitations of sand-casting and die-casting? Compare and contrast these processes, and the quality and properties of the castings produced.
                    (The Polytechnic, London W.1., Diploma Course.)
3. Which metals are normally pressure die-cast? Describe ONE method of pressure die-casting.   (C. & G., Paper No. 293/7.)
4. Compare and contrast the effects of cold-working and hot-working on metals.   (Brunel College, Dip. Tech.)
5. (a) Give a brief description, with sketches, of the type of rolling mill used in the initial treatment of a large steel ingot.
    (b) Why is steel hot-rolled in the production of slabs, sheet, bars and strip?   (C. & G., Paper No. 293/6.)
6. Compare forging and casting as processes of production; with especial reference to metallurgical effects.
              (University of Aston in Birmingham, Inter B.Sc.(Eng.).)
7. (a) Describe the effect of forging on the structure and properties of a metal.
    (b) Why is grain flow important in a forged material?
    (c) Make a sketch to illustrate the correct flow lines in a horizontal section of a crankshaft.   (C. & G., Paper No. 293/8.)
8. Write short accounts on TWO of the following working processes: (a) cold-rolling of sheet and strip; (b) wire-drawing; (c) drop-forging.
                    (Nottingham Regional College of Technology.)

9. Give an account of the formation of the "extrusion defect" in the direct extrusion process. What methods are available for minimising this defect?
(Derby and District College of Technology, H.N.D. Course.)

10. (a) Draw diagrams to indicate the difference between the "direct" and the "inverted" extrusion processes.

(b) What effects have these differences on the pressure required for extrusion, assuming that the conditions are otherwise identical?

(c) Sketch the flow pattern expected in the billets in the two processes.
(Institution of Production Engineers Associate Membership Examination, Part II, Group A.)

11. Explain fully the terms "cupping", "re-drawing" and "ironing" as applied to deep-drawing operations. Give simple sketches of the tools and dies used in these processes.   (U.L.C.I., Paper No. C111–2–4.)

12. Briefly review the commercial working processes used to produce semi-finished or final products.
(Nottingham Regional College of Technology.)

13. Describe, briefly, suitable fabrication techniques for the following: (a) a shell or cartridge case; (b) an aluminium saucepan; (c) a carburetter body; (d) a metal H-channel for a curtain runner.

Indicate the type of alloy to be used in each case.
(The Polytechnic, London W.1.)

14. Describe the various stages involved in the production of a component by the powder-metallurgy process. Under what circumstances would this process be used in preference to any other?
(Derby and District College of Technology, H.N.D. Course.)

15. (a) State the essential stages in the process of manufacture by powder-metallurgy technique.

(b) Using as examples typical components, explain the advantages and practical limitations of the process.
(Institution of Production Engineers Associate Membership Examination, Part II, Group B, Production Metallurgy.)

16. Write notes on FOUR of the following processes: (a) rolling; (b) extrusion; (c) powder metallurgy; (d) wire-drawing; (e) forging.
(Constantine College of Technology, H.N.D. Course.)

CHAPTER VII

PLAIN CARBON STEELS

7.10. It is impossible adequately to study the structure of steel, or, for that matter, any other alloy, without reference to what is known as a thermal-equilibrium diagram. Many readers will have been introduced to the iron–carbon thermal-equilibrium diagram during their earlier studies of Workshop Technology. The purpose of this chapter, therefore, is to clarify such ideas as those readers have formulated on the subject and, at the same time to introduce other readers to this very important branch of physical metallurgy. Moreover, it has been considered desirable at this stage to make a preliminary study of the structures and properties of plain carbon steels in preparation for a more exhaustive study later in the book. But first we will examine the construction and interpretation of a simple equilibrium diagram by reference to a series of alloys, of which two are well known.

7.20. Most engineers are aware that there are two main varieties of tin–lead solder. Best-quality tinman's solder contains 62% tin and 38% lead* and its solidification begins and ends at the same temperature (183° C). Plumber's solder, however, contains 33% tin and 67% lead, and whilst it begins to solidify at about 265° C, solidification is not complete until 183° C. Between 265 and 183° C, then, plumber's solder is in a pasty, partly solid state which enables the plumber to "wipe" a joint with the aid of his "cloth".

From observations such as these it can be concluded that the temperature range over which a tin–lead alloy solidifies depends upon its composition. On further investigation it will be found that an alloy containing 50% tin and 50% lead will begin to solidify at 220° C, and be completely solid at 183° C; whilst one containing 80% tin and 20% lead will begin to solidify at 200° C and finish solidifying at 183° C.

From the data accumulated above we can draw a diagram which will indicate the state in which any given tin–lead alloy (within the range of compositions investigated) will exist at any given temperature (Fig. 7.1). This diagram has been obtained by plotting the temperatures at which the alloys mentioned above begin and finish solidifying, on a

---

* Whilst for reasons of economy tinman's solder often contains less than 62% tin, the latter composition is ideal, since the solder will melt and freeze quickly at a fixed temperature.

temperature–composition diagram. All points—a_1, a_2, a_3, a_4—at which the various alloys begin to solidify, are joined, as are the points—b_1, b_2, b_3, b_4—where solidification is complete.

Any alloy represented in composition and temperature by a point above *AEB* will be in a completely molten state, whilst any alloy similarly represented by a point below *CED* will be completely solid. Like-

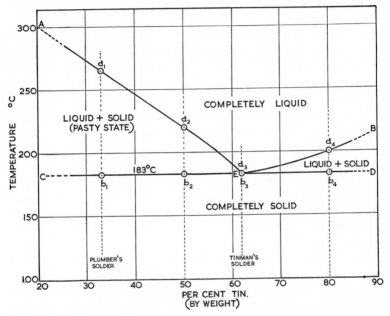

FIG. 7.1.—Diagram Showing the Relationship between Composition, Temperature and Physical State for a Series of Tin–Lead Alloys.

wise, any alloy whose temperature and composition are represented by a point between *AE* and *CE* or between *EB* and *ED* will be in a part liquid–part solid state.

7.21. Such a diagram is of great use to the metallurgist, and is called a thermal-equilibrium diagram. The meaning of the term "equilibrium" in this context will become apparent as a result of later studies in this book, but for the moment we will consider an everyday example which goes some way to illustrate its meaning.

On a hot summer's day we can produce a delightfully refreshing drink by putting a cube of ice into a glass of lager. The contents of the glass, however, are not in thermal equilibrium with the surroundings, and as heat-transfer takes place into the lager the ice ultimately melts and the liquid warms up, so that the whole becomes more homogeneous if less palatable. Rapid cooling, as we shall see later, often produces an alloy structure which, like the ice and lager, is not in thermal equilibrium at room temperature. The basic difference between the ice–lager mixture and the non-equilibrium metallic structure is that the former is able to reach "structural" equilibrium with ease, but in the case of the metallic structure rearrangement of the atoms is more difficult, since they are packed in geometrical layers in a crystal lattice. A non-equilibrium metallic structure produced by rapid cooling may therefore be retained permanently at room temperature.

7.22. If we assume that a series of alloys has been cooled slowly enough for structural equilibrium to obtain, then the thermal-equilibrium diagram will indicate the relationship which exists between composition, temperature and microstructure of the alloys concerned. By reference to the diagram, we can, for an alloy of any composition in the series, find exactly what its structure or physical condition will be at any given temperature. We can also in many cases forecast with a fair degree of accuracy the effect of a particular heat-treatment on the alloy; for in modern metallurgy heat-treatment is not a process confined to steels, but is applied also to many non-ferrous alloys. These are two of the more important uses of the thermal-equilibrium diagram as a metallurgical tool. Let us now proceed with our preliminary study of the iron–carbon alloys, with particular reference to their equilibrium diagram.

7.30. Plain carbon steels are usually regarded as being those alloys of iron and carbon which contain up to 1·7% carbon. In practice, most ordinary steels also contain appreciable amounts of manganese, residual from the deoxidation process (2.83). For the present, however, we can neglect the effects of this manganese and consider steels as being straight iron–carbon alloys.

7.31. As we have seen (4.14), the pure metal iron, at temperatures below 910° C, has a body-centred cubic structure, and if we heat it to above this temperature the structure will change to one which is face-centred cubic. On cooling, the change is reversed and a body-centred cubic structure is once more formed. The importance of this reversible transformation lies in the fact that up to 1·7% carbon can dissolve in face-centred cubic iron, forming what is known as a "solid solution",* whilst in body-centred cubic iron no more than 0·03% carbon can dissolve.

7.32. As a piece of steel in its face-centred cubic form cools slowly and changes to its body-centred cubic form, any dissolved carbon

---

* We shall deal more fully with the nature of solid solutions in the next chapter, and for the present it will be sufficient to regard a solid solution as being very much like a liquid solution in that particles of the added metal are absorbed without visible trace, even under a high-power microscope, into the structure of the parent metal.

present in excess of 0·03% will be precipitated, whilst if it is cooled rapidly enough such precipitation is prevented. Upon this fact depends our ability to heat-treat steels.

7.33. The solid solution formed when carbon dissolves in face-centred cubic iron is called *Austenite*, and the very weak solid solution formed in the body-centred cubic structure is called *Ferrite*. For most practical purposes we can regard ferrite as having the same properties as pure iron. In most text-books on metallurgy the reader will find that the symbol γ ("gamma") is used to denote both the face-centred cubic form of iron and the solid-solution austenite, whilst the symbol α ("alpha") is used to denote both the body-centred cubic form of iron existing below 910° C and the solid-solution ferrite. The same nomenclature will be used in this book.

When carbon is precipitated from austenite it is not in the form of elemental carbon (graphite), but as the compound iron carbide, Fe_3C, usually called *Cementite*. This substance, like most other metallic carbides, is very hard, so that, as the amount of carbon (and hence, of cementite) increases, the hardness of the slowly cooled steel will also increase.

7.34. Fig. 7.1. indicates the temperatures at which *solidification* begins and ends for any homogeneous *liquid* solution of tin and lead. In the same way Fig. 7.2 shows us the temperatures at which *transformation* begins and ends for any *solid* solution (austenite) of carbon and face-centred cubic iron. Just as the melting point of either tin or lead is lowered by adding each to the other, so is the allotropic transformation temperature of face-centred cubic iron altered by adding carbon. Fig. 7.2 includes only a part of the whole iron–carbon equilibrium diagram, but it is the section which we make use of in the heat-treatment of carbon steels. On the extreme left of this diagram is an area labelled "ferrite". This indicates the range of temperatures and compositions over which carbon can dissolve in body-centred cubic (α) iron. On the left of the sloping line AB all carbon present is dissolved in the body-centred cubic iron, forming the solid-solution ferrite, whilst any point representing a composition and temperature to the right of AB indicates that the solid-solution α is saturated, so that some of the carbon contained in the steel will be present as cementite. The significance of the slope of AB is that the solubility of carbon in body-centred cubic iron increases from 0·006% at room temperature to 0·03% at 723° C. Temperature governs the degree of solubility of solids in liquids in exactly the same way. If the reader will refer to 1.60 he will find an exact parallel between the solubility of potassium chlorate in water and that of carbon in body-centred cubic iron.

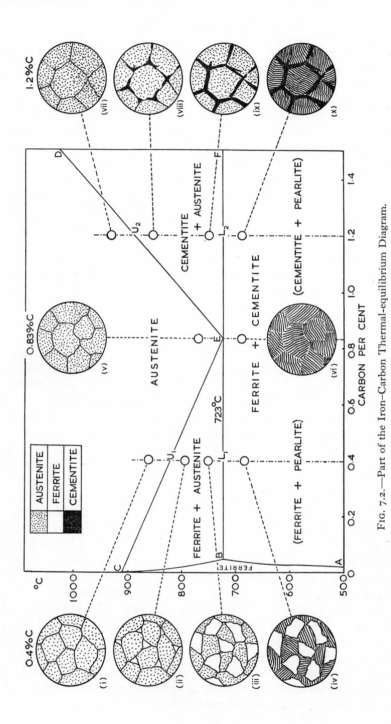

FIG. 7.2.—Part of the Iron–Carbon Thermal-equilibrium Diagram.

7.35. We will now study the transformations which take place in the structures of three representative steels which have been heated to a temperature high enough to make them austenitic and then allowed to cool slowly. If a steel containing 0·40% carbon is heated to some temperature above U_1 it will become completely austenitic (Fig. 7.2 (i)). On cooling again to just below U_1 (which is called the "upper critical temperature" of the steel), the structure begins to change from one which is face-centred cubic to one which is body-centred cubic. Consequently, small crystals of body-centred cubic iron begin to separate out from the austenite. These body-centred cubic crystals (Fig. 7.2 (ii)) retain a small amount of carbon (less than 0·03%), so we shall refer to them as crystals of ferrite. As the temperature continues to fall the crystals of ferrite grow in size at the expense of the austenite (Fig. 7.2 (iii)), and since ferrite is almost pure iron, it follows that most of the carbon present accumulates in the shrinking crystals of austenite. Thus, by the time our piece of steel has reached L_1 (which is called its "lower critical temperature") it is composed of approximately half ferrite (containing only 0·03% carbon) and half austenite, which now contains 0·83% carbon. The composition of the austenite at this stage is represented by E. Austenite can hold no more than 0·83% carbon in solid solution at this temperature (723° C.), therefore, as the temperature falls still farther, the carbon begins to precipitate as cementite. At the same time ferrite is still separating out and we find that these two substances, ferrite and cementite, form as alternate layers until all the remaining austenite is used up (Fig. 7.2 (iv)). This laminated structure of ferrite and cementite, then, will contain exactly 0·83% carbon, so that it will account for approximately half the volume of our 0·4% carbon steel. It is an example of what, in metallurgy, we call a *eutectoid*. This particular eutectoid is known as *Pearlite* because when present on the etched surface of steel it acts as a "diffraction grating", splitting up white light into its component spectrum colours and giving the surface a "mother of pearl" sheen. In order to be able to see these alternate layers of ferrite and cementite of which pearlite is composed, a metallurgical microscope capable of a magnification in the region of 500 diameters is necessary.

Any steel containing less than 0·83% carbon will transform from austenite to a mixture of ferrite and pearlite in a similar way when cooled from its austenitic state. Transformation will begin at the appropriate upper critical temperature (given by a point on CE which corresponds with the composition of the steel) and end at the lower critical temperature of 723° C. The relative amounts of ferrite and pearlite will depend upon the carbon content of the steel (Fig. 7.3), but in every case

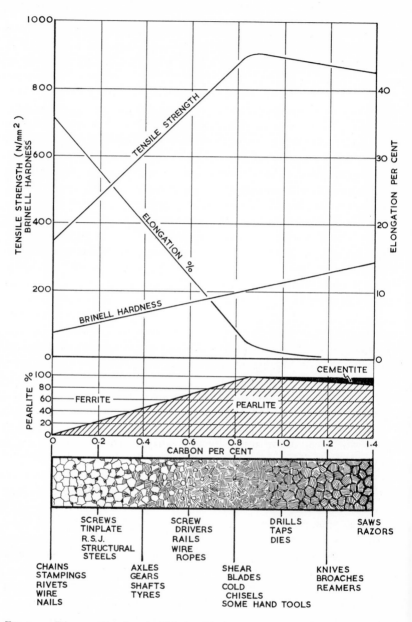

Fig. 7.3.—Diagram Showing the Relationship between Carbon Content, Microstructure and Mechanical Properties of Plain Carbon Steels in the Normalised Condition.

Typical uses of such steels are also indicated.

the ferrite will be almost pure iron and the pearlite will contain exactly 0·83% carbon.

7.36. A steel containing 0·83% carbon will not begin to transform from austenite on cooling until the point E is reached. Then *transformation* will begin and end at the same temperature (723° C), just as tinman's solder *solidifies* at a single temperature (183° C). Since the steel under consideration contained 0·83% carbon initially, it follows that the final structure will be entirely of pearlite (Fig. 7.2 (vi)).

7.37. A steel which contains, say, 1·2% carbon will begin to transform from austenite when the temperature falls to its upper critical at U_2. Since the carbon is this time in excess of the eutectoid composition, it will begin to precipitate first; not as pure carbon but as needle-shaped crystals of cementite round the austenite grain boundaries (Fig. 7.2 (viii)). This will cause the austenite to become progressively less rich in carbon, and by the time a temperature of 723° C has been reached the remaining austenite will contain only 0·83% carbon. This remaining austenite will then transform to pearlite (Fig. 7.2 (x)), as in the two cases already dealt with.

Any steel containing more than 0·83% carbon will have a structure consisting of cementite and pearlite if it is allowed to cool slowly from its austenitic state. Since the pearlite part of the structure always contains alternate layers of ferrite and cementite in the correct proportions to give an overall carbon content of 0·83% for the pearlite, it follows that any variation in the total carbon content of the steel above 0·83% will cause a corresponding variation in the amount of primary cementite. (The terms "primary cementite" and "primary ferrite" are used to denote that cementite or ferrite which forms first, before the transformation of the residual austenite to pearlite takes place.)

A plain carbon steel which contains less than 0·83% carbon is generally referred to as a *hypo-eutectoid* steel, whilst one containing more than 0·83% carbon is known as a *hyper-eutectoid* steel. Naturally enough, a plain carbon steel containing exactly 0·83% carbon is called a *eutectoid* steel.

7.40. So far we have been dealing only with structures produced in plain carbon steels by cooling them slowly from an austenitic condition. Such conditions prevail during industrial processes such as normalising and annealing. By very rapid cooling from the austenitic condition, such as would be obtained by water-quenching, another structure, called *Martensite*, is produced. This does not appear on the equilibrium diagram simply because it is not an equilibrium structure. Rapid cooling has prevented structural equilibrium from being reached.

7.41. As most readers will already know, martensite is very hard indeed. Unfortunately it is also rather brittle, and the steel is used in this condition only when extreme hardness is required. To increase the steel's toughness after quenching, at the expense of a fall in hardness, the steel can be tempered. A modification in the structure will take place, depending upon the tempering temperature. At tempering temperatures in the region of 400° C *Troostite* will form, whilst in the region of 600° C *Sorbite* will be produced. In either case tempering has allowed the steel to proceed in some measure back towards structural equilibrium, with the precipitation of microscopical particles of cementite in varying amounts from the martensitic structure. We shall deal more fully with the heat-treatment of steel, however, in Chapters XI and XII.

The Uses of Plain Carbon Steels

7.50. As shown in Fig. 7.3, the hardness of plain carbon steel increases progressively with increase in carbon content, so that generally the low- and medium-carbon steels are used for constructional work,

TABLE 7.1

Type of steel	Percentage carbon	Uses
Dead mild .	0·05–0·15	Chain, stampings, rivets, wire, nails, seam-welded pipes, mattresses, hot- and cold-rolled strip for many purposes
Mild . .	0·10–0·20	Structural steels, R.S.J., screws, machine parts, tin-plate, case-hardening, drop-forgings, stampings
	0·20–0·30	Machine and structural work, gears, free-cutting steels, shafting, levers, forgings
Medium carbon.	0·30–0·40	Connecting-rods, shafting, wire, axles, fish-plates, crane hooks, high-tensile tubes, forgings
	0·40–0·50	Crankshafts, axles, gears, shafts, die-blocks, rotors, tyres, heat-treated machine parts
	0·50–0·60	Loco tyres, rails, laminated springs, wire ropes
High carbon .	0·60–0·70	Drop-hammer dies, set-screws, screw-drivers, saws, mandrels, caulking tools, hollow drills
	0·70–0·80	Band saws, anvil faces, hammers, wrenches, laminated springs, car bumpers, small forgings, cable wire, dies, large dies for cold presses
	0·80–0·90	Cold chisels, shear blades, cold setts, punches, rock drills, some hand tools
Tool steels .	0·90–1·00	Springs, high-tensile wire, axes, knives, dies, picks
	1·00–1·10	Drills, taps, milling cutters, knives, screwing dies
	1·10–1·20	Ball bearings, dies, drills, lathe tools, woodworking tools
	1·20–1·30	Files, reamers, knives, broaches, lathe and woodworking tools
	1·30–1·40	Saws, razors, boring and finishing tools, machine parts where resistance to wear is essential

whilst the high-carbon steels are used for the manufacture of tools and other components where hardness and wear-resistance are necessary. Typical uses of steels of various carbon content are shown in Table 7.1 and also in Fig. 7.3.

Brittle Fracture in Steels

7.60. It has been shown in earlier chapters that fracture in metals is usually preceded by measurable slip. Under some circumstances, however, fracture may occur in a metal which has undergone little or no previous plastic deformation. Such fracture is termed "cleavage" fracture or "brittle" fracture. Failure of this type was experienced in the welded "liberty ships" manufactured during the Second World War for carrying supplies from America to Europe and was unexpected and dangerous.

Under normal conditions the stress required to cause cleavage is higher than that necessary to cause slip, but if, by some circumstances, slip is suppressed, brittle fracture will occur when the internal tensile stress increases to the value necessary to cause failure. This situation can arise under the action of bi-axial or tri-axial stresses within the material. Such stresses may be residual from some previous treatment, and the presence of points of stress concentration may aggravate the situation. The liberty ships mentioned above were fabricated by welding plates together to form a continuous body. Cracks usually started at sharp corners or arc-weld spots and propagated right round the hull, so that, finally, the ship broke in half. Had the hull been riveted, the crack would have been arrested at the first rivet hole it encountered. Some riveted joints are now in fact incorporated in such structures to act as crack arresters.

7.61. The relationship between mechanical properties and the method of stress application has already been mentioned (3.52), with particular reference to the Izod impact test. Plastic flow depends upon the movement of dislocations and this occurs in some finite time. If the load is applied very rapidly it is possible for stress to increase so quickly that it cannot be relieved by slip. A momentary increase of stress to a value above the yield stress will produce fracture.

7.62. As temperature decreases, the movement of dislocations becomes more difficult and this increases the possibility of internal stress exceeding the yield stress at some instant. Brittle fracture is therefore more common at low temperatures. This is supported by the fact that the liberty ships were in service in the cold North Atlantic.

7.63. Those metals with a F.C.C. structure maintain ductility at low temperatures, whilst some metals with structures other than F.C.C.

tend to exhibit brittleness. B.C.C. ferrite is particularly susceptible to brittle fracture, which follows a transcrystalline path along the (100) planes (see 4.15). This occurs at low temperatures as indicated in Fig. 7.4, and the temperature at which brittleness suddenly increases is

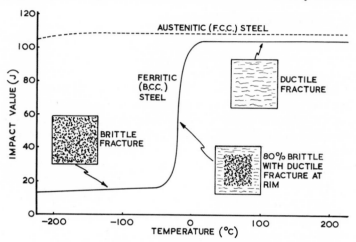

FIG. 7.4.—The Relationship Between Brittle Fracture and Temperature for Ferritic and Austenitic Steels.

known as the transition temperature. For applications involving atmospheric temperatures the transition temperature can be reduced to safe limits by increasing the manganese–carbon ratio of the steel, whilst at the same time controlling the grain size by small additions of aluminium. A suitable steel contains 0·14% carbon and 1·3% manganese. Where lower temperatures are involved it is necessary to use low-nickel steels.

RECENT EXAMINATION QUESTIONS

1. Give an account of the changes of structure and properties of normalised steels as the carbon content is raised from 0·1 to 0·8%. Comment on the choice of carbon content for some typical engineering applications.
   (University of Strathclyde, B.Sc.(Eng.).)

2. Draw and fully label the portion of the iron–carbon thermal-equilibrium diagram which applies to the heat-treatment of steel.
   Show how three mechanical properties of normalised plain carbon steels vary with the carbon content up to 1·4% C.
   (Derby and District College of Technology.)

3. Sketch the iron–carbon thermal-equilibrium diagram and label all the phase fields.
   Use the diagram to describe the structural changes which take place in a 1·0% carbon steel as it solidifies and cools slowly to room temperature.
   (U.L.C.I., Paper No. C111–1–4.)

4. Describe the slow cooling of the following plain carbon steels from 1 000° C to room temperature: 0·4% carbon; 1·2% carbon.

Compare the mechanical properties of these steels, and show how they are related to microstructure.

(Constantine College of Technology, H.N.D. Course.)

5. Discuss the importance of the carbon content in steels and explain how you would choose the carbon content for some typical engineering applications. (University of Strathclyde, B.Sc.(Eng.).)

6. Show how three mechanical properties of normalised plain carbon steels vary with the carbon content up to 1·4% C.

Sketch and fully label the microstructures of the following normalised steels: (a) 0·2% C; (b) 0·8% C; (c) 1·4% C.

(Derby and District College of Technology, H.N.D. Course.)

7. Give an account of the phenomenon known as "low-temperature notch brittleness". What methods are available for minimising this effect in ferritic steels?

(Derby and District College of Technology, H.N.D. Course.)

8. Discuss the relationship between impact value and the service temperature for a 0·2% carbon steel.

How may the properties of such a steel be improved for service at sub-zero temperatures? (U.L.C.I., Paper No. 237–2–8.)

CHAPTER VIII

THE FORMATION OF ALLOYS

8.10. In engineering practice we do not often use metals in their pure form except where high electrical conductivity, high ductility or good corrosion-resistance is required. These properities are generally at a maximum value in pure metals, but such mechanical properties as tensile strength, yield point and hardness are improved by alloying. Let us, then, consider what can happen when two metals are mixed together in the liquid state and allowed to solidify.

8.11. Broadly speaking, we are only likely to obtain a metallurgically useful alloy when the two metals in question dissolve in each other in the liquid state to form a completely homogeneous liquid solution. There are exceptions to this rule, as, for example, the suspension of undissolved lead particles in a free-cutting alloy (20.24). The globules of lead are undissolved in the molten, as well as in the solidified, alloy, and serve to improve machinability by rendering the swarf discontinuous and also by causing local brittleness. In general, however, a useful alloy is formed only when the metals in question are mutually soluble in the liquid state.

8.12. When such a liquid solution solidifies it will be found that one of the following conditions obtains:

(*a*) The solubility ends and the two metals which were mutually soluble as liquids become totally insoluble in the solid state. They therefore separate out as particles of the two pure metals.

(*b*) The solubility prevailing in the liquid state may be retained either completely or partially in the solid state. In the former case a single "solid solution" will be formed, whilst in the latter case a mixture of two different solid solutions will result.

(*c*) The two metals react chemically as solidification proceeds and form an "intermetallic compound".

8.13. Any one of the substances referred to above is called a "phase". Thus a solid phase may be either a pure metal or a solid solution or an intermetallic compound, of two or more metals, so long as it exists in the microstructure of the alloy as a separate entity.

A rule, based on thermodynamic considerations, and known as the Phase Rule, was formulated originally in 1876 by Josiah W. Gibbs,

Professor of Mathematical Physics at Yale University. Whilst otherwise this rule has little direct application to a study of physical metallurgy, it embodies an important statement that in a solid binary* alloy not more than two phases can exist together. Therefore a binary alloy may be built up of:

(a) Two pure metals existing as separate entities in the structure. This is a very rare occurrence, since in most cases there is a slight solubility of one metal in the other.

(b) A single complete solid solution of one metal dissolved in the other.

(c) A mixture of two solid solutions if the metals are only partially soluble in each other.

(d) A pure intermetallic compound.

(e) A mixture of two different intermetallic compounds.

(f) A solid solution and an intermetallic compound mixed together—and so on.

8.14. The sizes of the crystals comprising any of these phases in a binary alloy vary considerably. Sometimes the crystals are so large that they can be seen easily with the naked eye, whilst in other cases particles of a phase are so small that they can be seen clearly only with the aid of a metallurgical microscope capable of magnifications of 500 diameters or more. Practical microscopical metallography is generally the craft of a skilled laboratory technician, but it is hoped to give an introduction to this aspect of physical metallurgy in Chapter X.

We will now consider in some detail the nature of the phases and structures mentioned above.

The Eutectic

8.20. When two metals which are completely soluble in the liquid state but completely insoluble in the solid state solidify, they do so by crystallising out as alternate layers of the two pure metals. Thus a laminated type of structure, something like plywood, is obtained and is termed a "eutectic". The mechanism of this type of crystallisation is discussed in the next chapter (9.31), so that here we shall confine our studies to the nature of the eutectic.

8.21. In a eutectic of two pure metals there is no question whatsoever of solution, since the layers of the pure metals forming the eutectic can be seen quite clearly under the microscope, at magnifications usually between 100 and 500, as definite separate entities. The formation of the eutectic is therefore a result of insolubility being introduced when the alloy solidifies.

---

* Containing *two* metals.

If the two metals in question are partially soluble in each other in the solid state, we may still obtain a eutectic on solidification, but this time it will be a eutectic composed of alternate layers of two saturated solid solutions. One layer will consist of metal A saturated with metal B, and the other a layer of metal B saturated with metal A. This statement may at first cause some confusion in the mind of the reader, but it must be understood that the two layers are not one and the same as far as composition is concerned, since one layer is rich in metal A and the other is rich in metal B. Consider as a parallel case in liquid solutions the substances water and ether. If a few spots of ether are added to a testtube nearly filled with water and the tube shaken, a single solution of ether in water will result. If, on the other hand, a few spots of water are added to a test-tube nearly filled with ether, and the tube shaken, a single solution of water in ether will result. When, however, we put into the test-tube equal volumes of ether and water and again shake, we find that we are left with two layers. The upper layer will be ether saturated with water, and the lower layer will be water saturated with ether. The upper layer is a solution containing most of the ether, and the lower layer a different solution containing most of the water. We have a similar situation in the case of metallic solid solutions, in that one phase of a eutectic may be a solid solution rich in metal A and the other phase a solid solution rich in metal B.

8.22. Not all eutectics, whether of pure metals or solid solutions, appear under the microscope in the well-defined laminated form mentioned above. Usually layers of one of the phases are embedded in a matrix of the other phase, and quite often these layers are broken or disjointed. Again, surface tension may influence the crystallisation, and the final eutectic may consist of globules of one phase embedded in a matrix of the other. The effect of surface tension will be accentuated by slow rates of cooling which allow layers of one phase to break up; first, into smaller, thicker plates, and ultimately, into rounded globules (Fig. 11.8).

8.23. Sometimes a solid solution which has already formed in an alloy transforms at a lower temperature to a eutectic type of structure. When this happens, the eutectic type of structure produced is called a *eutectoid*, since it was not formed from a liquid solution like a eutectic, but from a solid solution. The transformation of the solid solution, austenite, to the eutectoid, pearlite (7.35), is an example of this type of change. Cementite nuclei form at random at the crystal boundaries of the austenite, and these nuclei initiate the growth of cementite plates in the direction in which the concentration of carbon in the austenite is highest. As a result of the extraction of the carbon from the surrounding austenite to form cementite, ferrite will nucleate alongside the cementite. In this way cementite and ferrite plates will develop alongside each other (Fig. 8.1).

If an alloy is water-quenched from a temperature above that at which the eutectoid begins to form, the solid solution (or possibly a modified form of it, such as martensite in steels) is often retained. This demonstrates that the eutectoid had not formed from the liquid but had been produced at a later stage.

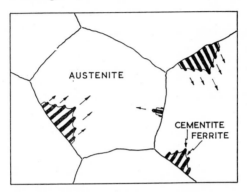

FIG. 8.1.—Mechanism of the Austenite ⟶ Pearlite Transformation.

The Solid Solution

8.30. A solid solution is formed when two metals which are mutually soluble in the liquid state remain dissolved in each other after crystallisation. If the resulting structure is examined, even with the aid of a high-power electron microscope, no trace can be seen of the parent metals as separate entities. Instead one sees crystals consisting of a homogeneous solid solution of one metal in the other, and if the metals are completely soluble in each other there will be crystals of one type only.

8.31. It will be clear that in order that one metal may dissolve in another to form a solid solution, its atoms must fit in some way into the crystal lattice of the other metal. This may be achieved by the formation of either a "substitutional" or an "interstitial" solid solution (Fig. 8.2).

8.32. Interstitial solid solutions can be formed only when the atoms of the added element are very small compared with those of the parent metal, thus enabling them to fit into the interstices or spaces in the crystal lattice of the parent metal. Not only can this occur during solidification but, also in many cases, when the parent metal is *already solid*. Thus, carbon can form an interstitial solid solution with F.C.C. iron during the solidification of steel, but it can also be absorbed by solid iron provided the latter is heated to a temperature at which the structure is face-centred cubic. This is the basis of carburising steel (19.20). Nitrogen is also able to dissolve interstitially in solid steel, making the nitriding process (19.40) possible; whilst hydrogen, the atoms of which are very small indeed, is able to dissolve in this way in a number of solid metals, usually producing mechanical brittleness as a result.

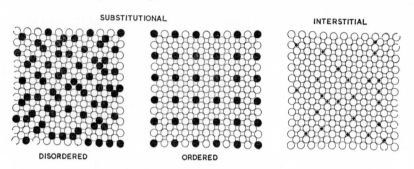

FIG. 8.2.—Ways in Which Solid Solution Can Occur.

8.33. In a substitutional solid solution the atoms of the parent metal are replaced in the crystal lattice by atoms of the added metal. This substitution may be either "ordered" or "disordered", as shown in Fig. 8.2. In the disordered solid solution replacement is quite indiscriminate, so that the concentration of atoms of the added element can vary considerably throughout the structure of a single crystal. In fact, when a solid solution of this type crystallises from the liquid state there is a natural tendency for the core of the dendrite to contain rather more atoms of the metal with the higher melting point, whilst the outer fringes of the crystal will contain correspondingly more atoms of the metal of the lower melting point. Nevertheless, within such a "cored" crystal, once formed, diffusion does take place; and this diffusion tends to produce uniform distribution of each kind of atom. Diffusion is only appreciable at relatively high temperatures, for which reason a rapidly cooled alloy will be appreciably cored and a slowly cooled alloy only slightly cored. Prolonged annealing will remove coring completely by allowing diffusion to take place and produce a uniform solid solution. The effects of the phenomena of coring and diffusion are illustrated in Fig. 8.3.

8.34. In an interstitial solid solution diffusion can take place with relative ease, since the solute atoms, being small, can move readily through the crystal lattice of the parent metal, but in substitutional solid solutions the mechanism of diffusion is still open to speculation. It was stated earlier (5.15) that faults can occur in crystals, giving rise to dislocations which facilitate slip. Similarly, it is suggested that what are

THE CORING OF SOLID SOLUTIONS.

DIAGRAMS INDICATING THE RELATIVE DISTRIBUTIONS OF ATOMS OF METALS Ⓐ & Ⓑ AT VARIOUS STAGES DURING THE SOLIDIFICATION AND ANNEALING OF A SOLID SOLUTION CONTAINING 50% OF EACH METAL.

● ATOMS OF Ⓐ – THE METAL WITH THE HIGHER MELTING POINT.

○ ATOMS OF Ⓑ – THE METAL WITH THE LOWER MELTING POINT

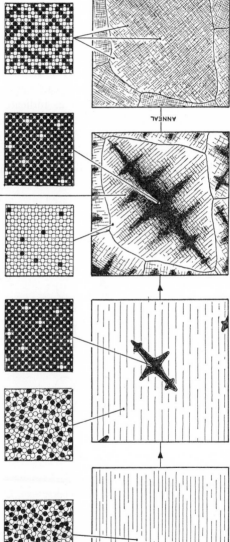

1. A HOMOGENEOUS LIQUID SOLUTION OF METALS Ⓐ & Ⓑ.

2. DENDRITES RICH IN METAL Ⓐ BEGIN TO GROW MAKING THE REMAINING LIQUID CORRESPONDINGLY RICH IN METAL Ⓑ.

3. COMPLETELY SOLID — THE CORE OF THE CRYSTAL IS RICH IN METAL Ⓐ AND THE OUTER FRINGES RICH IN METAL Ⓑ.

ANNEAL

4. ANNEALING HAS CAUSED DIFFUSION TO TAKE PLACE RESULTING IN A HOMOGENEOUS SOLID SOLUTION OF METALS Ⓐ & Ⓑ

Fig. 8.3.—The Variations in Composition Which Are Possible in a Solid Solution.

called "vacant sites" can exist in a metal crystal. These are positions in the crystal lattice which are not occupied by an atom (Fig. 8.4).

Since there is always some difference in size between the solute atoms and the atoms of the parent metal, the presence of a solute atom will

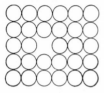

cause some distortion in the lattice of the parent metal. This distortion will be minimised by the association of a solute atom with a vacant site as shown in Fig. 8.5A. The associated pair of vacant site with solute atom can migrate easily through the crystal in stages as indicated in Fig. 8.5B–F.

FIG. 8.4.—A "Vacant Site" in a Crystal Lattice.

The rate of diffusion will depend to some extent upon the number of vacant sites, but also upon the concentration gradient of the solute atoms within the structure of the solvent metal (Fig. 8.6); and upon the temperature at which diffusion is taking place.

As long ago as 1858 Adolph Fick enunciated laws governing diffusion in substances in general. Here we are concerned with their application to diffusion in metallic alloys.

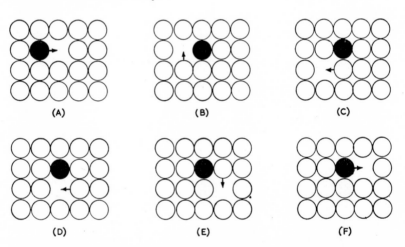

FIG. 8.5.—The Diffusion of a Solute Atom Which is Associated with a Vacant Site.

Fick's First Law states that the movement of atoms ($\mathcal{J}$) across unit area of a plane at any given instant is proportional to the concentration gradient, $\left(\dfrac{\partial c}{\partial x}\right)_t$, at the same instant, but of opposite sign, i.e.—

$$\mathcal{J} = -D\left(\frac{\partial c}{\partial x}\right) \qquad . \quad . \quad . \quad . \quad (1)$$

(assuming that the X-axis is parallel to the direction in which the concentration gradient is operating). Here D is a constant (the diffusion coefficient) which will vary with different compositions. It is in units $\dfrac{(\text{length})^2}{\text{time}}$ (usually $10^{-4} \text{m}^2/\text{s}$.). When $\dfrac{\partial c}{\partial x} = 0$, $\mathcal{J} = 0$, which satisfies the requirement that diffusion ceases when the concentration gradient is zero (line CD in Fig. 8.6).

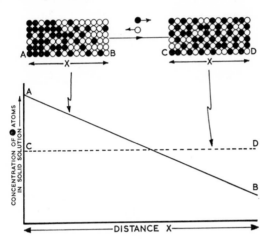

FIG. 8.6.—AB is the Initial Concentration Gradient and CD the Concentration Gradient After a Long Time During Which Diffusion Has Been Taking Place.

Fick's Second Law is derived from the first and the fact that matter is conserved. It can be shown that—

$$\frac{\partial c}{\partial t} = \frac{\partial}{\partial x}\left(D \frac{\partial c}{\partial x} \right) \qquad . \quad . \quad . \quad (2)$$

This takes into account the change in concentration which occurs at a given point as a function of time.

In the above it has been assumed that diffusion is taking place in one direction only (in the direction of the X-axis). In practice, it will occur along the Y and Z axes also.

8.35. In an ordered solid solution the atoms of the added metal take up certain fixed positions in the crystal lattice of the parent metal, as shown in Fig. 8.2. Solid solutions of the ordered type tend to be hard and brittle, whereas those of the disordered type are both tough and ductile, and are therefore the most useful of metallurgical phases. Fortunately most solid solutions are of the disordered type.

The Intermetallic Compound

8.40. Frequently two metals will combine chemically to form what is called an intermetallic compound. An intermetallic compound is often easy to detect in a microstructure, as it may be of a colour totally different from that of either of its parent metals. For example, the metal antimony, which is white, will combine with copper to form the intermetallic compound Cu_2Sb, which is bright purple in colour. Solid solutions, on the other hand, more often than not, present a blend of the colours of the two parent metals. Moreover, intermetallic compounds, being of fixed atomic constitutions, do not exhibit coring in the cast condition, as do solid solutions.

8.41. Intermetallic compounds may be of two types:

(*a*) *Normal Valency Compounds*, in which the laws of chemical valency are apparently obeyed, as in such compounds as Mg_3Sb_2, Mg_2Sn and Mg_3Bi_2. These valency compounds are generally formed when one metal (such as magnesium) has chemical properties which are strongly metallic, and the other metal (such as antimony, tin or bismuth) chemical properties which are only weakly metallic and, in fact, bordering on those of non-metals. Frequently such a compound has a melting point which is higher than that of either of the parent metals. For example, the intermetallic compound Mg_2Sn melts at 780° C, whereas the parent metals magnesium and tin melt at 650 and 232° C respectively.

(*b*) *"Electron Compounds"*. As was shown earlier (1.80), the chemical valency of a metal is closely related to the number of electrons in the outer "shell" of the atom, whilst the nature of the metallic bond is such that wholesale sharing of numbers of electrons takes place in the crystal structure of a pure metal.

In these "electron compounds" the normal valency laws are not obeyed, but in many instances there is a fixed ratio between the total number of valency bonds of all the atoms involved and the total number of atoms in a "molecule" of the intermetallic compound in question. There are three such ratios, commonly referred to as Hume-Rothery ratios:

(i) Ratio 3/2 (21/14)—β structures, such as $CuZn$, Cu_3Al, Cu_5Sn, Ag_3Al, etc.

(ii) Ratio 21/13—γ structures, such as Cu_5Zn_8, Cu_9Al_4, $Cu_{31}Sn_8$, Ag_5Zn_8, $Na_{31}Pb_8$, etc.

(iii) Ratio 7/4 (21/12)—ε structures, such as $CuZn_3$, Cu_3Sn, $AgCd_3$, Ag_5Al_3, etc.

Thus, in the compound CuZn, copper has a valency of 1 and zinc a valency of 2, giving a total of 3 valencies, and hence a ratio of 3 valencies to 2 atoms. In the compound $Cu_{31}Sn_8$ copper has a valency of 1 and tin a valency of 4. Therefore 31 valencies are donated by the copper atoms and 32 (i.e., 4×8) by the tin atoms, making a total of 63 valencies. In all, 39 atoms are present in a molecule of $Cu_{31}Sn_8$, therefore the ratio—

$$\frac{\text{Total number of valencies}}{\text{Total number of atoms}} = \frac{63}{39} \text{ or } \frac{21}{13}$$

These Hume-Rothery ratios have been valuable in relating structures which were apparently unrelated. There are, however, many intermetallic compounds which do not fall into either of the three groups mentioned above, nor are they valency compounds.

8.42. Intermetallic compounds are of only limited use in engineering alloys, since in most cases, though extremely hard, they are also very brittle, and quite often they can be crushed to a powder by gentle pressure in the jaws of a vice. There are occasions when these compounds can be utilised in small amounts, as, for example, in bearing alloys, in which small particles of the intermetallic compound are embedded in a matrix of solid solution. The intermetallic compound is hard wearing and of low frictional properties, whilst the solid solution provides the necessary tough matrix capable of withstanding mechanical shock and compressive forces.

8.50. Thus our useful metallurgical alloys are built up in varying proportions from solid solutions and intermetallic compounds. As we have seen, these two phases possess very widely contrasting properties, for whilst the solid solution is strong, tough and ductile, the intermetallic compound is brittle, hard and weak. It is the job of the metallurgist, by skilful adjustment of his alloy compositions coupled with suitable mechanical and thermal treatment, to produce alloys which will utilise both types of structure to produce a set of physical properties which fulfil the requirements of the engineer.

Strengthening Mechanisms

8.60. As was mentioned earlier in this chapter, the most common reason for alloying is to increase the yield strength of a metal. This involves impeding the movement of dislocations by making alterations to the structure on approximately the atomic scale.

8.61. Solid-solution Hardening. In a cold-worked metal the presence of a dislocation causes distortion of that part of the lattice structure near to it. This distortion, and the energy associated with it, can be reduced by the presence of solute atoms. *Large* substitutional solute atoms will reduce distortion if they take up positions where the lattice structure is being *stretched* due to the presence of a dislocation; whilst *small* substitutional solute atoms will have a similar effect if they replace solvent atoms in regions where the lattice is being *compressed*. When these so-called "atmospheres" of solute atoms are produced, the movement of dislocations will be impeded and a greater stress must be applied to move them. That is to say, the yield point has been raised.

In interstitial solid solutions the relatively small solute atoms will tend to occupy positions where the lattice is being extended, since inter-stitial gaps will be larger in these regions. This hypothesis affords an explanation of the yield-point effect in mild steel (3.31), and was a significant triumph in support of the dislocation theory in its earlier days.

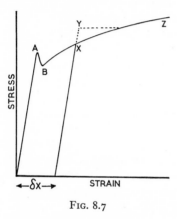

FIG. 8.7

In Fig. 8.7 the yield point, *A*, corresponds to the stress necessary to move dislocations away from their atmosphere of interstitial carbon atoms by which they have become anchored, whilst the point *B* indicates the stress necessary to keep the dislocations moving once movement has begun.

This yield-point effect in mild steel causes non-uniform localised deformation, which leads to the formation of "stretcher strains" (or "Lüder's lines") on the surface of such products as deep-drawn motor-car bodies. Such markings disfigure the surface, and their formation is avoided by applying slight initial deformation to the stock material, for example, by means of a light rolling pass. As this deformation is taking place in compression instead of in tension, stretcher strains will not form.

We will assume that, during this treatment, the stock material was stressed to the point *X* (Fig. 8.7). Since the material has now yielded, no further yield point will be encountered during the subsequent deep-drawing process. Thus, the stress–strain curve will follow *XZ*. If, however, the pre-stressed steel is allowed to remain at room temperature for a few months, or if it is heated to 300° C for a few hours, carbon atoms are able to move back to their initial positions relative to the dislocations and form their original "atmosphere", so that the yield-point effect returns. Since there are now more dislocations than were present in the original unstressed material, the new yield point will be

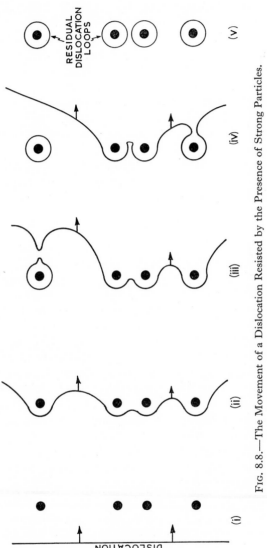

FIG. 8.8.—The Movement of a Dislocation Resisted by the Presence of Strong Particles.

at Y. This effect, which is known as "strain-ageing", can be caused by the presence of as little as 0·003% carbon or nitrogen. It is sometimes used to strengthen springs after coiling, but its effect in material destined for deep-drawing must be avoided. Consequently, such material is usually deep-drawn within twenty-four hours of the pre-stressing operation mentioned above.

8.62. Particle Hardening. The presence of small separate particles in the microstructure can impede the movement of dislocations provided that these particles are stronger than the matrix in which they are embedded. The degree of strengthening produced also depends on the size of particles, their distance apart and the tenacity of the bond between particles and matrix. The passage of a dislocation through an alloy containing isolated particles is represented in Fig. 8.8. Assuming that the particles are stronger than the matrix, the dislocation cannot pass through them, but if the stress is high enough, the dislocation can by-pass them leaving a "dislocation loop" around each particle. This will make the passage of a second dislocation much more difficult, particularly since dislocations have greater difficulty in passing between particles which are near to each other.

Particles of a separate phase are often precipitated from a solid solution during cooling. The cementite platelets in pearlite are an example, and their presence effectively strengthens steel. Such particles which have separated to form their own individual crystal structures are referred to as non-coherent precipitates. Prior to the actual precipitation of separate particles, however, an intermediate state often exists where atoms associated with the *solute* material begin to form groups *within the crystal lattice of the solvent metal*. These are small regions of less than one hundred atoms on an edge, and are called "coherent precipitates". Although still part of the continuous crystal structure of the matrix, they are attempting to form their own separate crystals, which will have a different crystal pattern. The presence of these coherent precipitates produces severe elastic distortions which impede the movement of dislocations. They are generally more effective than non-coherent precipitates. The nature of coherent precipitates, which give rise to the "precipitation hardening" of some aluminium alloys, beryllium bronzes and a number of other alloys, will be dealt with more fully later (9.82).

RECENT EXAMINATION QUESTIONS

1. Discuss the factors that may limit or prevent the formation of a substitutional solid solution in a binary alloy.
(Derby and District College of Technology.)

2. Describe, with the aid of sketches, the various types of solid solution which may be formed in metallic alloys.
Show, in each case, how the structure of the solid solution affects its behaviour during heat-treatment, giving typical examples among important engineering alloys.
(West Bromwich Technical College, A1 year.)

3. Describe, with the aid of sketches, how a cored structure develops in a substitutional solid solution.
Outline a theory which seeks to explain how the coring can be dispersed when such a solid solution is annealed at a sufficiently high temperature.
(U.L.C.I., Paper No. C111-1-4.)

4. Give an account of "solid-state diffusion", with specific reference to the eutectoid reaction in plain carbon steels and the heat-treatment of alluminium alloys.
(Aston Technical College, H.N.D. Course.)

5. State Fick's Law and show how it applies to the diffusion which takes place in a solid solution.
What other factors influence the rate of such diffusion?
(West Bromwich Technical College, A1 year.)

6. For what reasons may it be necessary to apply a heat-treatment process to a non-ferrous metal? Indicate the principles of each of the treatments you mention.
(University of Aston in Birmingham, Inter B.Sc.(Eng.).)

7. (a) Give a concise account of the nature of alloys.
(b) What is the difference between an interstitial and a substitutional solid solution?
(c) How do solid solutions differ from intermetallic compounds?
(Institution of Production Engineers Associate Membership Examination, Part II, Group A.)

8. What three factors in the constitution of an alloy are most important in determining its mechanical properties?
Illustrate the relevance of these factors in the case of any three industrial alloys with different applications.
(University of Strathclyde, B.Sc.(Eng.).)

9. Explain the following terms: (a) a substitutional solid solution; (b) an interstitial solid solution; (c) an electron compound; (d) a valency compound.
What factors affect the formation of solid solutions?
(Brunel College, Dip. Tech.)

10. Explain the following terms:

    (i) substitutional solid solution;
    (ii) electron compound;
    (iii) interstitial solid solution;
    (iv) eutectoid.

        (Derby and District College of Technology, H.N.D. Course.)

11. Give an account of how an element may alloy with another metal in the solid state.   (Brunel College, Dip. Tech.)

12. Write an account of the effect of the constitution of alloys on the industrial properties of them.

        (West Bromwich Technical College, A1 year.)

13. (a) State three reasons why *pure* metals are sometimes used in engineering. Give examples.

    (b) State four purposes of alloying. Illustrate your answer by reference to two well-known alloys.   (C. & G., Paper No. 293/6.)

CHAPTER IX

THERMAL-EQUILIBRIUM DIAGRAMS

9.10. As we saw earlier in this book (7.21), the thermal-equilibrium diagram is one of the metallurgist's most important "tools". With its aid he can find precisely what the structure of a given alloy will be at any given temperature, provided that the alloy has been allowed to reach equilibrium. In the metallurgical sense, "equilibrium" refers to the state of balance which exists, or which tends to be attained, between the phases in the structure of an alloy after a chemical or physical change has taken place. In some cases equilibrium may not be reached for long periods after the change has begun. In fact, rapid cooling (or "quenching") to room temperature may suppress the chemical or physical change involved to such an extent that equilibrium will never be reached so long as the alloy remains at room temperature. Reheating (or "tempering") followed by slow cooling to room temperature will allow the physical or chemical change to proceed in some degree towards completion; the extent to which equilibrium is thus attained being dependent upon the temperature reached, the time the alloy remains at that temperature and the rate of cooling to room temperature or at whatever other temperature equilibrium is being studied.

9.11. We have already seen that physical and chemical changes which take place after solidification in carbon steel do so with relative ease because the small carbon atoms can move quickly through the crystal lattice of face-centred cubic iron. Thus, during the process of normalising, a steel will attain structural equilibrium. In some alloy systems, however, physical changes which take place after solidification are so sluggish that it is doubtful if equilibrium is ever attained. Often we have no means of telling whether or not equilibrium has been reached, even after extremely slow rates of cooling involving periods of days or even weeks. There is therefore a tendency these days to refer to these charts as "constitutional diagrams" instead of "thermal-equilibrium diagrams". The latter term, however, is still in general use, and we shall employ it throughout this book.

9.12. The thermal-equilibrium diagram is in reality a chart which shows the relationship between the composition, temperature and structure of any alloy in a series. Let us consider briefly how these charts are constructed.

161

A pure metal will complete its solidification without change of temperature (4.12), whilst an alloy will solidify over a range of temperature which will depend upon the composition of the alloy (7.20). Consider, for example, a number of alloys of different composition containing the two metals A and B which form a series of solid solutions (Fig. 9.1).

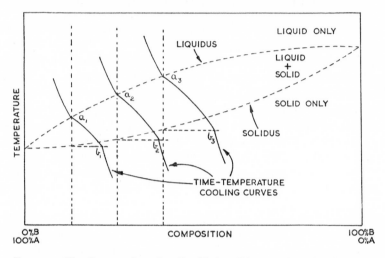

FIG. 9.1.—The Construction of an Equilibrium Diagram of the Solid-solution Type Using Cooling Curves.

For successive compositions containing diminishing amounts of the metal A, freezing commences at a_1, a_2, a_3, etc., and ends at b_1, b_2, b_3, etc. Thus, if we join all points a_1, a_2, a_3, etc., we shall obtain a line called the *liquidus*, indicating the temperature at which any given alloy in the series will commence to solidify. Similarly, if we join the points b_1, b_2, b_3, etc., we have a line, called the *solidus*, showing the temperature at which any alloy in the series will become completely solid. Hence the liquidus can be defined as the line above which all indicated alloy compositions represent completely homogeneous liquids, whilst the solidus can be defined as the line below which all represented alloy compositions of A and B are completely solid. For temperatures and compositions corresponding to the co-ordinates of points between the two lines both liquid solutions and solid solutions can co-exist in equilibrium.

9.13. Such a system as this occurs when the two metals are soluble in each other in all proportions in both the liquid and solid states, as in the case of copper and nickel or gold and silver. When, on the other hand, the metals are completely insoluble in the solid state a eutectic type of equilibrium diagram represents the system (Fig. 9.2). Since the eutectic part of the structure of any alloy in the series solidifies at the same temperature, and is the portion of the alloy with the lowest melting point,* the line $ACEDB$ will be the solidus whilst AEB is the liquidus.

9.14. Diagrams of this type can be constructed by using the appropriate points (obtained from time–temperature cooling curves) which indicate where freezing began and where it was complete. Thus the liquidus and solidus temperatures for a number of alloys in the series between 100% A and 100% B can be obtained and then plotted on a temperature–composition diagram as indicated in Figs. 9.1 and 9.2.

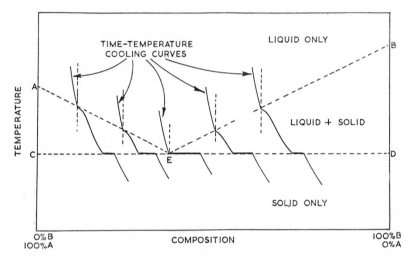

FIG. 9.2.—The Construction of an Equilibrium Diagram of the Eutectic Type Using Cooling Curves.

Such cooling curves will determine the positions of the liquidus and solidus; and if the alloy system is one in which further structural changes occur after the alloy has solidified, then the metallurgist must resort to other methods of investigation to determine the phase boundary lines. These methods include the use of X-rays, electrical-conductivity measurements and exhaustive microscopical examination following the quenching of representative specimens from different temperatures.

9.15. The construction of equilibrium diagrams, then, is the result of much tedious experimental work, and, since the conditions under which the work is carried out can be so variable, it is not surprising that the values assigned to compositions and temperatures at which phase changes occur are under constant review by research metallurgists. The

---

* "Eutectic" is derived from the Greek "eutektikos", which means "able to be melted easily".

reader will, therefore, find that the equilibrium diagrams printed in books on metallurgy often differ in small detail. In general, however, the variations are so small as not to affect the treatment of most of the industrially important alloys. As an example, the accepted value for the carbon content of a completely eutectoid carbon steel (7.37) has varied between 0·80 and 0·89% during recent years, whilst the eutectoid temperature (the lower critical temperature of plain carbon steels) has varied between 698 and 732° C.

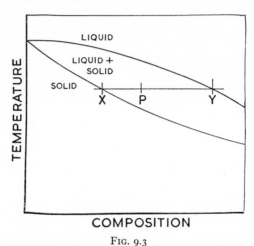

FIG. 9.3

9.20. In the interpretation of thermal-equilibrium diagrams the following definitions and rules* will be found helpful:

(*a*) The areas of the diagram are called "phase fields", and on crossing any sloping boundary line from one field to the next, the number of phases will always change by one, i.e. two single-phase fields will always be separated by a double-phase field containing both of the phases. In a binary system three phases can only exist together at a point, such as the eutectic point E in Fig. 9.2.

(*b*) At a point P (Fig. 9.3) in a two-phase field both liquid and solid can exist together. If a temperature horizontal is drawn through P, the composition of the solid is given by X, and the composition of the liquid solution in equilibrium with it by Y. P itself represents the overall composition of the mixture.

(*c*) The relative amounts of both liquid and solid at P are given by the relative lengths PX and PY—

---

* These rules are a modified combination of Gibbs' Phase Rule (8.13) and rules enunciated by the French metallurgist Portevin.

Weight of solid solution $\times PX =$ Weight of liquid solution $\times PY$

or $\qquad \dfrac{\text{Weight of solid solution}}{\text{Weight of liquid solution}} = \dfrac{PY}{PX}$

This is often referred to as the "lever rule", for reasons which will be apparent to engineers.

(d) A phase which does not occupy a field by itself, but appears only in a two-phase field, is either a pure metal or an intermetallic compound of invariable composition.

(e) If a vertical line, representing the composition of a given alloy, crosses some line of the diagram, it means that a change in the number of phases will occur at that point, i.e. a phase will be precipitated or absorbed.

We will now consider the basic types of equilibrium system, which are as follows:

In Which Two Metals, Mutually Soluble in all Proportions in the Liquid State, Become Completely Insoluble in the Solid State

9.30. Such cases are rare, since there is nearly always a small degree of solid solubility of one metal in another. One of the few examples in which no solubility has so far been detected in the solid state is supplied by the alloy system of the metals cadmium and bismuth. Incidentally, bismuth is a heavy, lustrous metal with a pinkish hue, and should not be confused with medicinal "bismuth" used for the relief of indigestion. This medicinal "bismuth" is actually bismuth carbonate, which is a white powder.

The bismuth–cadmium thermal-equilibrium diagram is shown in Fig. 9.4.

9.31. Let us consider the solidification of an alloy of composition x, i.e. containing about 80% cadmium and 20% bismuth. When the temperature falls to T, crystal nuclei of pure cadmium begin to form in accordance with rule (b) stated above. (The temperature horizontal, T, cuts the liquidus at the chosen composition, x, and the other phase boundary is the 100% cadmium ordinate.) Since pure cadmium is deposited, it follows that the liquid which remains becomes correspondingly richer in bismuth. Therefore the composition of the liquid moves to the left—say, to x_1—and, as indicated by the diagram, no further deposition of cadmium takes place until the temperature has fallen to T_1. When this happens more cadmium is deposited, and dendrites begin to develop from the nuclei which have already formed.

The growth of the cadmium dendrites, on the one hand, and the consequent enrichment of the remaining liquid in bismuth, on the other,
continues until the temperature has fallen to 140° C. The remaining
liquid then contains 40% cadmium and 60% bismuth, i.e. the eutectic
point E has been reached.

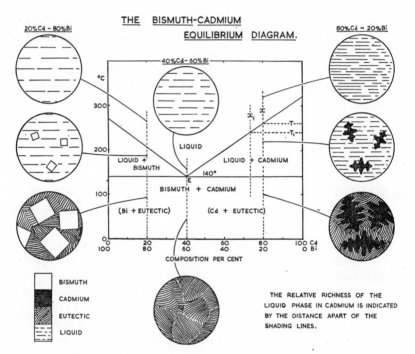

FIG. 9.4.—The Bismuth–Cadmium Equilibrium Diagram.

At this point the two metals are in equilibrium in the liquid, but, due
to the momentum of crystallisation, the composition swings a little too
far past the point E, resulting in the deposition of a little too much
cadmium. In order that equilibrium shall be maintained, a swing back
in composition across the eutectic point takes place by the deposition
of a layer of bismuth. In this way the composition of the liquid oscillates
about E by depositing alternate layers of cadmium and bismuth, whilst
the temperature remains at 140° C until the remaining liquid has
solidified. Thus the final structure will consist of primary crystals of
cadmium which formed between the temperature T and 140° C, and a
eutectic consisting of alternate layers of cadmium and bismuth which
formed at 140° C.

9.32. Had the original liquid contained less than 40% cadmium, then crystals of pure bismuth would have formed first, causing the composition of the remaining liquid to move to the right until ultimately the point E was reached as before, and the final liquid contained 40% cadmium and 60% bismuth. This remaining liquid would solidify as eutectic in the manner already described.

9.33. If the original liquid contained exactly 40% cadmium and 60% bismuth at the outset, then no solidification whatever would occur until the temperature had fallen to 140° C. Then a structure composed entirely of eutectic would be formed as outlined above.

9.34. In all three cases mentioned, the eutectic part of the structure will be of constant composition and will always contain 40% cadmium and 60% bismuth. Any variation either side of this in the *overall* composition of the alloy will be compensated for by first depositing appropriate amounts of either primary cadmium or primary bismuth, whichever is in excess of the eutectic composition. It is important to realise that there is no question of solid solubility existing in any way in the final structure, whatever its composition. With the aid of a microscope, we can see the two pure metals cadmium and bismuth as separate constituents in the microstructure. In other words, this is a case of complete insolubility in the solid state.

In which Two Metals, Mutually Soluble in all Proportions in the Liquid State, Remain Mutually Soluble in the Solid State

9.40. A number of pairs of metals fulfil these conditions, and are usually those which form disordered substitutional types of solid solution, since then the atoms of the second metal are able to replace those of the parent metal in all proportions. Moreover, the atoms are usually of a similar size in each of the metals, and the metals themselves often bear other physical and chemical resemblances to each other. Such is the case in the alloy systems of gold–silver, antimony–bismuth and copper–nickel, each of these pairs forming continuous series of solid solutions. Since the copper–nickel alloys are the only ones of the three groups mentioned which are commercially useful, it is proposed to deal with that system. The copper–nickel thermal-equilibrium diagram is shown in Fig. 9.5.

Again this is a simple type of equilibrium diagram, and since, as in the cadmium–bismuth system, no transformations take place in the solid, the diagram consists of two lines only—the liquidus and solidus. Above the liquidus we have a uniform liquid solution for any alloy in the series, whilst below the solidus we have a single solid solution for any alloy, though in the cast condition, as we shall see, the solid solution may vary in concentration due to coring. Between the liquidus and solidus both liquid and solid solutions co-exist.

9.41. Consider the freezing of an alloy of composition A (Fig. 9.5), i.e. containing 60% nickel and 40% copper. Assume that cooling is taking place rapidly, such as would be the case during the industrial

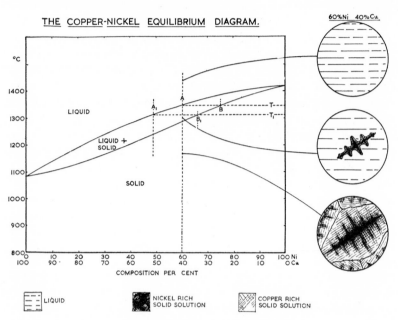

FIG. 9.5.—The Copper–Nickel Equilibrium Diagram.

The microstructures indicated are those obtained under *non-equilibrium* conditions of solidification.

casting of slab ingots into cast-iron moulds. Above $T°$ C the alloy will exist as a uniform liquid solution, but as the temperature falls to T, dendrites of solid solution will begin to form. They will not, however, be dendrites of composition A but dendrites of composition B; indicated by drawing a "temperature horizontal" from the point where the vertical line through composition A cuts the liquidus to the solidus. Thus the dendrites which form will contain approximately 75% nickel, and, since the original liquid contained only 60% nickel, it follows that the remaining liquid will contain an even lower proportion of nickel. Hence its composition will move to the left, say to A_1.

Solidification will continue when the temperature falls farther to T_1, and this time a layer of solid of composition B_1 will be deposited. This is less rich in nickel than the original seed crystals and, as crystallisation proceeds, successive layers will contain less and less nickel, and consequently more and more copper, until ultimately the liquid is used up.

Clearly, then, a non-uniform solid solution is formed, and whilst its *overall* composition will be 60% nickel and 40% copper, due to the coring effect the initial skeleton of a crystal will contain about 75% nickel and its outside fringes only about 50% nickel.

9.42. The situation is further influenced, however, by diffusion, which is taking place simultaneously with crystallisation. Due to the fact that successive layers of the alloy which deposit are richer in nickel than the remaining liquid, a concentration gradient is set up which tends to make copper atoms diffuse inwards towards the centre of the dendrite, whilst nickel atoms move outwards towards the copper-rich liquid. In the above case we have assumed very rapid cooling so that little or no diffusion has been possible, and that under these conditions the structure has been unable to reach equilibrium. Since we are dealing with non-equilibrium conditions, only a limited amount of information can be obtained from the equilibrium diagram—in this case a guide as to how the structure develops. If cooling were slow, diffusion could take place extensively so that coring would be slight in the final crystals. Annealing at a sufficiently high temperature will also permit diffusion to proceed (8.34), resulting in the formation of almost completely uniform solid-solution crystals.

9.43. We will now consider the cooling of the same alloy under conditions which are ideally slow so that complete equilibrium is attained at each stage of solidification. The liquid (composition A, Fig. 9.6) will begin to solidify at temperature T by depositing nuclei of composition B. This causes the composition of the remaining liquid to move to the left, but, due to the prevailing slow cooling, *diffusion is able to keep pace with solidification* so that,

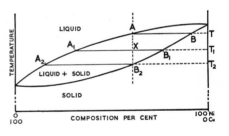

FIG. 9.6

as the composition of the liquid follows the liquidus from A to A_2, the composition of the solid follows the solidus from B to B_2.

Thus, at some temperature T_1, the composition of the *uniform* solid solution is given by B_1, whilst that of the remaining homogeneous liquid in equilibrium with it is given by A_1. Since the *overall* composition of the alloy is indicated by $X(A)$, then—

$$\frac{\text{Weight of solid solution (composition } B_1)}{\text{Weight of remaining liquid solution (composition } A_1)} = \frac{XA_1}{XB_1}$$

Under conditions of perfect equlibrium the thermal-equilibrium diagram therefore gives us complete information about the system.

At temperature T_2 the last trace of liquid (composition A_2) has just disappeared and, by the process of diffusion, its composition has been absorbed by the solid which is now of uniform composition B_2. A and B_2, of course, represent the same composition, which is obvious, since a uniform liquid has been replaced by a uniform solid.

In Which the Two Metals, Mutually Soluble in All Proportions in the Liquid State, Are only Partially Soluble in the Solid State

9.50. This case is, in effect, intermediate between those already dealt with, and the thermal-equilibrium diagram is consequently a sort of "hybrid" of the other two. As in the cadmium–bismuth system, a eutectic is formed, but in this case it is a eutectic of two solid solutions instead of two pure metals. It is in fact the "general" case and purists may argue that, as such, it should have been dealt with before either of the preceding particular cases. The systems have, however, been dealt with in order of complexity rather than logical classification in order to assist the student who may be encountering the subject for the first time.

The only new feature of this system, as compared with those foregoing, is that we have phase boundaries occurring below the solidus, indicating that phase changes can take place in the solid. On the tin–lead thermal-equilibrium diagram (Fig. 9.7), which we shall use as an example of this type of system, such phase boundaries as these are indicated by the lines *BC* and *FG*. A line such as *BC* or *FG* is often termed a SOLVUS. Part of the tin–lead system was used in Chapter VII as an introduction to equilibrium diagrams. We will now consider it in its entirety.

9.51. As the diagram shows, we have two solid solutions in the system, since:

(*a*) Tin will dissolve up to a maximum of 2·6% lead at the eutectic temperature, forming the solid solution α.

(*b*) Lead will dissolve up to a maximum of 19·5% tin at the eutectic temperature, giving the solid solution β.

(The different phases present in an alloy series are generally lettered from left to right across the diagram, using letters of the Greek alphabet for convenient reference.)

The slope of the phase boundaries *BC* and *FG* indicates that both the solubility of lead in tin (α) and the solubility of tin in lead (β) decrease as the temperature falls. This is a normal phenomenon with

THE TIN-LEAD EQUILIBRIUM DIAGRAM.

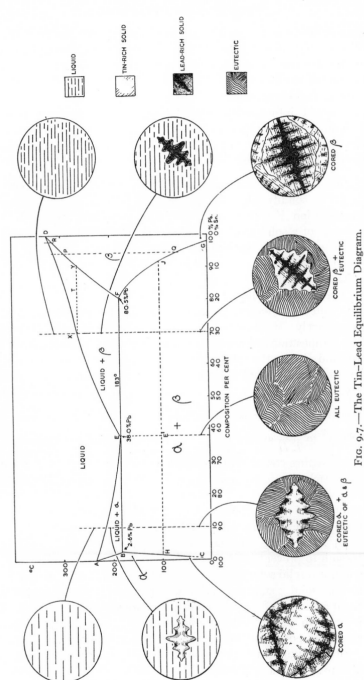

FIG. 9.7.—The Tin-Lead Equilibrium Diagram.

The microstructures indicated are those obtained under *non-equilibrium* conditions of solidification.

liquid solutions as well as with solid solutions, as was shown in the case of the solution of potassium chlorate in water (1.60). Similarly, cold tea will dissolve less sugar than will hot tea, though normally the amount of sugar we use is less than that required to form a saturated solution, whatever the temperature of the tea.

In the case of the tin–lead alloys the sloping phase boundaries BC and FG indicate changes taking place in the solid. Several different structures may be formed, depending upon the alloy composition. For example, let us consider:

(a) An alloy of composition X (70% lead–30% tin). This will begin to solidify when the temperature falls to T and dendrites of composition Y will deposit. The alloy continues to solidify in the manner of a solid solution until at 183° C the last layer of solid to form will be of composition F (80·5% lead–19·5% tin) and the remaining liquid will be of composition E (the eutectic composition with 38% lead and 62% tin).

The remaining liquid now solidifies by depositing, in the form of a eutectic, alternate layers of α and β, of compositions B and F respectively. If this structure now cools *slowly* to room temperature the compositions of the solid solutions α and β will follow the lines BC and FG, i.e the solid solution α will become progressively poorer in lead and the solid solution β will become poorer in tin, until at, say, 100° C α will contain less than 1% lead and β will contain less than 10% tin. The proportions of α and β will also vary from $\dfrac{EF}{EB}$ at 183° C to $\dfrac{E'\mathcal{J}}{E'H}$ at 100° C. In the same way, provided we assume that the structure cooled slowly enough to allow the primary dendrites of β to reach a uniform composition F, these dendrites will now alter in composition to $\mathcal{J}$ at 100° C, and in so doing precipitate some α of composition H in order to adjust their own composition. The α precipitated will join that already present in the eutectic.

(b) If the alloy contained, say, 95% lead, and was cooled slowly enough to prevent coring, solidification would be complete at P and a uniform solid solution β would result. On continuing to cool slowly, further solid changes would begin at Q with the precipitation of small amounts of α at the grain boundaries of the β. This α would increase in amount as the temperature fell, and the β became progressively poorer in tin. Hence the final structure would consist of crystals of uniform β containing about 98% lead with small amounts of α precipitated at the crystal boundaries.

(c) If the original alloy contained more than 98% lead and was cooled slowly, the structure would remain entirely β throughout, after its solidification had been completed at R.

9.52. In actual practice it is unlikely that cooling would be slow enough to prevent some coring from taking place. Since the initial β dendrites would then be relatively rich in lead, this could lead to the formation of small amounts of α at the β grain boundaries in (c) and more than the expected amount of α in (b). In case (c) this α would be absorbed on annealing. "Solution annealing" is an important process used in the treatment of many alloys to absorb some constituent which has been precipitated due to coring. After treatment the alloy will be soft and ductile and able to receive cold-work. Solution annealing is also an integral part of most strengthening processes based on precipitation hardening, the principles of which will be discussed later (9.82).

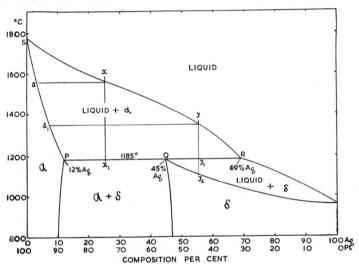

FIG. 9.8.—The Platinum–Silver Equilibrium Diagram.

In Which a Peritectic Reaction Takes Place

9.60. Sometimes in an alloy system, a solid phase which has already been formed will react with the remaining liquid to produce an entirely new phase. This is known as a peritectic reaction. The term is derived from the Greek "peri", which means "around", since, during the reaction, the original solid phase becomes surrounded or coated by the reaction product. Such a reaction occurs between the phase δ and the remaining liquid (producing austenite) in the iron–carbon system (11.20), but for the moment we will consider the peritectic reaction which takes place in the platinum–silver system shown in Fig. 9.8.

9.61. If we ignore coring effects, that is, assume very slow rates of cooling, the peritectic reaction will occur only in those platinum–silver alloys containing between 12 and 69% silver. Let us first consider a liquid of composition x, i.e. containing about 25% silver. This will begin to solidify by depositing dendrites of the solid solution α of composition a. If the rate of cooling is slow enough for diffusion to remove the effects of coring, by the time we reach 1 185° C the structure will consist of α dendrites containing 12% silver (composition P) and a remaining liquid containing 69% silver (composition R), since the composition of α will change as it follows the solidus SP. At this stage—

$$\frac{\text{Weight of } \alpha \text{ (composition } P)}{\text{Weight of remaining liquid (composition } R)} = \frac{x_1 R}{x_1 P}$$

9.62. At 1 185° C, the peritectic temperature, the α dendrites begin to react with the remaining liquid and form the solid solution δ, which contains 45% silver (composition Q). Since x_1 lies between P and Q this means that there is an excess of the solid solution α, so that the liquid is used up first. Hence the final structure will consist of α containing 12·0% silver (composition P), and δ, containing 45·0% silver (composition Q). These phases will be in the ratio—

$$\frac{\text{Weight of } \alpha \text{ (composition } P)}{\text{Weight of } \delta \text{ (composition } Q)} = \frac{x_1 Q}{x_1 P}$$

9.63. Let us now consider a liquid alloy of initial composition y. This will begin to solidify by depositing crystals of α of composition a_1, and if we again neglect coring, the crystals will gradually change in composition to P when the temperature has fallen to 1 185° C, and again the composition of the remaining liquid will be R. At this stage—

$$\frac{\text{Weight of } \alpha \text{ (composition } P)}{\text{Weight of remaining liquid (composition } R)} = \frac{y_1 R}{y_1 P}$$

When the peritectic reaction takes place, we shall find that this time there is an excess of liquid because y_1 lies between Q and R. The α will therefore be completely used up, and just below 1 185° C we shall be left with the new solid solution, δ, and some remaining liquid in the ratio—

$$\frac{\text{Weight of } \delta \text{ (composition } Q)}{\text{Weight of remaining liquid (composition } R)} = \frac{y_1 R}{y_1 Q}$$

As the temperature continues to fall slowly, this remaining liquid will solidify as δ, which will change in composition along Qy_2. At y_2 solidification will be complete and the structure will consist of crystals of uniform δ of the same composition as that of the original liquid. In actual practice we would probably find some α remaining in the final structure because it had become coated with a protective layer of δ when the reaction began.

In Which the System Contains One or More Intermediate Phases (Intermetallic Compounds)

9.70. Frequently two metals will not only form limited solid solutions with each other but, in other suitable proportions, will also combine to form intermetallic compounds. The formation of an intermetallic com-

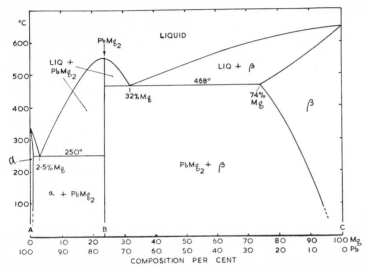

Fig. 9.9.—The Lead–Magnesium Equilibrium Diagram, Showing the Effect of the Introduction of a Single Intermetallic Compound on the System.

pound further complicates the thermal-equilibrium diagram by introducing another phase, and since some alloy systems contain a number of intermetallic compounds, formed at different compositions within the system, the resulting diagram may appear somewhat complex. Let us examine the lead–magnesium system (Fig. 9.9) as being a simple example of this combination type of diagram.

Here only one intermetallic compound, $PbMg_2$ (containing 23·4% magnesium), is formed, and it will be apparent that we really have two separate simple equilibrium diagrams joined at *B*. The section *AB* is a eutectiferous series (of the tin–lead type) involving α and $PbMg_2$; whilst the section *BC* is a similar eutectiferous series involving $PbMg_2$ and the solid solution β. If we consider the complete system between pure lead and pure magnesium, then, clearly, the two sections become joined as shown, forming one complete diagram.

Precipitation from a Solid Solution

9.80. When the temperature of a solid solution falls such that it reaches a state of saturation, any further fall in temperature will lead to the precipitation of some second phase. This phenomenon was mentioned in 9.51 and is similar in principle to the precipitation which takes place from saturated liquid solutions when these continue to cool (1.60). Precipitation from a solid solution, however, takes place much more sluggishly than from a liquid solution because of the greater difficulty of movement of the solute atoms in a solid solution, particularly if it is of the substitutional type. We must, therefore, consider carefully the relationship between the rate of cooling of a saturated solid solution and the extent to which precipitation can take place— and consequently equilibrium be attained—either during the cooling process or subsequently.

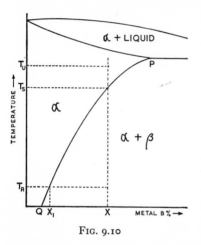

FIG. 9.10

9.81. Precipitation under Equilibrium Conditions. Fig. 9.10 represents part of a system such as was described in 9.50. Here metal B is partially soluble in metal A in the solid state forming the solid solution α. Any B in excess of solubility at any given temperature is precipitated, under equilibrium conditions, as the phase β, which in this case may be either a solid solution or an intermetallic compound.*

Consider the slow cooling of some alloy of composition X from the temperature T_U. At T_U the solid solution α is unsaturated and this state of affairs prevails until the temperature falls to T_S. Here the solid solution reaches saturation and when the temperature falls below T_S, precipitation may begin. Random nucleation takes place at the grain boundaries as well as on certain crystallographic planes in the α matrix, and nuclei of β begin to form. As the alloy cools, precipitation of β continues, and since β is rich in B, the composition of α changes progressively along the solvus line PQ. Because β is richer in B than is α, B must diffuse through α in order to reach the growing nuclei of β. This tends to reduce the concentration of B rather more rapidly in that

---

* Since the remainder of the diagram is not given, we cannot say whether β is a solid solution as indicated in 9.50 or an intermetallic compound as indicated in 9.70.

α near to the β nuclei than in those regions of α which are far away from any nuclei. The concentration gradient is, however, too small to cause rapid diffusion of the B atoms, and, since the temperature is falling, regions of α far from any β nuclei ultimately become supersaturated with excess B. Consequently more β nuclei will form and begin to grow.

Precipitates which form under equilibrium conditions in this manner are generally non-coherent. That is, the new phase which has formed has a crystal structure which is entirely its own and is completely separate from the surrounding matrix from which it was precipitated. Its strengthening effect on the alloy as a whole is somewhat limited (8.62) and generally much less than that resulting from the presence of a coherent precipitate, the formation of which will be dealt with in the next section.

Assuming that the temperature has fallen to T_R and that the precipitation of β has taken place under equilibrium conditions, the structure will consist of non-coherent particles of β (usually of such a size as can be seen easily using an ordinary optical microscope) in a matrix of α which is now of composition X_1, i.e. it contains much less B than did the original α (composition X).

9.82. Precipitation under Non-equilibrium Conditions.

We will now assume that the alloy X, which has been retained at temperature T_U, long enough for its structure to be completely homogeneous α, is cooled very rapidly to room temperature (T_R). This could be achieved by quenching it in cold water. In most cases treatment such as this will prevent any precipitation from taking place and we are left with α solid solution which, at T_R, is now super-saturated with B. The structure will not be in equilibrium and is said to be in a *metastable* state. It has an urge to return to a state of equilibrium by precipitating some β. At room temperature such precipitation will be unlikely to occur due to the extreme sluggishness of movement of B atoms, but if the temperature is increased, diffusion will begin and then accelerate as the temperature continues to rise.

The alloy is held at some selected temperature so that diffusion occurs at a low but definite rate. The temperature chosen is well below T_S in order to ensure that non-coherent precipitation does not occur. During this heat-treatment clusters of atoms of both A and B, but with the overall composition of the phase β, slowly form groups at many points in the α lattice. *These clusters have the important property that their lattice structures are continuous with that of the α matrix*, and there is no discontinuous interface as exists between α and β, which has precipitated during cooling under equilibrium conditions. This unbroken continuity of the two lattices is known as *coherency*. Since the cluster size is extremely small and the rate of diffusion very slow, a large number of these coherent nuclei will form and the chosen temperature will not be high enough to allow the formation of a separate β structure. Instead, this intermediate structure—we will call it β′—is produced, and the mismatching between β′ and the α matrix leads to distortion in the

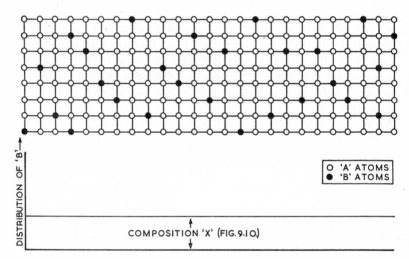

(A) The alloy has been heated to T_U (Fig. 9.10) and quenched. It consists of homogeneous α, which is supersaturated with B.

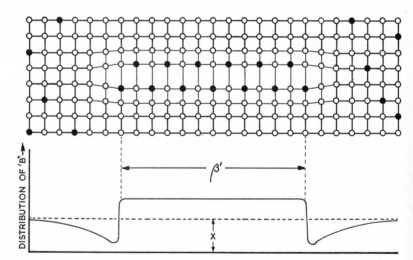

(B) Heating at some selected temperature causes the B atoms to migrate and form clusters within the α lattice. These clusters produce β′, which, although similar to β, is still continuous (coherent) with the original α lattice.

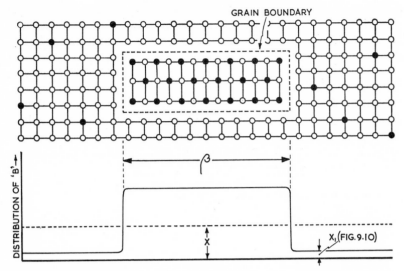

(C) Here equilibrium has been attained—the alloy has been heat-treated (9.81) so that β has been rejected from α as a non-coherent precipitate. The lattice parameters of β no longer "match up" to those of α as do those of the intermediate β′. Hence we have a crystal boundary as indicated by the broken line.

FIG. 9.11.—The Formation of Coherent and Non-coherent Precipitates.

α lattice in the neighbourhood of these nuclei. Such distortions will hinder the movement of dislocations and so the strength and hardness increase. The greater the number of nuclei and the larger they are, provided that coherency is retained, the greater the strength and hardness of the structure as a whole. Time and temperature will influence this "precipitation" procedure and, hence, also the ultimate mechanical properties which result. If the temperature is high (T_1, Fig. 9.12) a

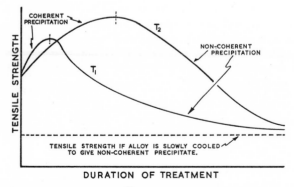

FIG. 9.12

high rate of diffusion prevails, and this in turn leads to the formation of relatively few nuclei, which, however, will grow to a large size. At a lower temperature (T_2) a larger number of nuclei will form and grow slowly so that strength increases slowly.

Alloys Containing More Than Two Metals

9.90. In this chapter we have dealt only with binary alloys, that is, alloys containing two different metals. In the case of ternary alloys, namely, those containing three different metals, a third variable is obviously introduced, so that the system can no longer be represented by

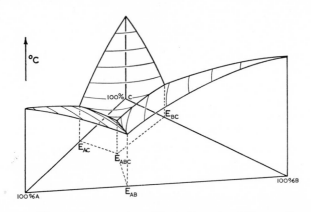

FIG. 9.13.—Diagram Representing a Three-dimensional Ternary System, of Three Metals A, B and C.

The curved isothermal contour lines meet binary eutectics which converge to a ternary eutectic point, E_{ABC}.

a two-dimensional diagram. Instead we must work in *three* dimensions in order adequately to represent the alloy system. The base of the three-dimensional diagram will be an equilateral triangle, each pure metal being represented by an apex of the triangle. Temperature will be represented by an ordinate perpendicular to the triangular base. Fig. 9.13 illustrates a ternary-alloy system of the metals A, B and C in which the binary systems A–B, B–C and A–C are all of the simple eutectic type. The three binary liquidus *lines* now become liquidus *surfaces* in the three-dimensional system, and these surfaces intersect in three "valleys" which drain down to a point of minimum temperature, vertically above E_{ABC}. This is the ternary eutectic point of the system, its temperature being lower than that of either of the binary eutectic points E_{AB}, E_{BC} or E_{AC}.

Equilibrium diagrams for alloys of four or more metals cannot be represented in a single diagram since more than three dimensions would be required. In practice such a system can often usefully be represented in the form of a pseudo-binary diagram in which the concentration of one component is varied whilst the others are kept constant—this is equivalent to taking a vertical "slice" through a ternary system, parallel to one of the end faces. One of these pseudo-binary systems is shown in Fig. 14.1. Such systems can be interpreted in the same way as an ordinary binary system.

We have dealt here only with equilibrium diagrams of the simple fundamental types. Many of those which the reader will encounter are far more complicated and often contain such a multitude of different phases as almost to exhaust the Greek alphabet. This is particularly true of equilibrium diagrams which represent those copper-base alloy systems in which a number of intermetallic compounds are formed and in which peritectic reactions are also common. In general, we are interested only in those parts of the diagram near to one end of the system, where we usually find a solid solution with, possibly, small amounts of an intermetallic compound. The interpretation then becomes much simpler, and the reader has been provided with sufficient information in this chapter to deal with most of the alloy systems likely to be encountered.

RECENT EXAMINATION QUESTIONS

1. Beryllium (m.pt 1 282° C) and silicon (m.pt 1 414° C) are completely soluble as liquids but completely insoluble as solids. They form a eutectic at 1 090° C containing 61% silicon.
    Draw the equilibrium diagram and explain, with the aid of diagrams, what happens when alloys containing (i) 10% silicon, (ii) 70% silicon, solidify completely.   (West Bromwich Technical College, A1 year.)

2. Explain the meaning of the terms *eutectic, eutectoid* and *solid solution*. Illustrate your answer by means of typical thermal-equilibrium diagrams and sketches of representative microstructures.
                            (U.L.C.I., Paper No. B111-1-4.)

3. Give an account of the factors which affect the formation of solid solutions between two metals. Show how these factors operate in the case of the metals copper and nickel and how the resulting properties of the alloys suit them for particular applications.
                            (University of Strathclyde, B.Sc.(Eng.).)

4. Draw the thermal-equilibrium diagram for two metals that show complete liquid and solid solubility.
    Describe the cooling of a typical alloy.
    What are the effects of non-equilibrium cooling?
                            (Nottingham Regional College of Technology).

5. With reference to a hypothetical equilibrium diagram, describe how coring occurs during solidification, under normal industrial conditions, of a solid-solution alloy.

Describe how coring is prevented or removed by (i) cooling under equilibrium conditions, and (ii) annealing after normal solidification.

(Derby and District College of Technology.)

6. Two metals, A and B, have melting points of 400° C and 700° C respectively. Cooling curves were plotted for several binary alloys of A and B and the following data was obtained:

% B in alloy	10	30	50	70	90
First arrest point ° C.	360	250	400	530	650
Second arrest point ° C.	250	—	250	250	400
Third arrest point ° C.	—	—	—	—	160

On squared paper draw accurately the equilibrium diagram for the binary alloy system of A and B, and fully label this. For an alloy containing 60% B calculate:

(i) the temperature at which solidification begins;
(ii) the ratio of phases present at 300° C (assuming equilibrium to have been attained).

(The Polytechnic, London W.1, Poly.Dip.)

7. The following results indicate the temperatures associated with discontinuities in the cooling curves of the alloys indicated:

% B. . .	0	10	20	35	50	55	60	75	90	95	100
° C. . .	600	545	495	410	330	300	315	365	415	430	450
° C. . .	—	450	300	300	300	—	300	300	300	370	—

If the maximum and minimum percentage solubilities of the two metals are 20B in A, 10B in A, and 10A in B, 5A in B, sketch and label the equilibrium diagram. (Assume solubility lines are straight.)

(a) Describe the cooling of an alloy contining 35% B and sketch typical microstructures.
(b) What proportions of α and β would you expect in the eutectic alloy at the eutectic temperature and at 0° C?
(c) Indicate a composition which might give a Widmanstätten* structure. Sketch the macrostructure.

(Constantine College of Technology, H.N.D. Course.)

* Discussed later in this book (10.10 and 11.53).

8. Draw a thermal-equilibrium diagram representing the system between two metals, X and Y, given the following data:

   (i) X melts at 1 000° C and Y at 800° C;
   (ii) X is soluble in Y in the solid state to the extent of 10·0% at 700° C and 2·0% at 0° C;
   (iii) Y is soluble in X in the solid state to the extent of 20·0% at 700° C and 8·0% at 0° C;
   (iv) a eutectic is formed at 700° C containing 40·0% X and 60·0% Y.

   Describe what happens when an alloy containing 70·0% X solidifies and cools slowly to 0° C.
   Sketch the microstructures of an alloy containing 15·0% Y (a) after it has cooled slowly to 0° C; (b) after it has been annealed at 700° C and then water-quenched. (West Bromwich Technical College, A1 year.)

9. Two metals, A and B, have melting points 750° C and 500° C respectively.
   They form a eutectic at 75% B which melts at 400° C. Their solubilities are, at eutectic temperature, 20% B in A and 10% A in B; and at 0° C, 5% B in A and 10% A in B.
   From the above information draw the equilibrium diagram for the system, clearly marking all phases present. From the diagram determine what structures would be obtained in slowly cooled alloys of the following compositions: 10% B; 40% B; 75% B; 85% B; 95%B.
   (Rutherford College of Technology, Newcastle upon Tyne.)

10. Draw the equilibrium diagram for two metals completely miscible in the liquid state and partially miscible in the solid state. Show how the diagram can be used to predict the equilibrium proportions of phases present for three representative alloys. Sketch the microstructures you would obtain in the three alloys chosen.
    (Nottingham Regional College of Technology.)

11. What is (a) a eutectic, (b) a peritectic? In each case name one commercial alloy system which exhibits the effect.
    Describe typical microstructures obtained with various compositions in a eutectic system with partial solid solubility.
    (The Polytechnic, London W.1.)

12. What is meant by a "peritectic reaction"? Sketch an equilibrum diagram of a system containing such a reaction and describe the equilibrium cooling of TWO typical alloys, at least one of which undergoes a peritectic reaction. What is the effect of faster cooling on the reaction?
    (Brunel College, Dip. Tech.)

13. Platinum (m.pt 1 773° C) dissolves 12·0% silver (m.pt 961° C) at 1 185° C forming a solid solution α. The liquid in equilibrium with α at this temperature contains 69·0% silver and 31·0% platinum. At 1 185° C a peritectic reaction takes place between α and the remaining liquid forming another solid solution, δ, which contains 45·0% silver and 55·0% platinum.

Draw as much of the thermal-equilibrium diagram as is possible from the details given above and explain what happens when liquids containing (*a*) 70% platinum/30% silver; (*b*) 50% platinum/50% silver, solidify slowly.   (U.L.C.I., Paper No. C111–1–4.)

14. Tungsten dissolves 5·0% platinum at 2 460° C to form a solid solution α. The composition of the liquid in equilibrium with α at this temperature is 61·0% tungsten/39·0% platinum. At 2 460° C a peritectic reaction takes place between α and the remaining liquid forming a solid solution, β, which contains 63·0% tungsten/37·0% platinum. At 1 700° C, α contains 4·8% platinum and β contains 37·1% platinum.

Draw the thermal-equilibrium diagram between 3 500° C and 1 700° C and explain what happens when liquids containing (i) 80·0% tungsten; (ii) 62·0% tungsten solidify slowly. (M.pts: Tungsten—3 400° C; Platinum—1 773° C).

(West Bromwich Technical College, A1 year.)

CHAPTER X

PRACTICAL METALLOGRAPHY

10.10. Aloys Beck von Widmanstätten lived to the venerable age of ninety-five, and when he died in 1849, had been, successively, owner of a printing works; an editor in Graz; manager of a spinning mill near Vienna; and, between 1806 and 1816, director of the State Technical Museum in Vienna. In 1808 he discovered that some meteorites, when cut and polished, developed a characteristic structure when subsequently oxidised by heating in air. Later, he found that etching with nitric acid gave better results and revealed the type of metallurgical structure which still bears his name. It may therefore be argued that von Widmanstätten originated metallographic examination. The microscope, however, was not used in this direction until 1841, when Paul Annosow used the instrument to examine the etched surfaces of Oriental steel blades.

In the early 1860s Professor Henry C. Sorby of Sheffield developed a technique for the systematic examination of metals under the microscope and can therefore lay claim to be the founder of that branch of metallurgy known as microscopical metallography. Some of the photomicrographs he produced are excellent even by modern standards.

It would not be possible in a book of this type to deal exhaustively with the subject of the microscopical examination of metals. However, it is hoped that what follows may help the reader to prepare representative microstructures for himself, and so equip him to be able, as a practising engineer, to trace many of the more common causes of failure which are attributed to microstructural defects.

Much useful work can be done with a minimum of equipment, and even at to-day's prices (1973) a sum of £450 would provide sufficient basic apparatus to enable a trained eye to evaluate most of the common microstructural defects encountered in commercial metals and alloys. Metallurgical knowledge of this type can only be accumulated as a result of practice and experience. Ability in this branch of metallurgical technique is, like beauty, in the eye of the beholder.

The Preparation of Specimens for Microscopical Examination

10.20. In preparing a specimen for microscopical examination it is first necessary to produce in it a surface which appears perfectly flat and scratch-free when viewed with the aid of a microscope. This involves first grinding the surface flat, and then polishing it to remove the marks left by grinding. The polishing process causes a very thin layer of amorphous metal to be burnished over the surface of the specimen, thus hiding the crystal structure. In order to reveal its crystal structure the specimen is "etched" in a suitable reagent. This etching reagent dissolves the "flowed" or amorphous layer of metal.

That, briefly, is the basis of the process employed in preparing a specimen for examination; but first it will be necessary to select a representative sample of the material under investigation.

10.21. Selecting the Specimen. The selection of a specimen for microscopical examination calls for a little thought, since a large body of metal may not be homogeneous either in composition or crystal structure. Sometimes more than one specimen will be necessary in order adequately to represent the material. In some alloys the structure may also exhibit "directionality", as, for example, in wrought iron. In a specimen of the latter, cut parallel to the direction of rolling, the slag will appear as fibres elongated in the direction of rolling, whilst a section cut at right angles to the direction of rolling will show the slag as apparent spherical inclusions, and give no hint to the fact that what is being observed is a cross-section *through* slag fibres.

For the examination of surface defects a specimen must be chosen so that a section through the surface layer is included in the face to be polished. Surface cracks and the like should be investigated by cutting a piece of metal containing the crack and mounting it in bakelite or a similar compound. The surface to be polished is then ground sufficiently so that a section through the crack is obtained.

A specimen approximately 20 mm diameter or 20 mm square is a convenient size to handle. It is difficult to grind a perfectly flat surface on smaller specimens, and these are best mounted as described below. The specimen should not be more than 12 mm thick, or it may rock during polishing, producing a bevelled surface.

10.22. When it is necessary to preserve an edge, or when a specimen is so small that it is difficult to hold it flat on the emery paper, the specimen may be mounted in a suitable compound. This can be done most satisfactorily by using one of the proprietary materials such as "N.H.P." plastic mounting compound. This consists of a powder which, when mixed with the liquids supplied with it, hardens to give a solid plastic substance which will retain a metal specimen during and after the polishing operation. Only simple apparatus of the type shown in Fig. 10.1 is required. The specimen is placed on a suitable flat surface and the two L-shaped retaining pieces arranged around it. (If possible, these retaining members should be held in position with a small clamp.)

The specimen is covered with the powder, and this is then moistened
with the first liquid. The whole is then saturated with the liquid

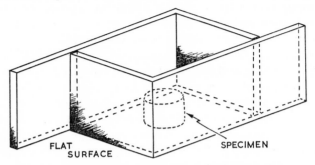

FIG. 10.1.—Method of Mounting Specimen in Plastic Material
where no pressure is required.

"hardener". In about twenty minutes the mass will have set hard and
the L-shaped members can be detached.

Specimens can be mounted more quickly by using some thermosetting substance, such as bakelite or, alternatively, a transparent

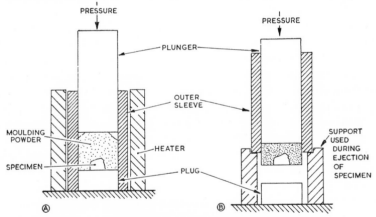

FIG. 10.2.—Mould for Mounting Specimens in Plastic Materials
when pressure is necessary.
(A) Moulding the mount.
(B) Ejecting the finished mount.

thermoplastic material. These substances mould at about 150° C,
which is usually too low a temperature to cause any structural change
in the specimen. They can be ground and polished easily and do not
promote any electrolytic action during etching. A small mould (Fig.
10.2) is required in conjunction with a press capable of giving pressures
up to about 25 N/mm².

After placing the specimen, the powder and the plunger in the mould the latter is heated by means of a special electric heater which encircles it. If this is not available a bunsen burner will suffice. In either case a thermometer should be inserted in a hole provided in the plunger so that overheating of the mould is avoided. Some mounting powders decompose at high temperatures, with the formation of dangerously high pressures.

10.23. Grinding and Polishing the Specimen. It is first necessary to obtain a reasonably flat surface on the specimen. This can be done either by using a fairly coarse file or, preferably, by using a motor-driven emery belt. If a file is used it will be found easier to obtain a flat surface by rubbing the specimen on the file than by filing the vice-held specimen in the orthodox way. Skilled workshop technologists may wince at the thought of such a procedure, but it is guaranteed to produce a flat surface for those readers who, like the author, possess negligible skill in the use of a file. Whatever method is used, care must be taken to avoid overheating the specimen by rapid grinding methods, since this may lead to alterations in the microstructure. When the original hack-saw marks have been ground out, the specimen (and the operator's hands) should be thoroughly washed in order to prevent carry-over of filings and dirt to the polishing papers.

Intermediate and fine grinding is then carried out on emery papers of progressively finer grade. These must be of the very best quality, particularly in respect of uniformity of particle size. With modern materials not more than four grades are necessary (220, 320, 400 and 600 from coarse to fine), since by using a paper with a waterproof base wet grinding can be employed. Special grinding tables are also available in which 300 mm × 50 mm strips of emery paper are clamped, side by side, on a sloping glass plate, the surface of which is flushed by a current of water (Plate 10.1A). This not only acts as a lubricant but also carries away particles of grit and swarf which might damage the surface being ground. The specimen is drawn back and forth along the entire length of the No. 220 paper, so that scratches produced are roughly at right angles to those produced by the preliminary grinding operation. In this way it can easily be seen when the original scratches produced by the primary grinding operation have been completely removed. If the specimen were ground in the same direction so that the new scratches were parallel to the original ones this would be virtually impossible. Having removed the primary grinding marks, the specimen is washed free of No. 220 grit. Grinding is then continued on the No. 320 paper, again turning the specimen through 90° and polishing until the previous scratch marks have been removed. This process is repeated with the No. 400 and No. 600 papers.

In cases where dry grinding is still used, complete cleanliness must be maintained at all stages in order to avoid carry-over of coarse emery grit to the finer papers. Ideally, a new paper should be used for each specimen, but, in the interests of economy, used papers can be shaken free of grit by pulling them taut repeatedly. They can be stored effectively between the pages of a glossy magazine. Alternatively, a strip of each grade of paper can be attached permanently to its own polishing block. It is essential that the specimen be washed in running water before passing from one grade of paper to the next, and particularly before passing to the final polishing cloth.

[*Courtesy of Messrs. C. Reichert, Vienna.*]

10.1A.—A Modern Wet-grinding "Deck".

Four strips of emery paper (water proof base) are held on a sheet of glass by means of the upper perspex "platen". A stream of water flows down the papers into the sump and thence to a drain.

10.1B.—Rotary Polishing Machine for Finishing Metallographic Specimens.

PLATE 10.1

Normally, most of the steels and harder non-ferrous alloys can be ground dry, provided that care is taken not to overheat them. For the softer aluminium alloys, bearing metals, etc., the paper should be impregnated with a lubricant, such as paraffin. A lighter pressure can then be used, and there is much less chance of particles of grit becoming embedded in the soft metal surface. However, modern wet-grinding processes, as outlined earlier in this section, are far more satisfactory for all materials, and have almost completely replaced dry grinding methods.

So far the operation has been purely one of grinding, and if our efforts have been successful we shall have finished with a specimen whose surface is covered by a series of parallel grooves cut by the particles on the last emery paper to be employed. The final polishing operation is somewhat different in character and really removes the ridged surface layer by means of a burnishing operation. When polishing is complete the ridges have been completely removed, but the mechanism of polishing is such that it leaves a "flowed" or amorphous layer of metal on the surface (Fig. 10.3). This hides the crystal structure and must be dissolved by a suitable etching reagent.

Irons and steels are polished by means of a rotating cloth pad (Plate 10.1B) impregnated with a suitable polishing medium. "Selvyt" cloth is possibly the best-known material used to cover the polishing wheel, though special cloths, such as "Metron", are now available and are generally more suitable for this purpose. The cloth is thoroughly wetted with *distilled* water and a small quantity of the polishing powder

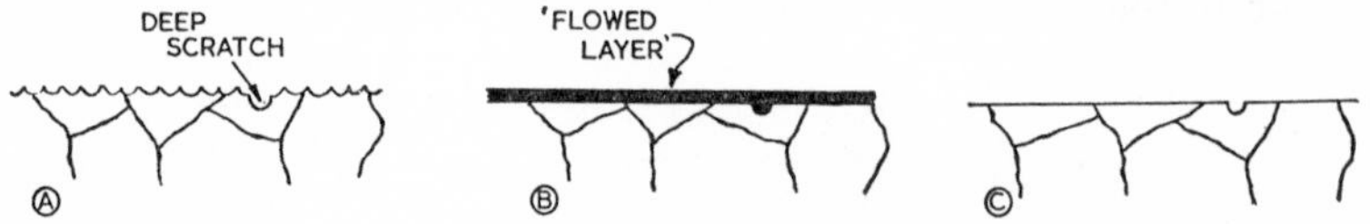

FIG. 10.3

(A) Grooves Produced in the Metal Surface by the Final Grinding Operation.
    A deep scratch produced by a particle of coarse grit is shown.
(B) Final Polishing Has Produced a "Flowed Layer".
    This may cover the deep scratch as shown, rendering it invisible.
(C) Etching Has Removed the "Flowed Layer", thus Revealing the Crystal Structure Beneath.
    Unfortunately the deep scratch is also visible again.

worked in with *clean* finger-tips. The most popular polishing powder is alumina (aluminium oxide) which has been prepared in a pure state by calcining ammonium alum. It is usually sold under the trade name of "Gamma Alumina". Finer grades of this material are supplied as a thick suspension in water. A constant drip of water should be applied to the wheel, which should be run at low speeds until the operator has acquired the necessary manipulative technique. Light pressures should be used, since heavy pressures are more likely to result in scratches being formed by grit particles embedded deep in the cloth. Moreover, the use of light pressure is less likely to result in the specimen being suddenly projected across the laboratory.

If specimens are to be polished in fairly large numbers it will be worth while using one of the proprietary diamond-dust polishing compounds such as "Dialap". In these materials the graded diamond particles are carried in a "cream" base which is soluble in both water and the special polishing fluid, a few spots of which are applied to the polishing pad in order to lubricate the work and promote even spreading of the compound. These compounds are graded and colour-coded according to particle size (in micrometres). For polishing irons and steels it is generally convenient to use a two-stage technique necessitating two polishing wheels. Preliminary polishing is carried out using a 6 μm particle size. The specimen is then washed and finished on the second wheel using a 1 μm material. Since these compounds are expensive, it is desirable that the operator should have some manipulative skill in order that frequent changing of the polishing pad is not made necessary due either to tearing of the cloth or lack of cleanliness in working. At the same time the polishing cloths have a much longer life than is the case with those polishing media which tend to dry on the cloth and render it unfit for further use. The oil-based lubricant used with diamond pastes keeps the cloth in good condition so that it can be used intermittently over long periods, provided that the pad is kept covered to exclude dust and grit. In this way the higher cost of the diamond paste can be offset.

Non-ferrous specimens are best polished by hand on a small piece of "Selvyt" cloth wetted with "Silvo". Polishing should be accomplished with a circular sweep of the hand, instead of the back-and-forth motion used in grinding. As at every other stage, absolute cleanliness should be observed if a reasonably scratch-free surface is to be obtained. Copper alloys can be polished quickly by passing from the No. 400 grade of emery paper to a piece of good-quality chamois leather wetted with "Brasso". The grinding marks are very quickly removed in this way, and final polishing is then accomplished with "Silvo" and "Selvyt" cloth. These polishing agents may be found unsuitable in a few cases because of an etching action on the alloy. It is then better to use magnesium oxide (magnesia).

When the specimen appears to be free from scratches it is thoroughly cleaned and examined under the microscope using a magnification of about 50 or 100. If satisfactorily free from scratches the specimen can be examined for inclusions, such as manganese sulphide (in steel) or slag fibres (in wrought iron), before being etched.

To summarise, the most important factors affecting a successful finish are:

(a) Care should be taken not to overheat the specimen during grinding. In steel this may have a tempering effect.

(b) Absolute cleanliness is essential at every stage.

(c) If a specimen has picked up deep scratches in the later stages of grinding it is useless to attempt to remove them on the polishing pad. If a specimen is polished for too long on the pad its surface may become rippled.

(d) Apply light pressure at all times during grinding and polishing.

10.24. Etching the Specimen. Before being etched the specimen must be absolutely clean, otherwise it will undoubtedly stain during etching. Nearly every case of failure in etching can be traced to inadequate cleaning of the specimen so that a film of grease still remains.

The specimen should first be washed free of any adhering polishing compound. The latter can be rubbed from the *sides* of the specimen with the fingers, but care must be exercised in touching the polished face. The best way to clean this is by very gently smearing the surface with a finger-tip dipped in grit-free soap solution, and washing under the tap. Even now the specimen may be slightly greasy, and the final film of grease is best removed by immersing the specimen in boiling alcohol ("white" industrial methylated spirit) for about two minutes. The alcohol should not be heated over a naked flame, but preferably by an electrically heated water-bath.

From this point onwards the specimen must not be touched by the fingers but handled with a pair of nickel crucible tongs. It is removed from the alcohol and cooled in running water before being etched. With specimens mounted in thermoplastic materials it may be found that the mount is dissolved by hot alcohol. In such cases swabbing with a piece of cotton wool soaked in caustic soda solution may be found effective for degreasing.

When thoroughly clean, the specimen is etched by being plunged into the etching reagent and agitated vigorously for several seconds. The specimen is then *quickly* transferred to running water to wash away the etching reagent, and then examined to see the extent to which etching has taken place. Such inspection is carried out with the naked eye. If successfully etched the surface will appear slightly dull, and in cast materials the individual crystals may actually be seen without the aid of the microscope. If the surface is still bright further etching will be necessary. The time required for etching varies with different alloys and etching reagents. Some alloys can be etched sufficiently in a few seconds, whilst some stainless steels, being resistant to attack by most reagents, require as much as thirty minutes.

After being etched the specimen is washed in running water and then dried by immersion for a minute or so in boiling alcohol. If it is withdrawn from the alcohol and shaken with a flick of the wrist to remove the surplus, it will dry almost instantaneously. For mounted specimens, the mounts of which are affected by boiling alcohol, it is better to spot a few drops of alcohol on the surface of the specimen. The surplus is then shaken off and the specimen held in a current of warm air from a hair drier. The specimen must be dried *evenly and quickly*, or it will stain.

A summary of the most useful etching reagents is given in Tables 10.1, 10.2, 10.3 and 10.4.

10.25. Both polishing and etching can be carried out electrolytically. This involves setting up an electrolytic cell in which the surface of the specimen acts as the anode. By choosing a suitable electrolyte and appropriate current conditions, the surface of the specimen can be selectively dissolved to the required finish. Not only can "difficult" metals and alloys be attacked in this way, but a high-quality, scratch-

TABLE 10.1—*Etching Reagents for Iron, Steels and Cast Irons*

Type of etchant	Composition	Characteristics and uses
Nital	2 cm³ nitric acid; 98 cm³ alcohol (industrial methylated spirit)	The best general etching reagent for irons and steels. Etches pearlite, martensite and troostite, and attacks the grain boundaries of ferrite. For pure iron and wrought iron the concentration of nitric acid may be raised to 5 cm³. To resolve pearlite, etching must be very light. Also suitable for ferritic grey cast irons and blackheart malleable irons.
Picral	4 g picric acid; 96 cm³ alcohol	Very good for etching pearlite and spheroidised structures, but does not attack the ferrite grain boundaries. It is the most suitable reagent for all cast irons, with the exception of alloy and completely ferritic cast irons.
Alkaline sodium picrate	2 g picric acid; 25 g sodium hydroxide; 100 cm³ water	The sodium hydroxide is dissolved in the water and the picric acid then added. The whole is heated on a boiling water-bath for 30 minutes and the clear liquid poured off. The specimen is etched for 5–15 minutes in the boiling solution. Its main use is to distinguish between ferrite and cementite. The latter is stained black, but ferrite is not attacked.
Mixed acids and glycerol	10 cm³ nitric acid; 20 cm³ hydrochloric acid; 20 cm³ glycerol; 10 cm³ hydrogen peroxide	Suitable for nickel–chromium alloys and iron–chromium-base austenitic steels. Also for other austenitic steels, high chromium–carbon steels and high-speed steel. Warm the specimen in boiling water before immersion.
Acid ammonium persulphate	10 cm³ hydrochloric acid; 10 g ammonium persulphate; 80 cm³ water	Particularly suitable for stainless steels. Must be freshly prepared for use.

TABLE 10.2—*Etching Reagents for Copper and its Alloys*

Type of etchant	Composition	Characteristics and uses
Ammoniacal ammonium persulphate	20 cm³ ammonium hydroxide (0·880); 10 g ammonium persulphate; 80 cm³ water	A good etchant to reveal the grain boundaries of pure copper, brasses and bronzes. Should be freshly made to give the best results.
Ammonia–hydrogen peroxide	50 cm³ ammonium hydroxide (0·880); 20–50 cm³ hydrogen peroxide (3% solution); 50 cm³ water	The best general etchant for copper, brasses and bronzes. Etches grain boundaries and gives moderate contrast. The hydrogen peroxide content can be varied to suit a particular alloy. Used for swabbing or immersion, and should be freshly made as the hydrogen peroxide deteriorates.
Acid ferric chloride	10 g ferric chloride; 30 cm³ hydrochloric acid; 120 cm³ water	Produces a very contrasty etch on brasses and bronzes. Darkens the β in brasses. Can be used following a grain-boundary etch with the ammonium persulphate etchant. Use at full strength for nickel-rich copper alloys. Dilute 1 part with 2 parts of water for copper-rich solid solutions in brass, bronze and aluminium bronzes.
Acid bichromate solution	2 g potassium bichromate; 8 cm³ sulphuric acid; 4 cm³ saturated sodium chloride solution; 100 cm³ water	Useful for aluminium bronze and complex brasses and bronzes. Also for copper alloys of beryllium, manganese and silicon, and for nickel silvers.

free finish can be produced. The description of detailed techniques is, however, beyond the scope of this book. The traditional methods of polishing and etching already described are adequate for the successful preparation of most metallurgical materials.

TABLE 10.3.—*Etching Reagents for Aluminium and its Alloys*

Type of etchant	Composition	Characteristics and uses
Dilute hydrofluoric acid	0·5 cm³ hydrofluoric acid; 99·5 cm³ water	The specimen is best swabbed with cotton wool soaked in the etchant. A good general etchant.
Caustic soda solution	1 g sodium hydroxide; 99 cm³ water	A good general etchant for swabbing.
Keller's reagent	1 cm³ hydrofluoric acid; 1·5 cm³ hydrochloric acid; 2·5 cm³ nitric acid; 95 cm³ water	Particularly useful for duralumin-type alloys. Etch by immersion for 10–20 seconds.

N.B. On no account should hydrofluoric acid or its fumes be allowed to come into contact with the skin or eyes. Care must be exercised with all strong acids.

TABLE 10.4.—*Etching Reagents for Miscellaneous Alloys*

Type of etchant	Composition	Characteristics and uses
Acetic and nitric acids	3 cm³ glacial acetic acid; 4 cm³ nitric acid; 16 cm³ water	Useful for *lead and its alloys* (use freshly prepared and etch for 4–30 minutes). 5% Nital is also useful for lead and its alloys.
Acetic acid and hydrogen peroxide	30 cm³ glacial acetic acid; 10 cm³ hydrogen peroxide (30% solution)	Suitable for *lead-antimony alloys.* Etch for 5–20 seconds.
Acid ferric chloride	10 g ferric chloride; 2 cm³ hydrochloric acid; 95 cm³ water	Suitable for *tin-rich bearing metals.* Other tin-rich alloys can be etched in 5% Nital.
Dilute hydrochloric acid in alcohol	1 cm³ hydrochloric acid; 99 cm³ alcohol (industrial methylated spirit)	For *zinc and its alloys.* 1% Nital is also useful.
Iodine solution	10 g iodine crystals; 30 g potassium iodide; 100 cm³ water	The best etchant for *cadmium–bismuth alloys.*
Mixed nitric and acetic acids	50 cm³ nitric acid; 50 cm³ glacial acetic acid	Suitable for *nickel and monel metal.* Should be freshly prepared.

The Metallurgical Microscope

10.30. The metallurgical microscope is similar in optical principles to any other microscope, but differs from some of them in the method by which the specimen is illuminated. Most biological specimens can be prepared as thin, transparent slices mounted between sheets of thin glass, so that illumination can be arranged simply, by having a source of light *behind* the specimen. Metals, however, are opaque substances, and since they must be illuminated by frontal lighting, it follows that the source of light must be *inside* the microscope tube itself. This is usually accomplished, as in Fig. 10.4, by means of a small plain-glass reflector, *R*, placed inside the tube. With this system of illumination much of the light is lost both by transmission when it first strikes the plate and, by reflection, when the reflected ray from the specimen strikes the inclined plate again. Nevertheless, a small 6-volt bulb is usually sufficient as a source of illumination. The width of the beam is controlled by the iris diaphragm, *D*. Generally speaking, this should be partly closed so that the beam of light is just sufficient to cover the back component of the objective lens. An excess of light, reflected from the sides of the microscope tube, will cause light-scatter and, consequently, "glare" in the field of view.

The optical system of the microscope consists of two lenses, the objective, O, and the eyepiece, E. The former is the more important and expensive of the two lenses, since it has to resolve the fine detail of the

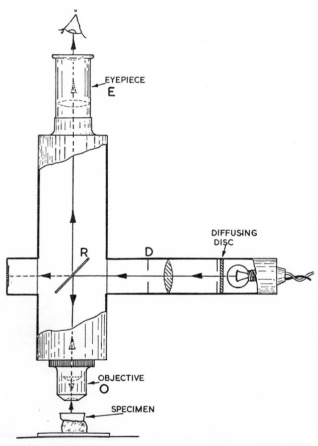

FIG. 10.4.—The Optical System of the Metallurgical Microscope.

object being examined. Good-quality objectives are corrected for chromatic and spherical aberrations, and hence, like camera lenses, are of compound construction. The magnification given by the objective depends upon its focal length—the shorter the focal length, the higher the magnification. In addition to magnification, resolving power is also important. This is defined as the ability of a lens to show clearly separated two lines which are very close together. In this way resolution

[*Courtesy of Messrs. C. Reichert, Vienna.*]

PLATE 10.2.—Simple Metallurgical Microscope Equipped for
Illumination by Incident Light.

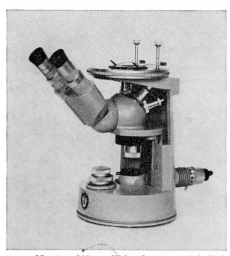

[*Courtesy of Messrs. Vickers Instruments Ltd., York.*]

PLATE. 10.3.—Most Modern Metallurgical
Microscopes Are of "Inverted" (Le
Chatelier) Design.

The specimen is placed face-down on the
upper platform whilst the illumination
system is built into the base. This instrument, Messrs. Vickers' "Metallette", is
equipped for binocular vision.

[*Courtesy of Messrs. C. Reichert, Vienna.*]

PLATE 10.4.—Inverted Type of Metallurgical Microscope with Built-in
Camera Which Can Produce Nega-
tives up to 120 mm × 165 mm
($\frac{1}{2}$-Plate) in Size.

can be expressed as a number of lines per mm. Thus resolving power depends upon the quality of the lens, and it is useless to increase the size of the image, either by extending the tube length of the microscope or by using a higher-power eyepiece, beyond a point where there is a falling off of resolution. A parallel example in photography is where a small 90 mm × 60 mm snapshot, on enlarging, fails to show any more *detail* than it did in its original size and in fact shows blurred outlines as a result.

[Courtesy of Messrs. Polaron Equipment Ltd., London N.3

PLATE 10.5.—An inverted type Metallurgical Microscope, equipped for binocular vision and Photography (using a standard 35 mm. single-lens reflex camera).

The eyepiece is so called because it is the lens nearest the eye. Its purpose is to magnify the image produced by the objective. Eyepieces are made in a number of powers, usually ×6, ×8 and ×10.

The overall approximate magnification of the complete system can be found from the expression:

$$\frac{T \cdot N}{F}$$

where T = the tube length of the microscope measured from the back
        component of the objective to the lower end of the eyepiece;
F = the focal length of the objective; and
N = the power of the eyepiece.

Thus for a microscope having a tube length of 200 mm and using a 16-mm focal-length objective and a ×10 eyepiece, the magnification would be $\frac{200 \times 10}{16}$, i.e. 125. Most metallurgical microscopes have a standard tube length of 200 mm. Assuming a tube length of this value, approximate magnifications with various objectives of standard focal lengths will be found in Table 10.5.

TABLE 10.5.—*Approximate Magnifications with Different Combinations of Objective and Eyepiece* (*Assuming a Tube Length of* 200 *mm*)

Focal length of objective (mm)	Power of eyepiece	Approximate magnification
32	× 6 × 8 × 10	37·5 50 62·5
16	× 6 × 8 × 10	75 100 125
8	× 6 × 8 × 10	150 200 250
4	× 6 × 8 × 10	300 400 500
2 (oil-immersion objective)	× 6 × 8 × 10	600 800 1000

10.31. Using the Microscope. The specimen must first be mounted so that its surface is normal to the axis of the instrument. This is most easily achieved by fixing the specimen to a microscope slide by means of a piece of plasticine. Normality is assured by using a mounting ring, as shown in Fig. 10.5. Obviously, the mounting ring must have perfectly

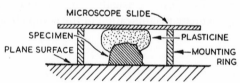

FIG. 10.5.—Mounting a Specimen for Examination under the Microscope.

parallel end faces. Mounting may not be necessary for specimens which have been set in bakelite, since the top and bottom faces of the bakelite mount are usually parallel, so that it can be placed directly on to the table of the microscope.

The specimen is brought into focus by using first the coarse adjustment and then the fine adjustment. It should be noted that the lenses supplied are generally designed to work at a fixed tube length (usually

200 mm), under which conditions they give optimum results. Therefore, the tube carrying the eyepiece should be drawn out the appropriate amount (a scale is usually engraved on the side of the tube). Slight adjustments in tube length may then be made to suit the individual eye.

Finally, the iris in the illumination system should be closed to a point where illumination just begins to decrease. This will limit glare due to internal reflections in the tube.

It is a mistake to assume that high magnifications in the region of 500 or 1 000 are always the most useful. In fact, they will often give a completely meaningless impression of the structure, since the field under observation will be so small. Directional properties in wrought structures or dendritic formation in cast structures are best seen using low powers of ×40 to ×100. Even at ×40 a single crystal of, say, cast 70–30 brass may completely fill the field of view. The dendritic pattern, however, will be clearly apparent, whereas at ×500 only a small area between two dendrite arms would fill the field of view. In a similar way, very little of Paris can be seen whilst standing in the Champs Elysée, but from the top of the Eiffel Tower a general idea of the shape of the city can be obtained. Thus, as a matter of routine, a low-power objective should always be used first to gain a general impression of the structure before it is examined at a high magnification.

10.32. The Care of the Microscope. Care should be taken never to touch the surface of optical glass with the fingers, since even the most careful cleaning may damage the surface, particularly if it has been "bloomed" (coated with magnesium fluoride to increase light transmission). In normal use dust may settle on a lens, and this is best removed by sweeping gently with a high-quality camel-hair brush.

If a lens becomes accidentally finger-marked, this is best dealt with by wiping gently with a piece of soft, well-washed linen moistened with xylol. Note that the operative word is *wipe* and not *rub*. Excess xylol must be avoided, as this may penetrate into the mount of the lens and soften the cement holding the components together.

High-power objectives of the oil-immersion type should always be wiped clean of cedar-wood oil before the latter has a chance to harden. If hardening takes place due to the lens being left standing for some time, then the oil will need to be removed by xylol, but the use of the latter should be avoided when possible.

Soft, well-washed linen should always be used to clean lenses. It is superior to chamois leather, which is more likely to absorb particles of grit, and to silk, which has a tendency to scratch the surface of soft optical glass.

10.33. The Electron Microscope. Whilst much of the routine micro-examination of metals is carried out at low magnifications in the region of ×100, it is often necessary in metallurgical research to be able to examine structures at very high magnifications. Unfortunately the highest magnification possible with an ordinary optical microscope is in the region of ×2 000. Above this magnification the dimensions being dealt with are comparable with the wavelength of light itself. Indeed, since blue light is of shorter wavelength than red light, it is advantageous to view specimens by blue light when examining very fine detail at high magnifications.

For very high-power microscopy (between ×2 000 and several millions) light rays can be replaced by a beam of electrons. The bending or *refracting* of light rays in an optical microscope is achieved by using a suitable glass-lens system. A similar effect is produced in the electron miscroscope by using an electro-magnetic "lens" to refract the electron beam. This "lens" consists of coil systems which produce the necessary electro-magnetic field to focus the electron beam.

Whereas light rays are reflected to a very high degree from the surface of a metal, electrons are transmitted. They are able to pass through a thin metallic foil, but will be absorbed by greater thicknesses of metal (Fig. 3.12). We must therefore view the specimen by transmission instead of by reflection, and in order to examine the actual surface of a metal, the replica method is generally used. This involves first preparing the surface of the specimen and then coating it with some suitable plastic material, a very thin film of which will follow the contours of the metallic surface. The thin plastic mould, or replica of the surface, is then separated from the specimen and photographed by transmitted electrons using the electron microscope.

Alternatively a very thin foil of the metal can be examined by a transmitted beam. Such foils are of the order of 5×10^{-8} to 5×10^{-7}m thick and are generally prepared by anodic dissolution (10.25). By examining a foil in this way we are in fact investigating the *internal* structure rather than surface detail. This method is therefore regarded as complimentary rather than as an alternative to the replica method already mentioned.

Macro-Examination

10.40. Useful information about the structure of a piece of metal can often be obtained without the aid of a microscope. Such investigation is usually referred to as "macro-examination", and may be carried out with the naked eye or by using a small hand magnifying lens.

The method of manufacture of a component can often be revealed by an examination of this type (Plate 10.6B and Fig. 6.5), whilst defects, such as the segregation of antimony–tin cuboids in a bearing metal (Plate 18.1), can also be effectively demonstrated. Macro-examination is also a means of assessing crystal size, particularly in cast structures (Plate 10.6A), whilst various forms of heterogeneity, such as is shown in the distribution of sulphide globules in cast steel, can be similarly examined.

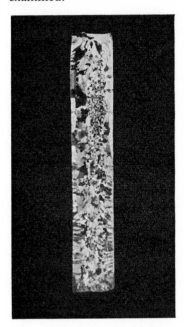

[*Courtesy of Messrs. Forgings and Presswork Ltd., Birmingham.*]

10.6A.—A Vertical Section Through a Small Ingot of Aluminium.

Normal size. Etched by swab-bing with a 20% solution of hy-drofluoric acid.

10.6B.—Section Through a "Close-form" Forging.

Deep-etched in boiling 50% hydrochloric acid for 30 minutes to reveal the flow lines. The strength of the teeth is increased by the continuous "fibre" produced by forging. × ½.

PLATE 10.6.

Usually, medium grinding is sufficient to produce a satisfactory surface for examination, and for large components an emery belt will be found almost indispensable. Polishing is not necessary, and for most specimens grinding can be finished at Grade 320 paper. Bearing in mind the rougher nature of the work, it will be realised that grinding should not be carried out on papers which are used for the preparation of specimens for examination under the microscope. Discarded papers from micro-polishing processes can be used for the preparation of macro-sections.

After being ground, the specimen should be washed to remove grit, but it will not generally be necessary to degrease it in view of the fact that macroscopic etching is often prolonged and removes more than the "flowed" layer. The fibrous structure of a forged-steel component, for example, is revealed by deep-etching the component in boiling 50% hydrochloric acid for up to fifteen minutes. After being etched, the specimen is washed and dried in the usual way, though fine detail is often more clearly seen when the component is wet. Suitable etchants for the macroscopic examination of various alloys are shown in Table 10.6.

TABLE 10.6.—*Etching Reagents for Macroscopic Examination*

Material to be etched	Composition of etchant	Working details
Steel	50 cm³ hydrochloric acid; up to 50 cm³ water	Use boiling for 5–15 minutes. Reveals flow lines; the structure of fusion welds; cracks; and porosity; also the depth of hardening in tool steels.
	25 cm³ nitric acid; 75 cm³ water	Similar uses to above, but can be used as a cold swabbing reagent for large components.
	1 g cupric chloride; 4 g magnesium chloride; 2 cm³ hydrochloric acid; 100 cm³ alcohol	*Stead's Reagent.* The salts are dissolved in the smallest possible amount of hot water along with the acid. Shows dendritic structure in cast steel. Also phosphorus segregation—copper deposits on areas *low* in phosphorus.
Copper and its alloys	25 g ferric chloride; 25 cm³ hydrochloric acid; 100 cm³ water	Useful for showing the dendritic structure of α solid solutions.
	50 cm³ ammonium hydroxide (0·880); 50 cm³ ammonium persulphate (5% solution); 50 cm³ water	Useful for alloys containing the β-phase.
Aluminium and its alloys	20 cm² hydrofluoric acid; 80 cm³ water	Degrease the specimen first in carbon tetrachloride and then wash in *hot* water. Swab with etchant.
	45 cm³ hydrochloric acid; 15 cm³ hydrofluoric acid; 15 cm³ nitric acid; 25 cm³ water	A much more active reagent—care should be taken to avoid contact with the skin.

Sulphur Printing

10.50. This affords a useful means of determining the distribution of sulphides in steel. The specimen should first be ground to Grade 320 emery paper and then thoroughly degreased and washed. Meanwhile a sheet of single-weight matt photographic bromide paper is soaked in a 2% solution of sulphuric acid for about five minutes. It is then removed from the solution and any surplus drops are wiped from the surface.

The emulsion side of the paper is then placed on the surface of the specimen and gently rolled with a squeegee to expel any air bubbles and surplus acid from between the surfaces. Care must be taken that the paper does not slide over the surface of the specimen, for which reason matt paper is preferable. For small specimens the paper can be laid emulsion side upwards on a flat surface and the specimen then pressed firmly into contact with it; care again being taken to prevent slipping between the paper and the specimen.

After about five minutes paper and specimen can be separated, and it will be found that the paper has been stained brown where it was in contact with particles of sulphide. The sulphuric acid reacts with the sulphides to produce the gas hydrogen sulphide, H_2S:

$$MnS + H_2SO_4 = MnSO_4 + H_2S$$
$$FeS + H_2SO_4 = FeSO_4 + H_2S$$

The liberated hydrogen sulphide then reacts with the silver bromide, $AgBr$, in the photographic emulsion to form a dark-brown deposit of silver sulphide, Ag_2S:

$$2AgBr + H_2S = 2HBr + Ag_2S$$

The print is rinsed in water and "fixed" for ten minutes in a solution containing 100 g of "hypo" (sodium thiosulphate) in 1 litre of water. The function of this treatment is to dissolve any surplus silver bromide, which would otherwise darken on exposure to light. Finally, the print is washed for thirty minutes in running water, and dried.

RECENT EXAMINATION QUESTIONS

1. Describe the essential steps in the preparation of micro-specimens and explain how etching reveals the microstructure. Explain why a single phase appears as different shades when examined under the metallurgical microscope. (Nottingham Regional College of Technology.)

2. Outline the procedure for the preparation of a specimen of plain carbon steel for examination under the microscope.
   Sketch and explain the structures of:

   (i) 0·3% carbon steel in the cast condition;
   (ii) 0·4% carbon steel in the fully hardened condition;
   (iii) 0·7% carbon steel in the annealed condition.
   (West Bromwich Technical College, A1 year.)

3. Describe in detail how you would prepare a specimen of cast 70–30 brass for microscopical examination.
   Sketch the type of structure typical of this alloy, and also the structure which should prevail after cold-working and annealing at 600° C.
   (U.L.C.I., Paper No. B111–1–4.)

4. Describe, with the aid of sketches, the optical system used in the metallurgical microscope.

What is meant by "resolving power" as applied to the objective used in such a microscope?

A metallurgical microscope employs a tube length of 200 mm. What ocular magnification will be obtained when this microscope is used in conjunction with a 4-mm objective and a × 10 eyepiece?

(West Bromwich Technical College, A1 year.)

5. What are the objects of (a) macro-examination of a metal sample, (b) micro-examination of a metal sample; and how are these accomplished?

(The Polytechnic, London W.1)

6. Describe in detail the preparation of a cross-section of a small forging for macro-examination.

Sketch a typical section and say what useful information could be obtained from such an examination.

(West Bromwich Technical College, A1 year.)

7. You are given a small steel bolt and asked to find out by metallographic means whether it has been forged or turned to shape.

Describe in detail the procedure adopted in the investigation.

Make sketches of the type of macrostructure expected in each case.

(U.L.C.I., Paper No. C111-1-4.)

CHAPTER XI

THE HEAT-TREATMENT OF PLAIN CARBON STEELS—(I)

11.10. In the first part of this chapter we shall consider the development of equilibrium structures in steels in greater detail than was possible in Chapter VII. We shall follow this with a study of those heat-treatment processes which depend upon equilibrium being reached in the structure of the steel under treatment.

11.11. The complete iron–carbon thermal-equilibrium diagram is shown in Fig. 11.1. This system is of the type dealt with in 9.50, that

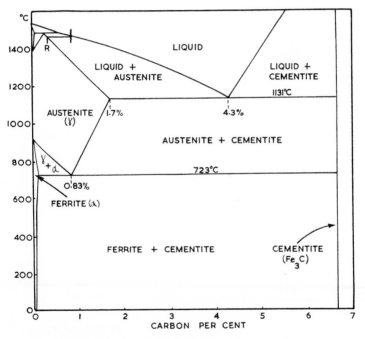

FIG. 11.1.—The Iron–Carbon Thermal-equilibrium Diagram.

is, where two substances are completely soluble in each other in the liquid state but are only partially soluble in the solid state. The diagram is modified somewhat by the structural changes which take place in the crystal lattice of iron at 910 and 1 400° C. However, despite the apparent complexity of the diagram, we have only three important phases to consider, namely, austenite (γ), ferrite (α) and cementite (Fe$_3$C). As we have seen (7.33), austenite is the solid solution which is formed when carbon dissolves in face-centred cubic iron, whilst ferrite is the very weak solid solution produced when carbon dissolves in body-centred cubic iron below 910° C. Cementite is a carbide, Fe$_3$C, formed

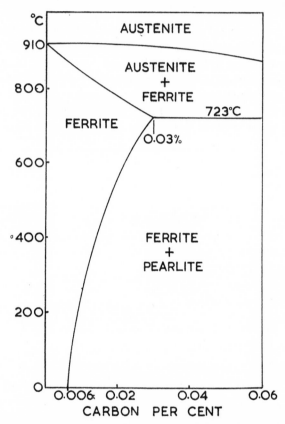

FIG. 11.2.—The "Ferrite Area" of the Iron–Carbon Equilibrium Diagram, Showing the Extent to Which Carbon is Soluble in α Iron.

by combination of the carbon with some of the iron. For the sake of clarity the important areas of the equilibrium diagram are shown in greater detail in Figs. 11.2, 11.3 and 11.4.

11.20. Let us now consider the type of structure likely to be produced in a large steel sand-casting, containing 0·3% carbon, as it solidifies and cools slowly to room temperature. Such an alloy will begin to solidify at about 1 515° C by forming dendrites of the solid solution δ

(Fig. 11.3). These dendrites will develop and change in composition, due to diffusion promoted by the slow rate of cooling, until at 1 492° C they will be of composition C (0·08% carbon). The remaining liquid will have become correspondingly enriched in carbon and will be of composition B (0·55% carbon). At 1 492° C a peritectic reaction takes

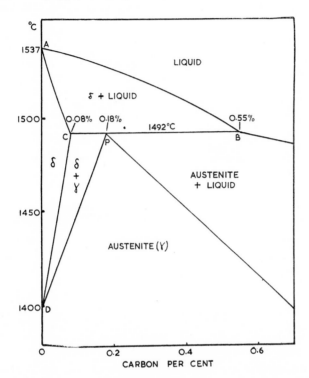

FIG. 11.3.—The Portion of the Iron–Carbon Diagram Including the Peritectic Reaction.

place between the remaining liquid and the δ-phase, resulting in the disappearance of the latter and the formation of austenite of composition P (0·18% carbon). As the temperature continues to fall the remaining liquid solidifies as austenite, and when solidification is complete the structure will consist of cored crystals of austenite of overall composition 0·3% carbon.

Since cooling is slow, diffusion of the carbon will be promoted, and by the time we reach the line FE (Fig. 11.4) the structure will consist entirely of crystals of austenite of practically uniform carbon content. Moreover, due to the considerable time that the casting remained at high temperatures, grain growth will have been excessive, leading to the formation of very large austenite crystals, particularly in the centre of the casting.

11.21. The Upper Critical (or A_3) Temperature of a steel* varies with its carbon content, and is that temperature below which austenite begins to transform to ferrite and cementite under conditions of slow cooling. It will thus be represented by a point on *FEG*. In the case of

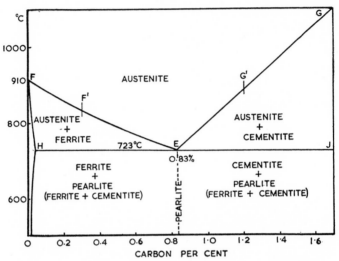

FIG. 11.4.—The "Steel" Portion of the Iron–Carbon Diagram.

the steel under consideration the upper critical temperature will be represented by F', and as the temperature falls below it, the face-centred cubic structure becomes unstable and crystals of ferrite begin to separate out from the austenite. Since this ferrite will contain rather less than 0·03% carbon, the carbon content of the remaining austenite will increase progressively as more and more ferrite is formed, until, at 723° C (the eutectoid temperature), the structure consists of ferrite, containing 0·03% carbon, and austenite, containing 0·83% carbon.

11.22. At 723° C the remaining austenite transforms to the eutectoid, pearlite, by depositing alternate layers of ferrite and cementite in the manner dealt with earlier in this book (7.35 and 8.23). The temperature, 723° C, at which pearlite is formed is called the Lower Critical (or A_1) Temperature,* and is the same for carbon steels of all compositions, since the eutectoid temperature is constant, i.e. *HEJ* is horizontal.

11.23. A 0·83% carbon steel will begin to solidify at approximately 1 470° C by depositing dendrites of austenite of composition R (Fig.

* See 13.11 for an explanation of the terms "A_1" and "A_3".

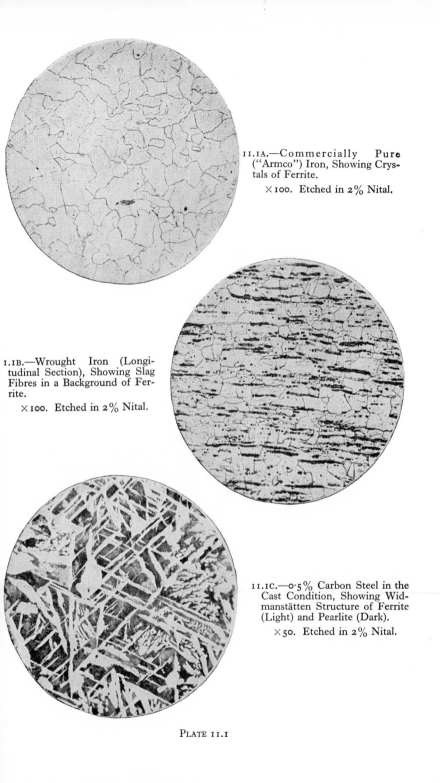

11.1A.—Commercially Pure ("Armco") Iron, Showing Crystals of Ferrite.

×100. Etched in 2% Nital.

1.1B.—Wrought Iron (Longitudinal Section), Showing Slag Fibres in a Background of Ferrite.

×100. Etched in 2% Nital.

11.1C.—0·5% Carbon Steel in the Cast Condition, Showing Widmanstätten Structure of Ferrite (Light) and Pearlite (Dark).

×50. Etched in 2% Nital.

PLATE 11.1

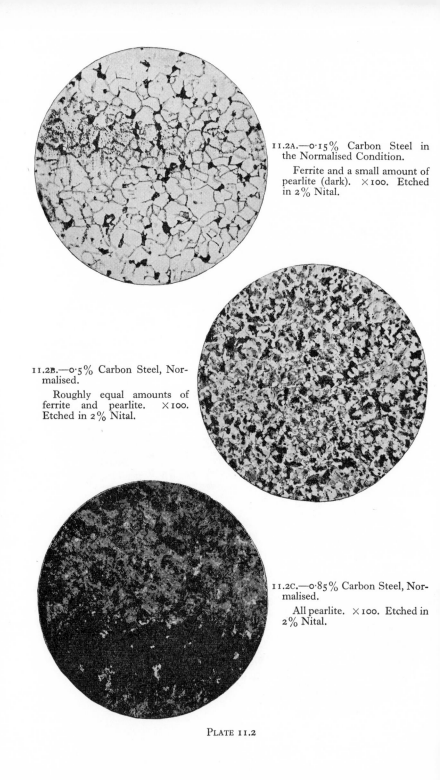

11.2A.—0·15% Carbon Steel in the Normalised Condition.

Ferrite and a small amount of pearlite (dark). ×100. Etched in 2% Nital.

11.2B.—0·5% Carbon Steel, Normalised.

Roughly equal amounts of ferrite and pearlite. ×100. Etched in 2% Nital.

11.2C.—0·85% Carbon Steel, Normalised.

All pearlite. ×100. Etched in 2% Nital.

PLATE 11.2

11.3A—The Structure of Lamellar Pearlite Revealed by a Micrograph Taken at ×1000 Etched Picral–Nital.

Courtesy of Messrs. United Steel Companies Ltd., Rotherham.

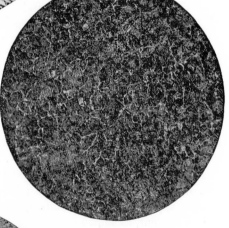

11.3B.—1·3% Carbon Steel, Normalised.

Network of free cementite around the patches of pearlite. ×100. Etched in 2% Nital.

Courtesy of Messrs. Hadfields Ltd., Sheffield.

11.3C.—Similar to Above, but Higher Magnification Reveals the Lamellar Nature of the Pearlite as Well as the Network of Free Cementite.

×750. Etched Picral–Nital.

Courtesy of Messrs. United Steel Companies Ltd., Rotherham

PLATE 11.3

11.1), and, when solidification is complete, the structure will consist of cored crystals of austenite of overall composition 0·83% carbon. As the steel cools slowly, the structure becomes uniform by diffusion, and no further structural change will take place until the point E (Fig. 11.4) is reached. For a steel of this composition the upper and lower critical temperatures coincide, and the austenitic structure transforms at this temperature to one which is totally pearlitic.

11.24. A 1·2% carbon steel will solidify in a similar way to the 0·83% alloy by forming austenite crystals of an overall carbon content of 1·2%. As the temperature falls to the upper critical for this alloy at G' (Fig. 11.4), needles of primary cementite begin to precipitate at the crystal boundaries of the austenite. Since cementite is being deposited, the remaining austenite will be rendered less rich in carbon, so that its composition will move to the left, and when the temperature has fallen to 723° C the remaining austenite will contain 0·83% carbon. As before, pearlite will now form, giving a final structure of primary cementite and pearlite.

11.25. Thus, in a steel which has been permitted to cool slowly enough to enable it to reach structural equilibrium, we shall find one of the following structures:

(a) With less than 0·006% carbon it will be entirely ferritic. In practice, such an alloy would be classed as commercially pure iron.

(b) With between 0·006% and 0·83% carbon the structure will contain ferrite and pearlite. The relative proportions of ferrite and pearlite appearing in the microstructure will vary according to the carbon content, as shown in Fig. 7.3.

(c) With exactly 0·83% carbon the structure will be entirely pearlitic.

(d) With between 0·83% and 1·7% carbon the structure will consist of cementite and pearlite, in relative amounts which depend upon the carbon content.

11.26. The composition of the pearlite area in the microstructure of any plain carbon steel is always the same, namely, 0·83% carbon, and if the overall carbon content is either greater or smaller than this, then it will be compensated for by variation in the amount of either primary ferrite or primary cementite. The hardness of a slowly cooled steel increases directly as the carbon content, whilst the tensile strength reaches a maximum at the eutectoid composition (Fig. 7.3). These properties can be modified by heat-treatment, as we shall see in this chapter and the next.

Impurities in Steel

11.30. Most ordinary steels contain appreciable amounts of manganese, residual from the deoxidation process. The reader will also have gathered, from a study of Chapter II, that impurities, such as silicon, sulphur and phosphorus, are liable to be present in the finished steel. The effect of such impurities on mechanical properties will depend largely upon the way in which these impurities are distributed throughout the structure of the steel. If the liquidus and solidus lines of the appropriate equilibrium diagram, governing the alloying of the impurity in question, are close together, as in Fig. 11.5B, the compositions of

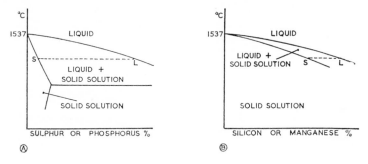

FIG. 11.5

liquid and solid will be maintained almost equal during solidification. This will lead to a relatively even distribution of the impurity throughout any crystal in the structure of the ingot. When, however, liquidus and solidus are widely separated, as in Fig. 11.5A, almost pure metal of composition S is first deposited from a liquid of composition L, thus causing the impurity to become concentrated in the last liquid to solidify, i.e. at the crystal boundaries. This effect, usually referred to as "coring", since the core of the crystal is of a different composition from the outer fringes at the crystal boundaries, was dealt with in 8.33.

The crystals in solid steel are never extensively cored with regard to silicon and manganese, and since these elements have high solid solubilities in steel, they will never appear as separate constituents in the microstructure. Sulphur and phosphorus, on the other hand, segregate appreciably and, if present in sufficient amounts, will precipitate during solidification, as their respective iron compounds, at the austenite grain boundaries. The effect will be aggravated by the low solid solubilities of these compounds in steel.

11.31. Manganese is not only soluble in austenite and ferrite but also forms a stable carbide, Mn_3C. In the nomenclature of the heat-treatment shop, manganese "increases the depth of hardening" of a steel, for reasons which will be discussed in Chapter XIII. It also improves strength and toughness. Manganese should not exceed 0·3% in high-carbon steels because of a tendency to induce quench cracks, particularly during water-quenching.

11.32. Silicon imparts fluidity to steels intended for the manufacture of castings, and is present in such steels in amounts up to 0·3%. In high-carbon steels silicon must be kept low, because of its tendency to render cementite unstable (as in pig iron—see 2.51) and liable to decompose into graphite (which precipitates) and ferrite.

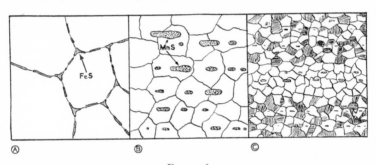

FIG. 11.6

(A) The Segregation of Ferrous Sulphide (FeS), at the Crystal Boundaries in Steel. × 750.
(B) The Formation of Isolated Manganese Sulphide (MnS) Globules When Manganese is Present in a Steel. × 200.
(C) "Ghost Bands", or Areas Lacking in Pearlite, Which Indicate the Presence of Phosphorus. × 75.

11.33. Phosphorus forms the brittle phosphide Fe_3P, which is soluble in steel. In solution, phosphorus has a considerable hardening effect on steel, but it must be rigidly controlled to amounts in the region of 0·05% or less because of the brittleness also introduced, particularly if Fe_3P should appear as a separate constituent in the microstructure.

In rolled or forged steel the presence of phosphorus is indicated by what are usually termed "ghost bands" (Fig. 11.6c). These are areas (which naturally become elongated during rolling) containing no pearlite, but instead, a high concentration of phosphorus. The presence of phosphorus and absence of pearlite will naturally make these ghost bands planes of weakness, particularly since, being areas of segregation, other impurities may be present in the ghosts.

11.34. Sulphur is the most deleterious impurity commonly present in steel. If precautions were not taken to render it harmless it would form the brittle sulphide, FeS. This is soluble in molten steel, but when solidification takes place the solid solubility falls to an equivalent of 0·03% sulphur. If the effects of extensive coring, referred to above, are also taken into account it will be clear that amounts as low as 0·01% sulphur may cause precipitation of the sulphide at the crystal boundaries. In this way the austenite crystals would become virtually coated with brittle films of ferrous sulphide. Since this sulphide has a fairly low

melting point, the steel would tend to crumble during hot-working. Being brittle at ordinary temperatures, ferrous sulphide would also render steel unsuitable for cold-working processes, or, indeed, for subsequent service of any type.

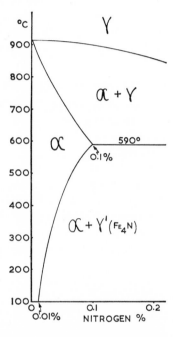

FIG. 11.7.—Part of the Iron–Nitrogen Thermal-equilibrium Diagram.

It would be very difficult, and certainly very expensive, to reduce the sulphur content to an amount less than 0·05% in the majority of steels. To nullify the effects of the sulphur present an excess of manganese is therefore added during deoxidation. Provided that about five times the theoretical manganese requirement is added, the sulphur then forms manganese sulphide, MnS, in preference to ferrous sulphide. The manganese sulphide so formed is *insoluble in the molten steel*, and some is lost in the slag. The remainder is present as fairly large globules, distributed throughout the steel, but since they are insoluble, they will not be associated with the structure when solidification takes place. Moreover, manganese sulphide is plastic at the forging temperature, so that the tendency of the steel to crumble is removed. The manganese sulphide globules become elongated into threads by the subsequent rolling operations (Fig. 11.6A and B).

11.35. Nitrogen. Atmospheric nitrogen is absorbed by molten steel during the manufacturing process. Whether this nitrogen combines with iron to form nitrides or remains dissolved interstitially after solidification (Fig. 11.7), it causes serious embrittlement and renders the steel unsuitable for severe cold-work. For this reason mild steel used for deep-drawing operations must have a low nitrogen content.

Due to the method of manufacture, Thomas steel is particularly suspect and may have a nitrogen content as high as 0·02%, probably leading to the presence of brittle Fe_4N in the structure (Fig. 11.7). This is more than four times the average nitrogen content of open-hearth steel adequate for deep-drawing operations. Naturally the relatively new "oxygen" processes (2.85) can produce mild steel with a very low nitrogen content (below 0·002%), since little or no nitrogen is present in the blast to the molten charge. Such steel is obviously ideal for deep-drawing. It is difficult, however, to prevent some atmospheric nitrogen from being absorbed, since the molten steel is in contact with the atmosphere during teeming.

The Heat-treatment of Steel

11.40. Because of the structural changes which take place in the solid, the iron–carbon alloys are among the relatively few engineering alloys which can be usefully heat-treated in order to vary their mechanical properties. This statement refers, of course, to heat-treatments other than simple stress-relief annealing processes.

Heat-treatments can be applied to steel not only to harden it but also to improve its strength, toughness or ductility. The type of heat-treatment used will be governed by the carbon content of the steel and its subsequent application.

11.41. The various heat-treatment processes can be classified as follows:

    (*a*) annealing;
    (*b*) normalising;
    (*c*) hardening;
    (*d*) tempering;
    (*e*) treatments which depend upon transformations taking place at a single predetermined temperature during a given period of time (isothermal transformations).

In all of these processes the steel is heated fairly slowly to some predetermined temperature, and then cooled, and it is the *rate of cooling* which determines the ultimate structure the steel will have. The final structure will be independent of the rate of heating, provided this has been slow enough for the steel to reach structural equilibrium at its maximum temperature. The subsequent rate of cooling, which determines the nature of the final structure, may vary between a drastic water-quench and slow cooling in the furnace.

Annealing

11.50. It has been suggested earlier in this book that an annealing process can be carried out for a number of different reasons, and this statement applies in the annealing of steels. Annealing processes for steels can be classified as follows:

11.51. Stress-relief Annealing. The recrystallisation temperature of mild steel is about 500° C, so that, during the normal hot-rolling process, recrystallisation will proceed automatically as the process is taking place. Thus, working stresses are relieved as they are set up.

Fig. 11.8.—The Spheroidisation of Pearlitic Cementite.

Frequently, however, we must apply a considerable amount of cold-work to mild steels, as, for example, in the drawing of wire. Stress-relief annealing then becomes necessary to soften the metal so that further drawing operations can be carried out. Such annealing is often referred to as "process" annealing, and is carried out at about 650° C. Since this temperature is well above the recrystallisation temperature of 500° C, recrystallisation will be accelerated so that it will be complete in a matter of minutes on attaining the maximum temperature. Prolonged annealing may in fact cause a deterioration in properties, since although ductility may increase further, there will be a loss in strength. A stage will be reached where grain growth becomes excessive, and where the layers of cementite in the patches of pearlite begin to coalesce and assume a globular form so that the identity of the eutectoid is lost (Fig. 11.8). In fact, the end-product would be isolated globular masses of cementite in a ferrite matrix. The result of this "balling-up" of the pearlitic cementite is usually called "deteriorated" pearlite.

It should be noted that process annealing is a *sub-critical* operation, that is, it takes place below the lower critical temperature (A_1). For this reason, although recrystallisation is promoted, there is no phase change and the constituents ferrite and cementite remain present in the structure throughout the process. The balling-up of the pearlitic cementite is purely a result of surface-tension effects which operate at the temperatures used.

Process annealing is generally carried out in either batch-type or continuous furnaces, usually with an inert atmosphere of burnt coal gas, though cast-iron annealing "pots" are still used, their lids being luted on with clay.

11.52. Spheroidising Anneals. "The spheroidisation of pearlitic cementite" may sound a somewhat ponderous phrase. In fact, it refers to the balling-up of the cementite part of pearlite mentioned above. This phenomenon is utilised in the softening of tool steels and some of the air-hardening alloy steels. When in this condition such steels can be drawn and will also machine relatively freely.

The spheroidised condition is produced by annealing the steel at a temperature between 650 and 700° C, that is, just *below* the lower critical temperature (A_1). Whilst no basic phase change takes place, surface tension causes the cementite to assume a globular form (Fig. 11.8) in a similar way to which droplets of mercury behave when mercury is spilled. If the layers of cementite are relatively coarse they take rather a long time to break up, and this would result in the formation of very large globules of cementite. This in turn would lead to tearing of the surface during machining. To obviate these effects it is better to give the steel some form of quenching treatment prior to annealing in order to refine the distribution of the cementite. It will then be spheroidised more quickly during annealing and will produce much smaller globules of cementite. These small globules will not only improve the surface finish during machining but will also be dissolved more quickly when the tool is ultimately heated for hardening.

11.53. Annealing of Castings. As stated earlier (11.20), the cast structure of a large body of steel is extremely coarse. This is due mainly to the slow rates of solidification and subsequent cooling through the austenitic range. Thus, a 0.35% carbon steel will be completely solid in the region of 1 450° C, but, if the casting is large, cooling, due to the lagging effect of the sand mould, will proceed very slowly down to the point (approximately 820° C) where transformation to ferrite and pearlite begins. By the time 820° C has been reached, therefore, the austenite crystals will be extremely large. Ferrite, which then begins to precipitate in accordance with the equilibrium diagram, deposits first at the grain boundaries of the austenite, thus revealing, in the final structure, the size of the original austenite grains. The remainder of the ferrite is then precipitated along certain crystallographic planes within the lattice of the austenite. This gives rise to a directional precipitation of the ferrite, as shown in Fig. 11.9 and Plate 11.1C, representing typically what is known as a Widmanstätten structure. This type of structure was first encountered by Widmanstätten in meteorites (10.10), which may be expected to exhibit a coarse structure in view of the extent to which they are overheated during their passage through the upper atmosphere. The mesh-like arrangement of ferrite in the Widmanstätten structure tends to isolate the stronger pearlite into separate patches, so that strength, and more particularly toughness, are impaired. The main characteristics of such a structure are, therefore, weakness and brittleness, and steps must be taken to remove it either by heat-treatment or by mechanical working. Hot-working will effectively break up this coarse as-cast structure and replace it by a fine-grained material, but in this instance we are concerned with retaining the actual shape of the casting. Heat-treatment must therefore be used to effect the necessary refinement of grain.

11.54. The most suitable treatment for a large casting involves heating it slowly up to a temperature about 40° C above its upper critical (thus the annealing temperature depends upon the carbon content of the steel, as shown in Fig. 11.10), holding it at that temperature only just long enough for a uniform temperature to be attained throughout the casting and then allowing it to cool slowly in the furnace. This treatment not only introduces the improvements in mechanical properties associated with fine grain but also removes mechanical strains set up during solidification.

As the lower critical temperature (723° C) is reached on heating, the patches of pearlite change to austenite, but these crystals of austenite are very small, since each grain of pearlite gives rise to a number of new austenite crystals. As the temperature rises, the Widmanstätten-type plates of ferrite are dissolved by the austenite until, when the upper critical temperature is reached, the structure consists entirely of fine-grained austenite. Cooling causes reprecipitation of the ferrite, but, since the new austenite crystals are small, the precipitated ferrite will also be distributed as small particles. Finally, as the lower critical temperature is reached, the remaining small patches of austenite will transform to pearlite. The structural changes taking place during annealing are illustrated diagrammatically in Fig. 11.9.

11.55. Whilst the tensile strength is not greatly affected by this treatment, both toughness and ductility are improved as shown by the following values for a cast carbon steel:

Condition	Tensile strength		Percentage elongation	Bend test
	N/mm²	tonf/in²		
Specimen "as cast" .	470	30·5	18	40°
Specimen annealed .	476	30·9	34	180° (without fracture)

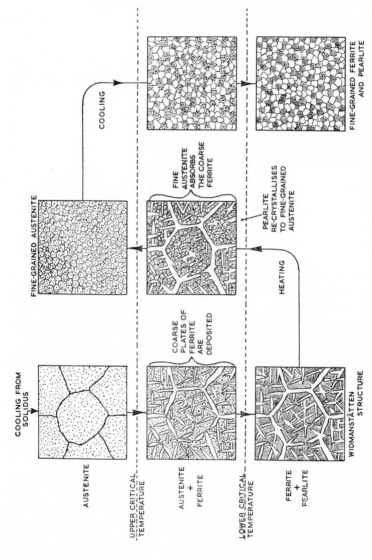

FIG. 11.9.—The Structural Effects of Heating a Steel Casting (Containing Approximately 0·35% Carbon) to a Temperature Just Above its Upper Critical, Followed by Cooling to Room Temperature.

11.56. Overheating during annealing, or heating for too long a period in the austenitic range, will obviously cause grain growth of the newly formed austenite crystals, leading to a structure almost as bad as the original Widmanstätten structure. For this reason the requisite annealing temperature should not be exceeded, and the casting should remain in the austenitic range only for as long as is necessary to make it completely austenitic. In fact, castings are sometimes air-cooled to about 650° C and then cooled more slowly to room temperature, by returning to a furnace to prevent stresses due to rapid cooling from being set up.

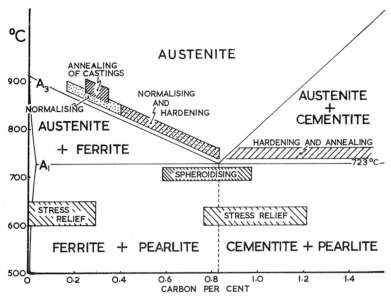

FIG. 11.10.—Heat-treatment Temperature Ranges of Classes of Carbon Steels in Relation to the Equilibrium Diagram.

11.57. Excessive overheating will probably cause oxidation, or "burning", of the surface, and the penetration by oxide films of the crystal boundaries following decarburisation of the surface. Such damage cannot be repaired by heat-treatment, and the castings can only be scrapped. To prevent "burning", castings are often annealed in cast-iron boxes into which they are packed with lime, sand, cast-iron turnings or carbonaceous material, according to the carbon content of the castings.

11.58. If annealing is carried out at too low a temperature, remnants of the as-cast structure will be apparent in the form of undissolved skeletons of Widmanstätten ferrite. As the temperature falls on cooling, the ferrite which did dissolve tends to reprecipitate on the existing ferrite skeleton so that the final structure resembles the unannealed.

Normalising

11.60. Normalising resembles annealing in that the maximum temperature attained is similar. It is in the method of cooling that the processes differ. Whilst, in annealing, cooling is retarded, in normalising the steel is removed from the furnace and allowed to cool in still air. This relatively rapid method of cooling limits grain growth in normalising, so that the mechanical properties are somewhat better than in an annealed component. Moreover, the surface finish of a normalised article is often superior to that of an annealed one when machined, since the high ductility of the latter often gives rise to local tearing of the surface.

11.61. The type of structure obtained by normalising will depend largely upon the thickness of cross-section, as this will affect the rate of cooling. Thin sections will give a much finer grain than thick sections, the latter often differing little in structure from an annealed section.

RECENT EXAMINATION QUESTIONS

1. Sketch the iron–carbon equilibrium diagram. With reference to this diagram, give a short account of the two main factors which affect the mechanical properties of alloys of iron and carbon.
   (University of Strathclyde, B.Sc.(Eng.).)
2. One particular plain carbon steel may be described as a "binary alloy" of exact "peritectic" composition and as being "hypo-eutectoid" in nature.

   (a) Define the three terms in inverted commas.
   (b) Sketch qualitatively the relevant portion of the equilibrium diagram and on it indicate clearly the composition of the steel.
   (c) Discuss the changes of structure that would occur if an alloy of this composition were cooled under equilibrium conditions from its molten state down to room temperature.

   (Institution of Production Engineers Associate Membership
   Examination, Part II, Group A.)
3. Draw and describe a Widmanstätten structure, also discuss how such a structure arises.   (Nottingham Regional College of Technology.)
4. Annealing treatments are commonly applied to both ferrous and non-ferrous metals. For what reasons may annealing be necessary? Explain the effects of the treatments you mention.
   (University of Aston in Birmingham, Inter B.Sc.(Eng.).)

5. Show, with the aid of sketches, and the appropriate thermal-equilibrium diagram, how it is possible to refine the grain structure of a sand-casting in steel by means of a heat-treatment process.

(U.L.C.I., Paper No. B111–1–4.)

6. Using the iron–carbon diagram of thermal equilibrium, explain the metallographic changes which occur in the annealing of a medium-carbon steel casting. Illustrate the answer with sketches where appropriate.

(Derby and District College of Technology, H.N.D. Course.)

7. (a) What is meant by the "normalising" process as applied to steel? Make special reference to the structural changes which occur.

(b) Mention two cases where the application of this process would be either necessary or advantageous.   (C. & G., Paper No. 293/6.)

8. Draw the iron–carbon thermal-equilibrium diagram up to 5% carbon and annotate it completely.

Use the diagram to explain what happens during (i) normalising a 0·5% C steel; (ii) spheroidising a 1·0% C steel.

(West Bromwich Technical College, A1 year.)

CHAPTER XII

THE HEAT-TREATMENT OF PLAIN CARBON STEELS—(II)

12.10. In the previous chapter those heat-treatment processes were dealt with in which the steel component was permitted to reach a state of structural equilibrium. That is, cooling took place sufficiently slowly to allow a pearlitic type of microstructure to form. Such treatments are normally only useful for improving the toughness and ductility of a steel component, and when increased hardness is required it is necessary to quench, or cool, the component sufficiently rapidly, in order to prevent the normal pearlitic structure from being formed.

12.11. Prior to the development of metallurgy as a science many of the processes associated with the hardening of steel were clothed in mystery. For example, it was thought that the water of Sheffield possessed certain magical properties, and it is said that an astute Yorkshire business man once exported it in barrels to Japan at considerable profit. In point of fact the high quality of Sheffield steel was a measure of the craftsmanship used in its production. Similarly, it is reported that Damascus steel swords were hardened by plunging the blade, whilst hot, into the newly decapitated body of a slave and stirring vigorously. Some metallurgists have suggested, possibly more out of cynicism than scientific accuracy, that hardening would be assisted by nitrogen absorption from the blood of the slave during this somewhat gruesome procedure. James Bowie, originator of the Bowie knife in the days of the "Wild West", is said to have quenched his knives *nine times* in succession in panther oil.

In this chapter, then, we shall deal with the production of structures, other than pearlite, in plain carbon steels.

Hardening

12.20. When a piece of steel, containing sufficient carbon, is cooled rapidly from above its upper critical temperature it becomes considerably harder than it would be if allowed to cool slowly. The degree of hardness produced can vary, and is dependent upon such factors as the initial quenching temperature; the size of the work; the constitution, properties and temperature of the quenching medium; and the degree of agitation and final temperature of the quenching medium.

12.21. Whenever a metallic alloy is quenched there is a tendency to suppress structural change or transformation. Frequently, therefore, it is possible to "trap" a metallic structure as it existed at a higher temperature and so preserve it at room temperature. This is usually an easy matter with alloys in which transformation is sluggish, but in iron–carbon alloys the reverse tends to be the case. Here, transformation, particularly that of austenite to pearlite, is rapid and is easily accom-

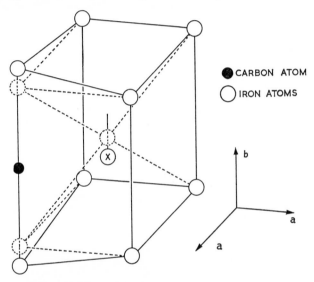

● CARBON ATOM

○ IRON ATOMS

FIG. 12.1.—The Lattice Structure of Martensite. (Atom X is displaced because of a carbon atom "vertically" above it in the lattice.)

plished during ordinary air-cooling to room temperature. This is due partly to the rapid diffusion of carbon in the face-centred cubic lattice, and partly to the allotropic transformation which takes place. (The rapid diffusion of carbon, due to the small size of its atoms relative to those of iron, also leads to the absence of coring, with respect to carbon, in a cast steel.)

Thus, when a plain carbon steel is quenched from its austenitic range it is not possible to trap austenite and so preserve it at room temperature. Instead, one or other phases is obtained intermediate between austenite, on the one hand, and pearlite, on the other. These phases vary in degree of hardness, but all are harder than either pearlite or austenite.

12.22. Even extremely rapid quenching rates are not sufficient to prevent the F.C.C. structure of the austenite from changing to one which is basically body-centred cubic. As a result of rapid cooling, however, the carbon remains in solution and a phase known as *Martensite* is obtained. The carbon atoms are retained in solution interstitially and this causes a distortion of the B.C.C. lattice to one which is tetragonal in nature (Fig. 12.1). In any one martensite crystal all of the carbon atoms occupy the same interstitial position on the *b*-axis. This elongates the lattice in the *b* direction and causes a contraction in the *a* directions (as compared with B.C.C. ferrite). Not every unit cell of the lattice will contain a carbon atom, since even in a eutectoid steel there is only one carbon atom for approximately every ten unit cells.

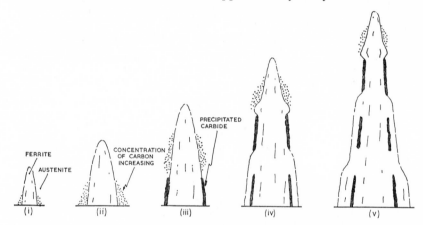

FIG. 12.2.—The Growth of Bainite.

As the ferrite crystals grow, so the concentration of carbon in the surrounding austenite increases until a point is reached where carbide is rejected.

Thus the presence of any carbon in excess of 0·03% will frustrate the formation of a simple B.C.C. structure when such a steel is quenched from the austenitic range. The degree of distortion existing in the resultant tetragonal martensite will effectively prevent the movement of dislocations from taking place. Consequently, since little or no slip is possible, martensite is an extremely hard, brittle phase as compared with austenite, which is soft and malleable. Under the microscope martensite appears to consist of a uniform mass of needle-shaped crystals (Plate 12.1A). These "needles" are in fact cross-sections through flat disc-shaped crystals. This is an instance of the sometimes misleading impression given by the two-dimensional image offered by the ordinary microscope.

Less severe quenching gives rise to a structure known as *Bainite*. This phase appears under the microscope, at magnifications in the region of ×100, as black patches (Plates 12.1B and C), but a higher magnification of ×1 000 shows that it is of a laminated nature something like pearlite. The growth of bainite (Fig. 12.2) differs from that of

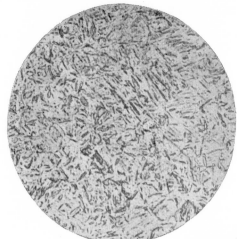

12.1A.—0·5% Carbon Steel, Water Quenched from 850° C.

Entirely martensite. ×100. Etched in 2% Nital.

12.1B.—0·5% Carbon Steel, Oil Quenched from 780° C.

Bainite (dark) and martensite. ×750. Etched in Picral–Nital.

Courtesy of Messrs. United Steel Companies Ltd., Rotherham

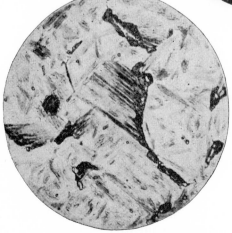

12.1C.—0·2% Carbon Steel, Water Quenched from 870° C on a Falling Gradient.

Acicular bainite (dark) and martensite. ×1000. Etched in Picral–Nital.

Courtesy of Messrs. United Steel Companies Ltd., Rotherham.

PLATE 12.1

pearlite in that ferrite nucleates first followed by carbide, whereas in pearlite it is the carbide which nucleates first (8.23). Bainite growth takes place quickly because the driving force is increased by a greater degree of non-equilibrium at the lower temperatures at which it is formed. Consequently particle size is too small to be seen by low-power microscopy.

Still slower rates of cooling produce normal pearlite, the coarseness of the ferrite and cementite laminations depending upon the rate of cooling. Thus, normalising leads to the formation of a fairly fine-grained structure whilst annealing produces coarse-grained structures.

12.23. In practice, factors such as composition, size and shape of the component to be hardened dictate the rate at which it shall be cooled. Large masses of steel of heavy section will obviously cool more slowly than small articles of thin section when quenched, so that whilst the surface skin may be martensitic, the core of a large section may be bainitic because it has cooled more slowly. If, however, small amounts of such elements as nickel, chromium or manganese are added to the steel, it will be found that the martensitic layer is much thicker than with a plain carbon steel of similar carbon content and dimensions which has been cooled at the same rate. Alloying elements therefore "increase the depth of hardening", and they do so by slowing down the transformation rates. This is a most important feature, since it enables an alloy steel to be hardened by much less drastic quenching methods than are necessary for a plain carbon steel. The liability to produce quench-cracks, which are often the result of water-quenching, is reduced in this way. Design also affects the susceptibility to quench-cracking. Sharp variations in cross-section and the presence of sharp angles, grooves, notches and rectangular holes are all likely to cause the formation of quench-cracks.

12.24. The quenching medium is chosen according to the rate at which it is desired to cool the steel. The following list of media is arranged in order of quenching speeds:

> 5% Caustic soda
> 5–20% Brine
> Cold water
> Warm water
> Mineral oil
> Animal oil
> Vegetable oil

The very drastic quench resulting from the use of caustic soda solution is used only when extreme hardness is required in components of simple shape. For more complicated shapes an oil-quenched alloy steel would give better results. Mineral oils are obtained during the refining of crude petroleum, whilst animal oils are produced by boiling the blubber of seal and whale, or by rendering down other animal tissue to obtain neatsfoot or lard oils. Vegetable oils include linseed, cottonseed and rape.

In addition to the rate of heat abstraction, such factors as flash point, viscosity and stability of oils are important. A high flash point is necessary to reduce fire risks, whilst a high viscosity will lead to a loss of oil by "drag-out", i.e. viscous oil clinging to the component as it is withdrawn from the quenching bath. Atmospheric oxidation and other chemical changes will often lead to a thickening of the oil and the formation of scum. On the other hand, some mineral oils "crack" or break down to simpler compounds of lower boiling point which will volatilise in use, leaving a thicker, more viscous mixture behind. All these factors govern the choice of a quenching medium.

12.25. To harden a piece of steel, then, it must be heated to between 30 and 50° C above its upper critical temperature and then quenched in some medium which will produce in it the desired rate of cooling. The medium used will depend upon the composition of the steel and the ultimate properties required. Symmetrically shaped components are best quenched "end-on", and all components should be agitated in the medium during quenching.

Tempering

12.30. A fully hardened carbon tool steel is relatively brittle, and the presence of stresses set up by quenching make its use, in this condition, inadvisable except in cases where extreme hardness is required. For this reason it is customary to reheat, or temper, the quenched component so that stresses will be relieved and, at the same time, brittleness reduced. Tempering causes a transformation of martensite to take place. The higher the tempering temperature, the more closely will the structure revert to the pearlitic structure stable at that temperature, since tempering is always carried out below the lower critical temperature.

12.31. Tempering at temperatures up to 200° C only relieves stress to some extent, but between 230 and 400° C the martensite changes to some degree to form a rapid-etching black constituent. Unlike bainite, which is started by ferrite separation, this new structure, called *Troostite*, is granular in form, and is, in fact, a mixture of extremely fine ferrite and cementite, rather like an emulsion of oil and water. Troostite is much tougher but somewhat softer than martensite, making it more useful in cases where strength and reliability are more important than extreme hardness.

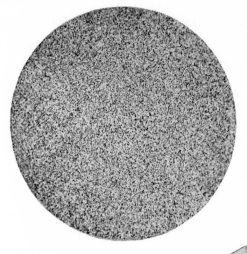

12.2A.—0·5% Carbon Steel, Water Quenched from 850° C and then Tempered at 600° C.

Sorbite. ×250. Etched in 2% Nital.

12.2B.—0·5% Carbon Steel, Normalised and then Annealed for 48 Hours at 670° C.

The pearlitic cementite has become spheroidised. ×750. Etched in Picral–Nital.

Courtesy of Messrs. United Steel Companies Ltd., Rotherham.

12.2C.—0·5% Carbon Steel, Water Quenched and then Tempered for 48 Hours at 670° C.

Spheroidised carbide has in this case been precipitated from martensite, making the distribution more even than in 12.2B. ×750. Etched in Picral–Nital.

Courtesy of Messrs. United Steel Companies Ltd., Rotherham

PLATE 12.2

12.32. Tempering to above 400° C causes the cementite particles to precipitate to such an extent that they are visible at magnifications of ×500 or so. This type of structure is known as *Sorbite*, but the difference between sorbite and troostite is only in the size of the particle and not in its fundamental nature. The point where troostite formation ends and sorbite formation begins is generally ill-defined. Consequently, there is a tendency among modern theoretical metallurgists not to use these terms, but instead to specify with accuracy the particular heat-treatment given to a steel (of stated composition) in order to produce the "tempered structure" in question. Nevertheless, the terms "troostite" and "sorbite" are still commonly used by the practical industrial metallurgist, and are therefore used in this book.

Due to increasing carbide precipitation, sorbite is weaker but more ductile than troostite, though as the tempering temperature rises above 550° C, strength falls rapidly, with little or no increase in ductility.

12.33. Tempering can be carried out in a number of ways, but, in all, the temperature needs to be fairly accurately controlled. As the steel is heated, the oxide film on the surface first assumes a pale-yellow colour and gradually thickens, with increase in temperature, until it is dark blue. This is a useful guide to the tempering of tools in small workshops where pyrometer-controlled tempering furnaces are not available. Table 12.1 shows typical colours obtained on clean surfaces when a variety of components are tempered to suitable temperatures. Such a colour–temperature relationship is only applicable to plain carbon steels. Stainless steels, for example, oxidise less easily, so that the colours obtained will bear no relationship to the temperatures indicated in the table. Moreover, the oxide film colour is only a reliable guide when the component has been progressively raised in temperature. It does not apply to one which has been maintained at a fixed temperature for some time, since here the oxide film will be thicker and darker in any case. In addition, the human element must also be taken into account, so that, in general, tempering in a pyrometer-controlled furnace is more successful.

12.34. Furnaces used for tempering are usually of the batch type (13.20—Part II). They employ either a circulating atmosphere or are of the liquid-bath type. Liquids transfer heat more uniformly and have a greater heat capacity, and this ensures an even temperature throughout the furnace. For low temperatures oils are often used, but higher temperatures demand the use of salt baths containing various mixtures of sodium nitrite and potassium nitrate. These baths can be used at about 500° C, but above that temperature either mixtures of chlorides or lead baths are necessary. Another popular furnace, in which the temperature can be varied easily and controlled thermostatically, is the circulating-air type. Here, uniform temperatures up to 650° C can be obtained by using fans to circulate the atmosphere, first over electric heaters, and then through a wire basket holding the charge.

TABLE 12.1.—*Tempering Colours for Plain-carbon-steel Tools*

Temperature (° C)	Colour	Type of component
220	Pale yellow	Scrapers; hack saws; light turning and parting tools
230	Straw	Screwing dies for brass; hammer faces; planing and slotting tools
240	Dark straw	Shear blades; milling cutters; paper cutters; drills; boring cutters and reamers; rock drills
250	Light brown	Penknife blades; taps; metal shears; punches; dies; woodworking tools for hard wood
260	Purplish-brown	Plane blades; stone-cutting tools; punches; reamers; twist drills for wood
270	Purple	Axes; gimlets; augers; surgical tools; press tools
280	Deeper purple	Cold chisels (for steel and cast iron); chisels for wood; plane cutters for soft woods
290	Bright blue	Cold chisels (for wrought iron); screw-drivers
300	Dark blue	Wood saws; springs

Failing a pyrometer-controlled furnace, temperature-indicating paints and crayons (23.30) are useful in determining the tempering temperature of small components, provided some method of uniform heating is available. Such indicators do indeed record the *actual temperature reached by the component*, which is more than can be said for a pyrometer controlling a furnace which is in the hands of an un-skilled operative.

Isothermal Transformations

12.40. As pointed out earlier in this chapter (12.22), the micro-structure and properties of a quenched steel are dependent upon the rate of cooling which prevails during quenching. This relationship, between structure and rate of cooling, can be studied for a given steel with the help of a set of isothermal transformation curves which are known as T.T.T. (Time–Temperature–Transformation) curves. The

T.T.T. curves for a steel of eutectoid composition are shown in Fig. 12.3. They indicate the time necessary for transformation to take place and the structure which will be produced when austenite is *supercooled* to any predetermined temperature.

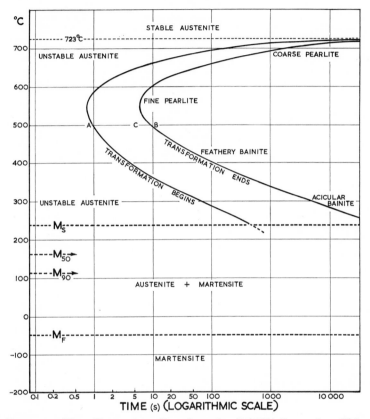

FIG. 12.3.—Time–Temperature–Transformation (T.T.T.) Curves for a Plain Carbon Steel of Eutectoid Composition.

Martensitic transformation is not complete until approximately $-50°$ C. Consequently a trace of "retained austenite" may be expected in a steel quenched to room temperature.

12.41. Such curves are constructed by taking a number of specimens of the steel in question, heating them into the austenitic range and then quenching them into baths at different temperatures. At predetermined time intervals individual specimens are removed from their baths and quenched in water. The microstructure is then examined to see the extent to which transformation had taken place at the holding temperature. Let us assume, for example, that we have heated a number of specimens of eutectoid steel to just above 723° C and have then quenched them into molten lead at 500° C (Figs. 12.3 and 12.4). Until one second has elapsed transformation has not begun, and if we remove a specimen from the bath in less than a second, and then quench it in water, we shall obtain a completely martensitic structure,* proving that at 500° C after one second (point *A* on Fig. 12.3) the steel was still completely austenitic. The production of martensite in the viewed

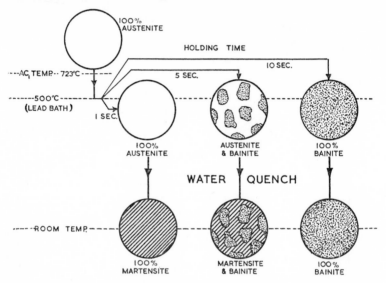

FIG. 12.4.—Diagrammatic Representation of the Method Used in Constructing T.T.T. Curves.

structure is entirely due to the final water-quench. If we allow the specimen to remain at 500° C for ten seconds (point *B*—Fig. 12.3) and then water-quench it, we shall find that the structure is composed entirely of bainite in feather-shaped patches, showing that after ten seconds at 500° C transformation to bainite was complete. If we quenched a specimen after it had been held at 500° C for five seconds (point *C*—Fig. 12.3) we would obtain a mixture of bainite and martensite (Fig. 12.4), showing that, at the holding temperature, the structure had contained bainite and austenite due to the incomplete transformation of the latter. By repeating such treatments at different holding temperatures we are able, by interpreting the resulting microstructures, to construct T.T.T. curves of the type shown in Fig. 12.3.

12.42. The horizontal line representing the temperature of 723° C is, of course, the lower critical temperature above which the structure

* Probably with traces of austenite for reasons which will be explained later.

of the eutectoid steel in question consists entirely of stable austenite. Below this line austenite is unstable, and the two approximately C-shaped curves indicate the time necessary for the austenite ⟶ ferrite + cementite transformation to begin and to be completed following rapid quenching to any predetermined temperature. Transformation is sluggish at temperatures just below the lower critical, but the delay in starting, and the time required for completion, decrease as the temperature falls towards 550° C. In this range the greater the degree of undercooling, the greater is the urge for the austenite to transform, and the rate of transformation reaches a maximum at 550° C. At temperatures just below 723° C, where transformation takes place slowly, the structure formed will be coarse pearlite, since there is plenty of time for diffusion to take place. In the region just above 550° C, however, rapid transformation results in the formation of very fine pearlite.

12.43. At temperatures between 550 and 220° C transformation becomes more sluggish as the temperature falls, for, although austenite becomes increasingly unstable, the slower rate of diffusion of carbon atoms in austenite at lower temperatures outstrips the increased urge of the austenite to transform. In this temperature range the transformation product is bainite. The appearance of this phase may vary between a feathery mass of fine cementite and ferrite for bainite formed around 450° C; and dark acicular (needle-shaped) crystals for bainite formed in the region of 250° C.

The horizontal lines at the foot of the diagram are, strictly speaking, not part of the T.T.T. curves, but represent the temperatures at which the formation of martensite will begin (M_s) and end (M_f) during cooling of austenite through this range. It will be noted that the M_f line corresponds approximately to −50° C. Consequently if the steel is quenched in water at room temperature, some "retained austenite" can be expected in the structure. This, however, will amount to less than 5% of the austenite which was present at the M_s temperature. In fact, at 110° C (Fig. 12.5) 90% of the austenite will have transformed to martensite.

12.44. These T.T.T. curves indicate structures which are produced by transformations which take place at a fixed temperature and specify a given "incubation" period which must elapse before transformation begins. There is no direct connection, therefore, between such isothermal transformations and transformations which take place under continuous cooling at a constant rate from 723° C to room temperature. Thus it is not possible to superimpose curves which represent continuous cooling on to a T.T.T. diagram. Modified T.T.T. curves which are related to continuous rates of cooling can, however, be produced.

These are similar in shape to the true T.T.T. curves, but are displaced to the right, as shown in Fig. 12.5. On this diagram are superimposed four curves, *A*, *B*, *C* and *D*, which represent different rates of cooling.

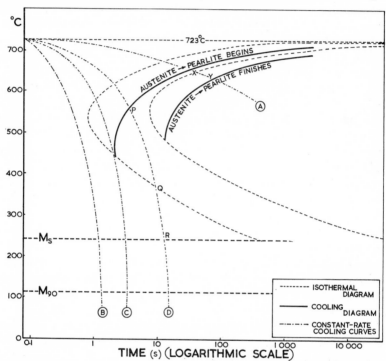

FIG. 12.5.—The Relationship between T.T.T. Curves and Curves Representing Continuous Cooling.

Curve *A* represents a rate of cooling of approximately 5° C per second, such as might be encountered during normalising. Here transformation will begin at *X* and be completed at *Y*, the final structure being one of fine pearlite. Curve *B*, on the other hand, represents very rapid cooling at a rate of approximately 400° C per second. This is typical of conditions prevailing during a water-quench, and transformation will not begin until 220° C, when martensite begins to form. The structure will consist of 90% martensite at 110° C and so contain a little retained austenite at room temperature. The lowest rate at which this steel (of eutectoid composition) can be quenched, in order to obtain a structure which is almost wholly martensitic, is represented by curve *C* (140 °C per second). This is called the critical cooling rate for the steel, and if a rate lower than this is used some fine pearlite will be formed. For example, in the case of the curve *D*, which represents a

cooling rate of about 50 °C per second, transformation would begin at
P with the formation of some fine pearlite. Transformation, however, is
interrupted in the region of Q and does not begin again until the M_s
line is reached at R, when the remaining austenite begins to transform
to martensite. Thus the final structure at room temperature is a mixture of pearlite, martensite and traces of retained austenite.

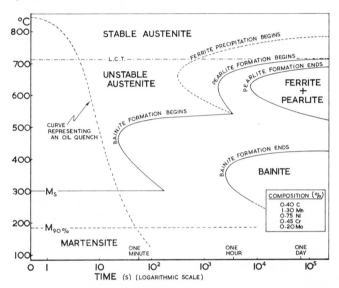

FIG. 12.6.—The Effects of Low-alloy Additions on the T.T.T. Curves for a
0·4% Carbon Steel.

The "nose" of the transformation begins curve is displaced well to the right and in
fact a "double nose" is formed. Even when a continuous-cooling curve (representing
an oil-quench) is superimposed on this isothermal diagram, it will be seen that there is
no transformation until the M_s line is reached and the structure will be wholly
martensitic. Since this diagram represents the T.T.T. curves for a hypo-eutectoid
steel, ferrite precipitation will begin before pearlite formation as indicated by the
broken line. This low alloy steel is covered by B.S. 970: 945M38. (See also Fig. 14.2.)

12.45. The T.T.T. curves illustrated in Fig. 12.3 are those for a steel
of eutectoid composition. If the carbon content is either above or below
this, the curves will be displaced to the left so that the critical cooling
rate necessary to produce a completely martensitic structure will be
greater. In order to obtain a structure which is entirely martensitic the
steel must be cooled at such a rate that the curve representing its rate of
cooling does not cut into the "nose" of the modified "transformation
begins" curve in the region of 550° C. Obviously, if the steel remains
in this region of the graph for more than one second, then transformation
to pearlite will begin. Hence the need for drastic water-quenches to
produce wholly martensitic structures in plain carbon steels.

Fortunately, the addition of alloying elements has the effect of slowing down transformation rates so that the T.T.T. curves are displaced to the right of the diagram. This means that much slower rates of cooling can be used, in the form of oil- or even air-quenches, and a martensitic structure still obtained. Small amounts of elements, such as nickel, chromium and manganese, are effective in this way (Fig. 12.6), and this is one of the most important effects of alloying.

12.46. We will now consider one or two practical applications arising from this study of modified isothermal transformation curves. Let us first examine the conditions under which a fairly large body of steel will cool, when quenched. The core will cool less quickly than the outside skin, and since its cooling curve B (Fig. 12.7A) cuts into the nose of the "transformation begins" curve, we can expect to find some fine pearlite in the core, whilst the surface layer is entirely martensitic. This feature is usually referred to as the "mass effect of heat-treatment" (12.50). Even if we are able to cool the component quickly enough to obtain a completely martensitic structure, as indicated in Fig. 12.7B, there will be such a considerable time interval CD between both core and surface reaching a martensitic condition that this will lead to quench-cracks being formed. These cracks will be caused by stresses set up due to the non-uniform volume change which takes place. Supposing, however, we cool the steel under conditions of the kind indicated in Fig. 12.7C. Here the steel is quenched into a bath at temperature E and left there until transformation is just about to commence. It is then removed from the bath and allowed to cool, so that martensite will begin to form at F. The net result is that, by allowing the core to attain the same temperature as the surface whilst in the bath at temperature E, we have prevented a big temperature gradient from being set up between the surface and the core of the specimen at the moment when martensite begins to form. The final air-cooling will not be rapid enough to allow a large temperature gradient to be set up, and both core and surface will become martensitic at the same time, thus minimising the tendency towards quench-cracking. The success of this treatment, which is usually known as "martempering", lies in cooling the steel quickly enough past the projecting "nose" of the modified "transformation begins" curve. Once safely past that point, relatively slow cooling will precipitate martensite. If we should cut into the "nose", fine pearlite will begin to form.

With suitable steels an "ausforming" operation can be combined with martempering. Whilst at the holding temperature the austenitic structure is heavily worked (up to 90% sectional reduction). The density of dislocations is increased by working and these dislocations are retained because the relatively low temperature does not permit recrystallisation. The steel is then allowed to cool and so transform to martensite which will *retain the high density of* dislocations, so that its strength is further increased. Moreover, the grain is fine due to the low forming tempera-

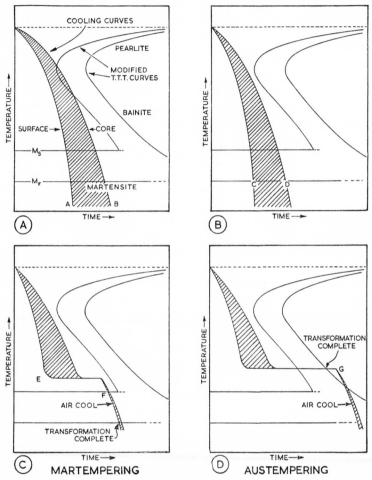

FIG. 12.7.—(A) and (B) Illustrate the Effects of Mass During Normal Quenching. (C) and (D) show how these effects may be largely overcome in martempering and austempering.

ture, whilst a fine dispersion of carbides, presumably deposited during forming, contributes some degree of particle hardening. Some low-alloy steels, having a deep "bay" between the pearlite and bainite "noses" of the T.T.T. curve (Fig. 12.6), can be treated in this way and will develop tensile strengths in the region of 3 000 N/mm².

12.47. Isothermal transformation offers a method by which we can obtain a tempered type of structure without the preliminary drastic water-quench. Such a treatment, known as "austempering", is illustrated in Fig. 12.7D. Here the steel is quenched into a bath at a temperature above that at which martensite can be formed and allowed to remain there long enough for transformation to be complete at G. Since transformation to bainite is complete at G, the steel can be cooled to room temperature at any desired rate, but air-cooling is preferable. We have thus succeeded in obtaining a structure which is similar in properties* to that of troostite, which is obtained by quenching and tempering. The drastic water-quench from above the upper critical temperature, however, has been avoided. Austempering is therefore a process of considerable importance when heat-treating components of intricate section. Such components might distort or crack if they were heat-treated by the more conventional methods of quenching and tempering.

12.48. Finally, mention must be made of "isothermal annealing". In this process the steel is heated into the austenitic range and then allowed to transform as completely as possible in the pearlitic range. The object of such treatment is generally to soften the steel sufficiently for subsequent cold-forming or machining operations.

The nature of the pearlite formed during transformation is influenced by the initial austenitising temperature. An austenitising temperature which is little above the upper critical for the steel promotes the formation of spheroidal pearlitic cementite during isothermal annealing, whilst a higher austenitising temperature favours the formation of lamellar pearlitic cementite. The pearlite structure is also influenced by the temperature at which isothermal transformation takes place, as would be expected. Transformation just below the lower critical temperature leads to the formation of spheroidal pearlitic cementite since precipitation is slow, whilst at lower temperatures transformation rates are higher and lamellar cementite tends to form. A structure containing spheroidal carbide is generally preferred for turning and cold-forming operations, whilst one with lamellar carbide is often used where milling or drilling are involved. It is claimed that isothermal annealing gives more uniform properties than does an ordinary annealing process.

Hardenability and Ruling Section

12.50. Brief mention of the "mass effect" in connection with the heat-treatment of steel has already been made. If a piece of steel is of heavy section, then it will be impossible to cool it as quickly as one of thin

* Though not in structure.

section, even by drastic quenching. The piece which is of heavy section will not, therefore, harden completely in its core, whilst that of thin section will be uniformly martensitic (Fig. 12.8). This difficulty may be remedied to some extent by adding alloying elements to the steel. These reduce the critical rates of transformation and make it possible to get a martensitic structure evenly throughout by oil-quenching.

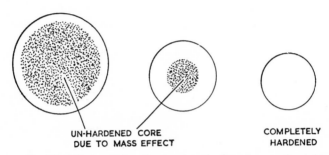

UN-HARDENED CORE
DUE TO MASS EFFECT

COMPLETELY
HARDENED

FIG. 12.8.—The Effects of Mass Produced in the Structure on Quenching. The heavier sections will cool too slowly to be entirely martensitic.

12.51. This is one of the most important functions of alloying, but in order to avoid the misuse of steels due to lack of appreciation of their properties, it has become necessary to specify the maximum diameter, or "ruling section", up to which the stated mechanical properties apply. If the section is exceeded, then the properties across the section cannot be expected to be uniform, since hardening of the core will not be complete.

12.52. The Jominy end-quench test is of great practical use in determining the hardenability of steel. Here a standard test piece is made (Fig. 12.9A) and heated up to its austenitic state. It is then dropped into position in a frame, as shown in Fig. 12.9B, and quenched at its end only, by means of a pre-set standard jet of water at 25° C. Thus different rates of cooling are obtained along the length of the bar. After the cooling, a "flat", 0·4 mm deep, is ground along the side of the bar and hardness determinations made every millimetre along the length from the quenched end. The results are then plotted as in Fig. 12.10.

These curves show that the depth of hardening of a nickel–chromium steel is greater than that of a plain carbon steel of similar carbon content, whilst the depth of hardening of a chromium–molybdenum steel is greater than that of the nickel–chromium steel.

With modifications, the results of the Jominy test can be used as a basis in estimating the "ruling section" of a particular steel. There is no direct simple relationship between the two, however, and it is often simpler to find by trial and error how a particular section will quench.

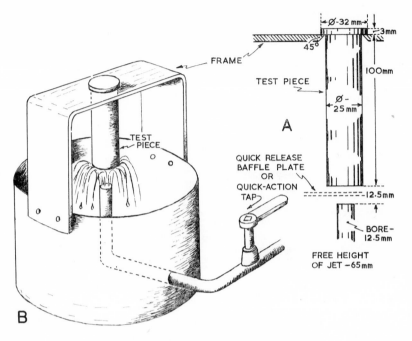

FIG. 12.9.—The Jominy End-quench Test.

(A) The standard form of test piece used. (B) A simple type of apparatus for use in the test.

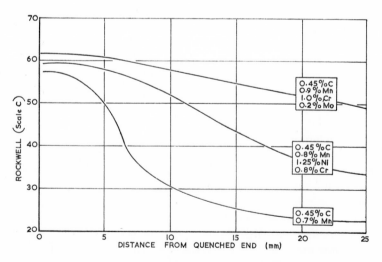

FIG. 12.10.—The Relative Depth of Hardening of Three Different Steels as Indicated by the Jominy Test.

RECENT EXAMINATION QUESTIONS

1. An annealed 0·4% C steel bar is cold-worked and placed with one end in a furnace at 900° C while the other end is maintained at room temperature. After a few hours the bar is quenched in cold water. Describe the structures you would expect to find along the length of the bar.
(Brunel College, Dip. Tech.)

2. Account for the hardness of martensite and discuss the changes in properties and applications of various typical tempering temperatures.
(University of Strathclyde, B.Sc.(Eng.).)

3. Describe, with the aid of a sketch, the type of atom arrangement which is thought to represent the crystal structure of martensite.
Show how an assumption of the existence of this type of structure helps to explain the mechanical properties of martensite.
Outline the microstructural changes which occur when martensite is tempered at various temperatures up to 600° C.
(West Bromwich Technical College.)

4. Draw the iron–carbon equilibrium diagram. By referring to it, distinguish between: (i) normalising; (ii) annealing; (iii) hardening; and (iv) tempering, as applied to carbon steels. (S.A.N.C.A.D.)

5. Sketch and label the "steel part" of the iron–iron carbide equilibrium diagram. With reference to the diagram describe the structural changes that occur when a cast 0·5% carbon steel is:
   (a) slowly heated to 900° C;
   (b) slowly cooled from 900° C;
   (c) quenched from 900° C;
   (d) Quenched from 900° C and reheated to 650° C.
Draw each microstructure, including that of the steel in the cast condition. (Constantine College of Technology, Middlesbrough.)

6. Both annealing and tempering are processes used to soften steel. Outline the conditions when these treatments would be used, and indicate any difficulties that may be encountered in practice.
(University of Aston in Birmingham, Inter B.Sc.(Eng.).)

7. Describe fully the nature of the following micro-constituents: (a) ferrite; (b) austenite; (c) martensite; (d) sorbite; (e) cementite.
Comment on the mechanical properties of each substance.
(U.L.C.I., Paper No. C111–1–4.)

8. Give an account of the principles involved in the quench-hardening of steel. Indicate the significance of critical cooling velocity.
(Derby and District College of Technology, H.N.D. Course.)

9. Give an account of how an isothermal transformation diagram for a steel may be compiled.
Describe the structures to be expected when austenite transforms at temperatures below the critical point.
Of what value are these transformation diagrams to the engineer?
(S.A.N.C.A.D.)

10. Sketch and label a typical time–temperature transformation curve (*S*-curve) for a plain carbon steel.

  With reference to this diagram describe the processes of: (*a*) martempering; (*b*) austempering.

  Indicate the advantages and limitations of each of these processes.

  (Carlisle Technical College, A1 year.)

11. (*a*) Explain what is meant by the "allotropy of iron" and show its importance in the heat treatment of ferrous materials with special reference to the martensite transformation. Mention any problems with these phenomena and suggest how "martempering" many minimise these problems.

  (*b*) A 0·6% carbon steel is water-quenched from 830° C and is then reheated slowly to 700° C.

  Describe in detail the changes in microstructure and mechanical properties that occur during the reheating period.

  (Aston Technical College.)

12. Give a brief account of each of the following heat-treatment processes: (i) martempering; (ii) austempering; (iii) process anneal; (iv) full anneal.

  (Derby and District College of Technology, H.N.D. Course.)

13. How are carbon steels heat-treated to develop their properties? What are the limitations of such steels and how may they be overcome? Explain your statements by reference to T.T.T. curves.

  (University of Aston in Birmingham, Inter B.Sc.(Eng.).)

14. What is meant by the term "Critical Cooling Velocity" as applied to steels?

  Describe, with the aid of diagrams, the structural changes which may take place when a plain carbon steel is cooled from the austenitic state, and how these changes are dependent upon the rate of cooling.

  (The Polytechnic, London W.1., Poly.Dip.)

15. Outline the following processes: (i) austempering; (ii) martempering; (iii) isothermal annealing.

  Contrast the advantages and limitations of these methods of heat-treatment with those of the more widely used procedures for obtaining comparable properties in steels.

  (West Bromwich Technical College.)

16. Four thin pieces of the same 0·8% carbon rolled-steel rod are heat-treated differently as follows:

  (i) heated to 680° C for twenty-four hours and cooled in still air;

  (ii) water-quenched from 750° C;

  (iii) heated to 730° C, quenched into molten lead at 400° C, allowed to remain there for five minutes and then cooled;

  (iv) heated to 1 200° C and cooled in still air.

  Sketch the type of microstructure produced in each specimen and explain the mechanism of its formation.

  (Institution of Production Engineers Associate Membership Examination, Part II, Group A.)

17. It is impossible to obtain maximum strength and hardness throughout the section of a plain carbon steel by quenching if the diameter is greater than about 25 mm. Why is this?

    Indicate how the effect may be overcome by alloying and describe a test which may be used to determine the depth of hardening.
    (Derby and District College of Technology.)

18. Using diagrammatic T.T.T. curves, explain the reasons for the addition of alloying elements to steels to overcome the limitations of carbon steels in heat-treatment.
    (University of Aston in Birmingham, Inter B.Sc.(Eng.).)

19. What is meant by the terms "hardenability", "mass effect" and "ruling section"?

    Describe a test for determining the hardenability of a steel.

    Sketch the profiles of the hardenability curves you would expect for the following steels:

    (a) 0.4% C, 0.6% Mn, 2.5% Ni, 0.6% Cr, 0.5% Mo;
    (b) 1.05% C, 0.6% Mn.

    (The Polytechnic, London W.1., Poly.Dip.)

20. (a) Distinguish carefully between "hardness" and "hardenability".

    (b) Name and briefly describe a test which gives some indication of hardenability.

    (c) Sketch the results which the test described in (b) would produce with (i) a plain 0.9% carbon steel; (ii) a nickel–chrome–molybdenum low-alloy steel. (C. & G., Paper No. 293/21.)

21. (a) Explain the term " hardenability" as applied to steel. State why it is important in engineering components and how improved hardenability may be obtained.

    (b) Explain what is meant by decarburisation and discuss its effects in engineering practice. (University of Strathclyde, B.Sc.(Eng.).)

22. Explain fully what is meant by the "ruling section" of a steel and discuss its significance in the choice of steels for engineering design.

    Outline ONE experimental procedure which is helpful in assessing the ruling section of a steel. (West Bromwich Technical College.)

23. (a) Describe how you would carry out a Jominy test on a steel sample.

    (b) What differences would be expected in the results obtained from such a test using:

    (i) a 0.3% plain carbon steel;
    (ii) a 0.3% carbon, 1½% nickel, 1% chromium?

    (c) Indicate clearly the practical importance of these differences in the production of *small* and *large* engineering components respectively.
    (Institution of Production Engineers Associate Membership Examination, Part II, Group A.)

CHAPTER XIII

ALLOY STEELS

13.10. Possibly the earliest attempt to produce an alloy steel was made in 1822 at the instigation of Michael Faraday. He was desirous of obtaining better cutting tools and non-corrodible metals for reflectors. He therefore attempted to alloy iron with a number of rare elements, including silver, gold, platinum and rhodium, and, incidentally, chromium; but it was not until ninety years later (in 1912) that Brearley discovered the stainless properties of high-chromium steel.

It may safely be said that the first *successful* attempt at producing an alloy steel was the result of the researches of the British metallurgist Robert Mushet, who had made Henry Bessemer's process a viable proposition by introducing deoxidation with manganese. Mushet's "self-hardening" tungsten–manganese steel produced in 1868 was indeed the forerunner of modern high-speed steel.

Systematic research into the properties of alloy steels dates from Sir Robert Hadfield's discovery of high-manganese steel in 1882. Some time afterward Dr. J. E. Stead said of this steel: "Hadfield had surprised the whole metallurgical world with the results obtained. The material produced was one of the most marvellous ever brought before the public." Large quantities of this alloy are produced to-day by the Sheffield firm which still carries the inventor's name.

So-called plain carbon steels contain up to 1.0% manganese, residual from deoxidation and desulphurisation processes, but an alloy steel is defined in Group A of the Trade and Navigation Accounts as: "Steel containing, by weight, at least 0.1% of molybdenum or vanadium; or 0.3% of tungsten or cobalt; or 0.5% of chromium or nickel; or 2.0% manganese".

By extending the usefulness of steel, the development of alloy steels has contributed greatly to the progress of those industries upon which modern civilisation largely depends. Certain elements are added to a plain carbon steel in order to improve its properties or, sometimes, to introduce entirely new properties. Thus, small amounts of nickel and chromium may be added to produce a general improvement in mechanical properties, whilst larger amounts of these metals will introduce such phenomena as the stabilisation of the austenite phase at room temperature, accompanied by a loss of ferromagnetism. The alloying elements added may either simply dissolve in the ferrite or they may combine with some of the carbon, forming carbides, which associate with the iron carbide already present.

The principal effects which these alloying elements have on the microstructure and properties of a steel can be classified as follows:

13.11. The Effect on the Allotropic Transformation Temperatures. The allotropic transformation temperatures referred to here are

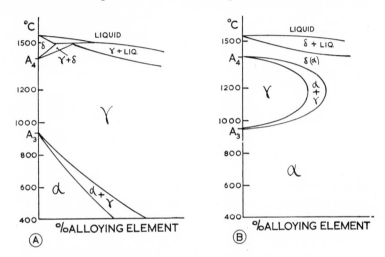

Fig. 13.1.—Relative Effects of the Addition of an Alloying Element on the Allotropic Transformation Temperatures at A_3 and A_4.
(A) Tending to stabilise γ, and (B) tending to stabilise α.

those at $910°$ C, where, on heating, the body-centred cubic (α) structure of pure iron changes to one which is face-centred cubic (γ); and at $1400°$ C, where the face-centred cubic structure changes back to body-centred cubic (δ) iron. These change points are designated A_3 and A_4 respectively. (The A_1 temperature is at $723°$ C, where the austenite $\longrightarrow$ pearlite change takes place in plain carbon steels; whilst the A_2 temperature is at $769°$ C, above which pure iron ceases to be magnetic. This point has no structural significance.)

Some elements, notably nickel, manganese, cobalt and copper, raise the A_4 temperature and lower the A_3 temperature, as shown in Fig. 13.1A. In this way these elements, when added to a carbon steel, tend to stabilise austenite (γ) and increase the range of temperature over which austenite can exist as a stable phase. Other elements, the most important of which include chromium, tungsten, vanadium, molybdenum, aluminium and silicon, have the reverse effect, in that they tend to stabilise ferrite (α) by raising the A_3 temperature and lowering the A_4, as indicated in Fig. 13.1B. Such elements restrict the field over which austenite may exist, and thus form what is often called a "gamma (γ) loop".

The elements of the γ-stabilising group generally have a face-centred cubic lattice. Since this is similar to that of austenite, these elements therefore retard the transformation of austenite to ferrite. At the same time these elements retard the precipitation of carbides, and again this has the effect of stabilising austenite. The α-stabilising elements are

ELEMENT	PROPORTION DISSOLVED IN FERRITE	PROPORTION PRESENT AS CARBIDE	ALSO PRESENT IN STEEL AS—
NICKEL	●	—	$NiAl_3$
SILICON	●	—	—
ALUMINIUM	●	—	SOME NITRIDES (SEE 19.41)
ZIRCONIUM	●	—	SOME NITRIDES (SEE 19.41)
MANGANESE	●	•	MnS GLOBULES (SEE 20.21)
CHROMIUM	●	•	—
TUNGSTEN	●	●	—
MOLYBDENUM	●	●	—
VANADIUM	•	●	SOME NITRIDES (SEE 19.41)
TITANIUM	•	●	SOME NITRIDES (SEE 19.41)
NIOBIUM	•	●	—
COPPER	○	—	Cu GLOBULES WHEN IN EXCESS OF 0.8 %
LEAD	—	—	Pb GLOBULES (SEE 20.22)

FIG. 13.2.—The Condition in Which Alloying Elements Are Present in Steel.

usually those with a body-centred cubic lattice. These will dissolve more readily in α-iron than in γ-iron, and at the same time diminish the solubility of carbon in austenite. In this way they stabilise ferrite. As shown in Fig. 13.1B, progressive increase in one or more of the α-stabilising elements will cause a point to be reached, beyond the confines of the γ-loop, where the γ-phase cannot exist at any temperature. Thus the addition of more than 30% chromium to a steel containing 0·4% carbon would lead to the complete suppression of the allotropic transformations, and such a steel would no longer be amenable to normal heat-treatment (Fig. 13.8).

13.12. The Effect on the Stability of the Carbides. Some of the alloying elements form very stable carbides when added to a plain carbon steel (Fig. 13.2). This generally has a hardening effect on the steel, particularly when the carbides formed are harder than iron carbide itself. Such elements include chromium, tungsten, vanadium, molybdenum, titanium and manganese. When more than one of these elements are present, a structure containing complex carbides is often formed.

Other elements have a graphitising effect on the iron carbide; that is, they tend to make it unstable so that it breaks up, releasing free graphitic carbon. This effect is more evident if no carbide stabilisers are present. Elements which tend to cause graphitisation include silicon, nickel and aluminium. Therefore, if it is necessary to add appreciable amounts of these elements to a steel, it can be done only when the carbon content

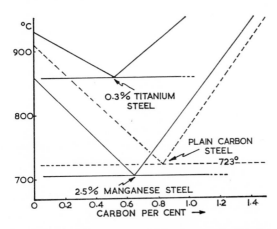

FIG. 13.3.—The Effects of Manganese and Titanium on the Displacement of the Eutectoid Point in Steel.

is extremely low. Alternatively, if the carbon content needs to be high, one or more of the elements of the first group, namely the carbide stabilisers, must be added in order to counteract the effects of the graphitising element.

13.13. The Effect on Grain Growth. The rate of crystal growth is accelerated, particularly at high temperatures, by the presence of some elements, notably chromium. Care must therefore be taken that steels containing elements in this category are not overheated or, indeed, kept for too long at an elevated temperature, or brittleness, which is usually associated with coarse grain, will result.

Fortunately, grain growth is retarded by other elements, notably nickel and vanadium, whose presence thus produce a steel which is less sensitive to the temperature conditions of heat-treatment.

13.14. The Displacement of the Eutectoid Point. The addition of an alloying element to carbon steel displaces the eutectoid point towards the left of the equilibrium diagram. That is, a steel can be completely pearlitic even though it contains less than 0.83% carbon. For example, the addition of 2.5% manganese to a steel containing 0.65% carbon produces a completely pearlitic structure in the normalised condition (Fig. 13.3). Similarly, although a high-speed steel may contain only 0.7% carbon, its microstructure exhibits masses of free carbide due to the displacement of the eutectoid point far to the left by the effects of the alloying elements (totalling 25%) which are present. The relative effects of various elements in displacing the eutectoid *composition* are shown in Fig. 13.4, whilst Fig. 13.5 shows the extent

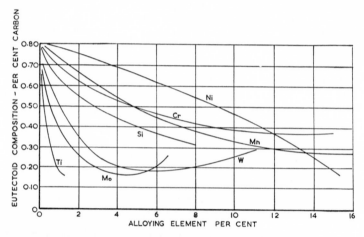

FIG. 13.4.—The Effect of Alloying Elements on the Eutectoid Composition.

to which some elements raise or lower the eutectoid *temperature*. This latter will move in sympathy with the A_3 point (13.11).

13.15. The Retardation of Transformation Rates. We have already seen (12.45) that the T.T.T. curves for a plain carbon steel are displaced to the right due to the effects produced by the addition of alloying elements. Thus, by adding alloying elements, we reduce the critical cooling rate which is necessary for the transformation of austenite to martensite to take place. This feature of the alloying of steels has obvious advantages and all alloying elements, with the exception of cobalt, will reduce transformation rates.

In order to obtain a completely martensitic structure in the case of a plain 0.83% carbon steel, we must cool it from above 723° C to room temperature in approximately one second. This treatment involves a very drastic quench, generally leading to distortion or cracking of the component. By adding small amounts of suitable alloying elements, such as nickel and chromium, we reduce this critical cooling rate to such an extent that a less drastic oil-quench is rapid enough to produce a totally martensitic structure. Further increases in the amounts of alloying elements will so reduce the rate of transformation that such a steel can be hardened by cooling in air. "Air-hardening" steels have the particular advantage that comparatively little distortion is produced

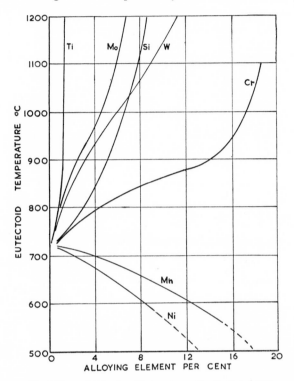

FIG. 13.5.—The Effect of Alloying Elements on the Eutectoid Temperature.

during hardening. Alternatively, such a steel, containing $4\frac{1}{4}\%$ nickel and $1\frac{1}{4}\%$ chromium and which will air-harden in thin sections, can be hardened completely through sections up to 0·15 m diameter by oil-quenching. This feature of alloying is one of the greatest significance.

13.16. The Improvement in Corrosion-resistance. The corrosion-resistance of steels is substantially improved by the addition of elements such as aluminium, silicon and chromium. These elements form thin but dense and adherent oxide films which protect the surface of the steel from further attack. Of the elements mentioned, chromium is the most useful when mechanical properties have to be considered. When nickel also is added in sufficient quantities, the austenitic structure is maintained at room temperature, so that, provided the carbon content is kept low, a completely solid-solution type of structure prevails. This also helps in maintaining a high corrosion-resistance by limiting the possibility of electrolytic attack (22.21).

13.17. Effects on the Mechanical Properties. One of the main reasons for alloying is to effect improvements in the mechanical proper-

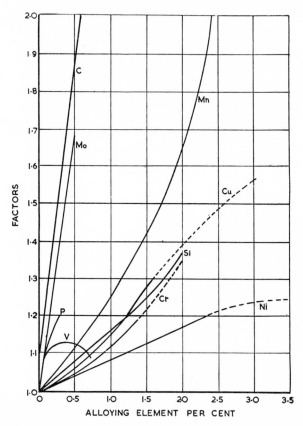

FIG. 13.6.—Walters' Factors for Estimating the Tensile Strength of Pearlitic Steels in the Normalised Condition.

ties of a steel. These improvements are generally the result of physical changes already referred to. For example, hardness is increased by stabilising the carbides; strength is increased when alloying elements dissolve in the ferrite; and toughness is improved due to refinement of grain. Many formulae have been devised, by the use of which the approximate tensile strength of an alloy steel may be estimated from its known composition. For example, Walters takes a basic tensile strength for pure iron of 250 N/mm² and multiplies this by factors for each alloying element present. The factor for each element changes as its quantity changes. Such factors for most of the principal alloying elements are shown in Fig. 13.6 and are true for pearlitic steels in the normalised condition. They give best results for steels with carbon contents below 0·25% and within the intermediate alloy range.

We will now examine the compositions and properties of some of the more important alloy steels.

Nickel Steels

13.20. Nickel is extensively used in alloy steels for engineering purposes, generally in quantities up to about 5·0%. When so used, its purpose is to increase tensile strength and toughness. It is also used in stainless steel, as mentioned later in this chapter. The main sources of nickel are the Sudbury mines in Northern Ontario, Canada; Cuba; the one-time cannibal island of New Caledonia in the Pacific; and the U.S.S.R.

13.21. The addition of nickel to a plain carbon steel tends to stabilise the austenite phase over an increasing temperature range, by raising the A_4 point and lowering the A_3 point. Thus, the addition of 25% nickel

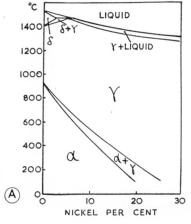
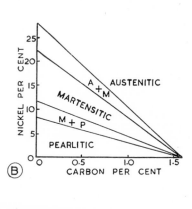

FIG. 13.7.—The Effects of Nickel as an Alloying Element.
(A) Its influence on the stability of γ.
(B) The Guillet diagram.

to pure iron renders it austenitic, and so non-magnetic, even after slow cooling to room temperature. The structure obtained after slow cooling to room temperature can be estimated for a nickel steel of known composition by referring to the type of diagram devised by Guillet (Fig. 13.7B). These diagrams show the approximate relationship

TABLE 13.1—*Nickel Steels and Alloys*

Type of steel	B.S. 970 designation	Proprietary trade names and codes	Composition (%)			Typical mechanical properties								Heat-treatment	Uses
			C	Mn	Ni	Condition	Yield point N/mm²	Yield point tonf/in²	Tensile strength N/mm²	Tensile strength tonf/in²	Elongation (%)	Redn. in area (%)	Izod (J)		
1% nickel	503M40	Dunford Hadfield's "Hecla 115"	0·4	1·50	1·0	Quenched and tempered at 600° C	494	32	695	45	25	55	91	Oil-quench from 850° C; temper between 550 and 660° C, and cool in oil or air.	Crankshafts, axles, connecting rods, other parts in the automobile industry, general engineering purposes.
3% nickel case-hardening	—	Firth Brown N3CH Carr's P153	0·12	0·45	3·0	38·1 mm dia. bar, hardened	510	33	772	50	20	60	83	After carburising, refine by oil-quenching from 860° C. Then harden by water-quenching from 770° C.	*Case-hardening;* crown wheels, gudgeon pins, differential pinions, camshafts.
5% nickel case-hardening	—	"Hecla 78" Firth Brown N5CH	0·12	0·40	5·0	28·5 mm dia. bar, hardened	602	39	849	55	22	47	68	After carburising, refine by oil-quenching from 850° C. Then harden by oil-quenching from 760° C.	*Heavy-duty case-hardened parts;* bevel pinions, gudgeon-pins, gear-box gears, worm shafts.
Thermal expansion alloy	—	Balfour Darwin's 36% Ni	0·05	—	36·0	—	—	—	—	—	—	—	—	Non-hardenable (except by cold-work).	Constant-coefficient, low-expansion nickel/iron alloy used for temperature-control equipment (thermostats, etc.).
Thermal expansion alloy	—	Balfour Darwin's 48% Ni	0·05	—	48·0	—	—	—	—	—	—	—	—	Non-hardenable (except by cold-work).	Higher-temperature thermostats and glass/metal sealing applications.

between microstructure, carbon content and the amount of alloying element added.

13.22. At the same time nickel makes the carbides unstable and tends to cause them to decompose to graphite. For this reason it is inadvisable to add nickel by itself to a high-carbon steel, and most nickel steels are low-carbon steels. If a higher carbon content is desired, then the manganese content is usually increased, since manganese acts as a stabiliser of carbides.

In addition to improving tensile strength and toughness, nickel has a grain-refining effect which makes the low-nickel, low-carbon steels very suitable for case-hardening (19.31), since grain growth will be limited during the prolonged period of heating in the region of 900° C.

13.23. The 3% and 5% nickel steels are the most widely employed. Those with the lower carbon contents are used mainly for case-hardening, whilst those with up to 0·4% carbon are used for structural purposes, shafting, gears, etc., as shown in Table 13.1.

Nickel lowers the coefficient of thermal expansion of steel progressively, until with about 35% nickel expansion is negligible. "Invar" is of approximately this composition (35% nickel; less than 0·1% carbon) and is useful in the manufacture of pendulum rods for master clocks, measuring tapes and delicate sliding mechanisms for use at varying temperatures.

Chromium Steels

13.30. The bulk of metallic chromium produced is used in the manufacture of alloy steels and in the electro-plating industry. The main producers of chromium are the U.S.S.R., South Africa, the Philippines, Jugoslavia, New Caledonia, Rhodesia, Cuba and Turkey, though not all of these countries are important exporters. Sierra Leone, a relatively small producer, is at present expanding its output.

13.31. The main function of chromium when added in relatively small amounts to a carbon steel is to cause a considerable increase in hardness. At the same time strength is raised with some loss in ductility, though this is not noticeable when less than 1·0% chromium is added. The increase in hardness is due mainly to the fact that chromium is a carbide stabiliser, and forms the hard carbides Cr_7C_3 or $Cr_{23}C_6$, or, alternatively, double carbides with iron carbide.

13.32. Chromium lowers the A_4 temperature and raises the A_3 temperature, forming the closed γ-loop already mentioned. In this way it stabilises the α-phase at the expense of the γ-phase. The latter is eliminated entirely, as shown in Fig. 13.8, if more than 11% chromium is added to pure iron, though with carbon steels a greater amount of chromium would be necessary to have this effect. Provided that the composition of the steel falls to the left of the γ-loop, chromium will give rise to a much greater depth of hardening, due to the retardation of the transformation rates which it also produces.

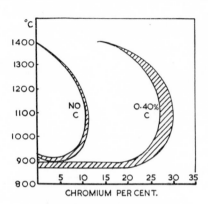

THE EFFECT OF C & Cr CONTENTS ON THE Y-LOOP IN Cr STEELS — Y LIES INSIDE THE LOOP; α+Y IN THE SHADED BAND; α OUTSIDE THE LOOP. A STEEL REPRESENTED BY A COMPOSITION TO THE RIGHT OF THE LOOP DOES NOT HARDEN ON QUENCHING.

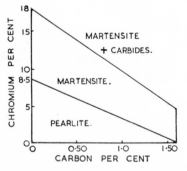

FIG. 13.8.—The Effects of Chromium as an Alloying Element.

13.33. The main disadvantage in the use of chromium as an alloying element is its tendency to promote grain growth, with the attendant brittleness that this involves. Care must therefore be taken to avoid overheating or holding for too long at the normal heat-treatment temperature.

13.34. Steels containing small amounts of chromium and up to 0·45% carbon are used for axle shafts, connecting-rods and gears; whilst those containing more than 1·0% carbon are extremely hard and are useful for the manufacture of ball-bearings, drawing dies and parts for grinding machines.

13.35. Chromium is also added in larger amounts—up to 21%— and has a pronounced effect in improving corrosion-resistance, due to the protective layer of oxide formed. This oxide layer is extremely thin, and these steels take a very high polish. They contain little or no carbon and are therefore completely ferritic and non-hardening (except by cold-work). They are used widely in the chemical-engineering industry; for domestic purposes, such as stainless-steel sinks; and in food containers, refrigerator parts, beer barrels, cutlery and table-ware. The best-known alloy in this group is "stainless iron", containing 13% chromium and usually less than 0·05% carbon.

If the carbon content exceeds 0·1% the alloy is a true stainless steel and is amenable to hardening by heat-treatment. The most common alloy in this group contains 13% chromium and approximately 0·3%

carbon. Due to displacement of the eutectoid point to the left, this steel is of approximately eutectoid composition. It is widely used in stainless-steel knives.

The stainless steels in this group are of the martensitic type, the structure being obtained by rapid cooling. If these steels are allowed to cool slowly, carbides will be precipitated, with consequent loss in corrosion-resistance. This is due partly to chromium coming out of solution in the form of carbides, thus leaving the matrix low in chromium, and partly to subsequent electrolytic action between the carbide particles and the matrix in the presence of an electrolyte.

The compositions and properties of the more important chromium steels are given in Tables 13.2a and b.

Nickel–Chromium Steels

13.40. Certain disadvantages attend the addition of either nickel or chromium, singly, to a carbon steel. Whilst nickel tends to prevent grain growth during heat-treatment, chromium accelerates it, producing the attendant brittleness under shock. On the other hand, whilst chromium tends to form stable carbides, making it possible to produce high-chromium, high-carbon steels, nickel has the reverse effect in assisting graphitisation. The deleterious effects of each element can be overcome, therefore, if we add them in conjunction with each other. Then, the tendency of chromium to cause grain growth is nullified by the grain-refining effect of the nickel, whilst the tendency of nickel to favour graphitisation of the carbides is counteracted by the strong carbide-forming tendency of the chromium.

13.41. At the same time other physical effects of each element are additive, so that they combine in increasing strength, corrosion-resistance and the retardation of transformation rates during heat-treatment. This latter effect is particularly useful in making drastic water-quenches avoidable.

13.42. In general, the low-nickel, low-chromium steels contain rather more than two parts of nickel to one part of chromium. Those with up to 3·5% nickel and 1·5% chromium are oil-hardened from temperatures between 810 and 850° C, followed by tempering at temperatures between 150 and 650° C according to the properties required. Some of these steels suffer from "temper-brittleness" when tempered in the range 250–400° C. This is shown by low resistance to shock in the Izod test. If tempered above 400° C, therefore, it is necessary to cool the steel quickly in oil through the range in which temper-brittleness is produced. The effect can be minimised by adding

TABLE 13.2a.—*Low-chromium Steels*

Type of Steel	B.S.970 designation	Proprietary trade names and codes	Composition (%)			Condition	Typical mechanical properties								Heat-treatment	Uses
			C	Mn	Cr		Yield point		Tensile strength		Elongation (%)	Reduction area (%)	Izod (J)	Hardness (Brinell)		
							N/mm²	tonf/in²	N/mm²	tonf/in²						
"60" C–Cr steel	526M60	Dunford Hadfield's "Hecla 104" Carr's P280 Thos. Andrews' "Monarch General Utility"	0·60	0·65	0·65	Oil-quenched and tempered at 200° C	—	—	—	—	—	—	—	700	Oil-quench from 800–850° C. Temper: (1) for cold-working tools at 200–300° C; (2) for hot-working tools at 400–600 °C.	General blacksmith's and boilermaker's tools and chisels. Hot and cold sates. Swages. Hot-stamping and forging dies. Builder's, mason's and miner's tools. Spring collets, chuck and vice jaws. Turning mandrels and lathe centres.
1% Cr steel	530M40	"Hecla 105" Firth Brown "Firthag" Carr's P609	0·45	0·90	1·00	28·5 mm diameter bar, oil-quenched and tempered at 550° C.	880	57	988	64	20	55	73	—	Oil-quench from 860° C; temper from 550–700° C.	Agricultural machine parts, machine-tool components. Lining plates, paddles and drums for concrete and tar mixers. Excavator cutting blades and teeth. Automobile axles, connecting-rods and steering arms. Spanners and small tools.
1% C–Cr steel	534A99	"Hecla 108" Firth Brown CCR2 Carr's P720 Jessop-Saville H2 Thos. Andrews' "Hardenite No. 5"	1·00	0·45	1·40	Hardened	—	—	—	—	—	—	—	850	Oil-quench from 810° C; temper at 150° C.	Ball- and roller-bearings. Roller- and ball-races. Instrument pivots and spindles. Cams. Small rolls.

TABLE 13.2b.—*Stainless High-chromium Steels and Irons*

Type of Steel	B.S. 970 (Pt. 4) designation	Proprietary trade names and codes	Composition (%)			Condition	Typical mechanical properties						Heat-treatment	Uses
			C	Mn	Cr		Yield point		Tensile strength		Elonga-tion (%)	Hard-ness Brinell		
							N/mm²	tonf/in²	N/mm²	tonf/in²				
Stainless iron	403S17	Br. Steel Corp. "Silver Fox 403"	0.04	0.45	14.0	Soft	340	22	510	33	31	—	Non-hardening except by coldwork.	Wide range of domestic articles—particularly forks and spoons. Can be pressed, drawn and spun.
Stainless iron	410S21	"Silver Fox 410" Firth Vickers' F1 Carr's P1000	0.10	0.50	13.0	Oil-quenched from 1000°C, and tempered at 750°C	371	24	571	37	33	170	Oil-quench, water-quench or air-cool from 950-1000°C; temper at 650-750°C.	Turbine-blade shrouding rivets, split pins, golf-club heads, solid-drawn tubes; structural and ornamental work.
Stainless steel	420S37	Firth Vickers' FG Carr's P1002	0.22	0.50	13.0	Oil-quenched from 960°C and tempered at 700°C	633	41	757	49	26	220	Oil-quench, water-quench or air-cool from 950-1000°C; temper at 500-750°C.	A general-purpose alloy—not in contact with non-ferrous metal parts or graphite packing. **Valve and pump parts.**
Stainless steel	420S45	"Silver Fox 420" Firth Vickers' FH Carr's P1003	0.3	0.50	13.0	Cutlery temper Spring temper	— —	— —	1 670 1 470	108 95	— —	534 450	Oil- or water-quench or air-cool from 950-1000°C; temper (for cutlery) at 150-180°C; temper (for springs) at 400-450°C.	Specially for cutlery and sharp-edged tools. Approximately pearlitic in structure when normalised.
Stainless iron	430S15	"Silver Fox 430" Firth Vickers' FI 17	0.05	0.50	17.0	Soft	370	24	525	34	27	—	Non-hardening—except by cold-work.	Can be deep-drawn or spun.
High-carbon chromium tool steel	BS 4659: BD3	Br. Steel Corp. CD Carr's 23S Jessop-Saville alloy C/WPS	2.10	0.30	12.5	Oil-quenched and tempered at 200°C	—	—	—	—	—	850	Heat slowly to 750-800°C, and then raise to 960-990°C, and oil-quench (small sections can be air-cooled); temper at between 150 and 400°C for 30-60 minutes.	Blanking punches, dies and shear-blades for hard, thin materials. Dies for moulding abrasive powders (ceramics). Master gauges, hobbing dies, thread-rolling dies.

small amounts of molybdenum (13.50), thus producing the range of well-known "nickel–chrome–molybdenum" steels.

13.43. The nickel–chromium steels are very adaptable and useful alloys. They forge well and also machine well in the softened condition, whilst their mechanical properties can be varied considerably by the treatment given. When the nickel content is increased to about 4·0% and chromium to about 1·5% an air-hardening steel is obtained. Such a steel is very useful for the manufacture of complex shapes which have to be hardened, and which would be likely to distort if water- or oil-quenching were attempted.

13.44. The high-nickel, high-chromium steels are all stainless alloys containing less than 0·1% carbon. The most popular is the "18–8" stainless steel containing 18% chromium and 8% nickel. It is an austenitic alloy in spite of the high proportion of chromium relative to that of nickel. Thus, the tendency of nickel to stabilise austenite is much stronger than the tendency of chromium to stabilise ferrite. 18–8 stainless steel takes a good polish and resists corrosion by many relatively corrosive organic and inorganic reagents.

The carbon content is kept low, since the presence of precipitated carbides in the microstructure reduces corrosion-resistance. Even with a carbon content below 0·1%, slow cooling of the steel to room temperature will cause some carbide precipitation to take place, and this considerably reduces corrosion-resistance, because the precipitation of chromium carbide particles leads to impoverishment of the surrounding austenite in chromium, so that electrolytic action (22.21) is induced between the chromium-depleted region and the adjoining chromium-rich regions. It is therefore necessary to heat the alloy to 1 050° C and then quench it, in order to retain the carbon in solid solution. Carbide precipitation leads to a defect usually known as "weld-decay" in fabricated articles. During welding, some part of the metal near to the weld will be maintained between 650 and 800° C long enough for carbides to begin to deposit there (Fig. 13.9). Subsequently corrosion will occur in this area near to the weld. The defect may be largely overcome by adding small amounts of titanium, molybdenum or niobium; elements which have a very strong carbide-forming tendency and thus "tie up" most of the carbon as carbides at a high temperature, so that this carbon is no longer available to form chromium carbide during slow-cooling following a welding process. In this way any variation in chromium content across the microstructure is avoided and post-welding treatment is unnecessary.

The compositions, properties and uses of some of the more important nickel–chromium steels are given in Tables 13.3a and b.

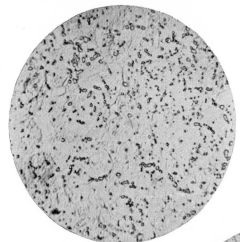

13.1A.—Stainless Cutlery Steel (13% Chromium; 0.3% Carbon) Oil-hardened from 950° C.

Particles of carbide in a martensite matrix. ×1500.

Courtesy of Messrs. Edgar Allen & Co. Ltd., Sheffield.

13.1B.—Stainless Steel (18% Chromium; 8% Nickel; 0.1% Carbon) Oil-quenched from 1100° C.

Twinned crystals of austenite. ×250. Etched in acidified ferric chloride.

Courtesy of Messrs. Hadfields Ltd., Sheffield.

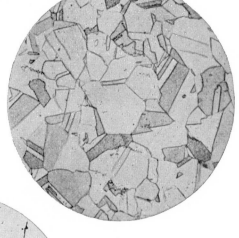

13.1C.—18/8 Stainless Steel Oil-quenched from 1100° C and then Reheated at 650° C for 1 Hour.

Carbides precipitated at the grain boundaries of the austenite. ×250. Etched in acidified ferric chloride.

Courtesy of Messrs. Hadfields Ltd., Sheffield.

PLATE 13.1

TABLE 13.3a.—Low-nickel, Low-chromium Steels

(See also low-nickel, low-chromium case-hardening steels (Table 19.1).)

B.S. 970 desig-nation	Proprietary trade names and codes	Composition (%)					Condition	Yield point		Tensile strength		Elonga-tion (%)	Reduc-tion area (%)	Izod (J)	Heat-treatment	Uses
		C	Mn	Ni	Cr	Other elements		N/mm²	tonf/in²	N/mm²	tonf/in²					
655M13	Firth Brown N3CCH Carr's P155	0·12	0·45	3·3	1·0	—	Carburised and double-quenched	850	55	925	60	13	40	39	After carburising, refine by heating to 850°C and then cool in oil or air. Harden by oil-quench from 770°C. Temper at 150°C.	A case-hardening steel of high core strength and hard-wearing surface.
653M31	Firth Brown NCR2 Dunford Hadfield's "Hecla 116" Carr's P564	0·3	0·6	3·0	0·8	Mo 0·65 (optional)	28·5 mm dia. bar, oil-quenched and tempered at 600°C	819	53	927	60	23	57	104	Oil-quench from 820–840°C; temper between 550 and 650°C. Cool in oil to avoid temper-brittleness if molybdenum is absent.	Highly-stressed parts in aero-, auto- and general engineering, e.g. differential shafts, stub axles, connecting-rods, high-tensile studs, pinion shafts. (For heavy sections the addition of molybdenum is advisable.)
640M40	"Hecla 98"	0·35	0·8	1·25	0·6	—	28·5 mm dia. bar, oil-quenched and tempered at 600°C	834	54	927	60	22	55	85	Oil-quench from 830–850°C; temper between 180 and 220°C, or between 550 and 650°C. Cool in oil. Suffers from temper-brittleness if tempered between 250 and 400°C.	Small and medium-sized sections for gears, differential pinions, etc. Automobile connecting rods, crankshafts, axles, bolts and screwed parts.
—	(largely obsolete)	0·3	0·45	4·25	1·25	—	Air-hardened and— *(a)* Tempered at 200°C *(b)* Tempered at 500°C (see H.T.) *(c)* Tempered at 630°C	1 440 / 1 036 / 819	93 / 67 / 53	1 810 / 1 170 / 973	111 / 76 / 63	12·5 / 17·5 / 21·5	34 / 47 / 57	17 / 13 / 61	Air-harden from 820–830°C; temper at 180–200°C for maximum hardness; temper at 600–650°C for maximum ductility. Cool in oil. Do not temper between 250 and 580°C, as this	Parts of complex shape where oil-quenching might cause distortion. However, because of the danger of temper-brittleness arising during heat-treatment, this steel has been superseded by the molybdenum-bearing type—835M30

Typical mechanical properties

B.S. 970 (Pt. 4) designation	Proprietary trade names and codes	C	Mn	Ni	Cr	Other elements	Condition	Yield point N/mm²	Yield point tonf/in²	Tensile strength N/mm²	Tensile strength tonf/in²	Elongation (%)	Reduction area (%)	Izod (J)	Brinell	Heat-treatment	Uses
302S25	Firth Vickers "Staybrite FST", Br. Steel Corp. "Silver Fox 302" Carr's P1010	0·05	0·8	8·5	18·0	—	Softened / Cold-rolled sheet	278 / 803	18 / 52	618 / 896	40 / 58	50 / 30	50 / —	143 / —	170 / —	Non-hardening except by cold-work. (Cool quickly from 1050° C to keep carbides in solution.)	Particularly suitable for domestic and decorative purposes.
321S20	"Staybrite FDP" "Silver Fox 321" Carr's P1011	0·05	0·8	8·5	18·0	Ti 0·6	Softened / Cold-rolled	278 / 402	18 / 26	649 / 803	42 / 52	45 / 30	50 / 42	104 / —	180 / 225		A "weld-decay" proof steel (fabrications may be used in the as-welded condition). Used extensively in plant for the manufacture of nitric acid. An austenitic steel.
—	"Staybrite DDQ" Carr's P1013	0·05	0·8	12·5	12·5	—	Softened	232	15	579	37·5	50	50	—	160		A deep-drawing austenitic alloy—suitable for table-ware and kitchen equipment.
304S15	"Staybrite FST(L)" "Silver Fox 304" Carr's P1014	0·03	0·8	10·0	18·5	—	Softened	232	15	618	40	60	65	104	160		A low-carbon alloy suitable for deep-drawing. Used mainly for domestic purposes, etc.
347S17	"Staybrite FCB" "Silver Fox 347" Carr's P1016	0·04	0·8	10·0	18·0	Nb 1·0	Softened	263	17	618	40	58	65	104	175		Immune from weld-decay, due to the stabilising effect of niobium. Used in welded plant where corrosive conditions are very severe, e.g, in nitric acid plant.
320S17	"Staybrite FMB(Ti)" "Silver Fox 320"	0·04	0·8	13·0	17·5	Mo 2·75 Ti 0·6	Softened	278	18	618	40	50	50	104	180		Molybdenum improves the corrosion-resistance to such reagents as sulphuric and phosphoric acids, chloride solutions and organic acids, such as acetic acid.
325S21	"Staybrite EMS"	0·05	1·5	8·5	18·0	Mo 0·3 Ti 0·6 S 0·25	Softened	232	15	587	38	53	54	65	175		A free-cutting stainless steel in which the machinability is increased by the presence of manganese and sulphur.

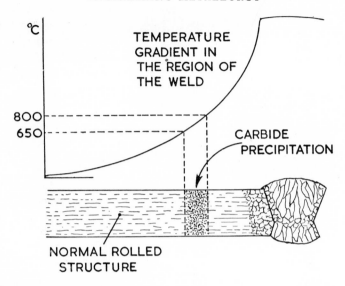

FIG. 13.9.—Structural Changes During Welding Which Lead to "Weld-decay" in Stainless Steels.

Steels Containing Molybdenum

13.50. The United States is the dominant producer of molybdenum, 35% of her output being a by-product of several copper-mines; Chile, Japan, Canada and Norway are other producers. At the end of last century practically the whole of the world production of molybdenum was used in making the chemical reagent ammonium molybdate for the analytical determination of phosphorus in iron, steel, and fertilisers. To-day the principal use of molybdenum is in the manufacture of alloy steels.

As mentioned earlier in this chapter (13.42), one of the main uses of molybdenum is to reduce the tendency to "temper-brittleness" in low-nickel, low-chromium steels. Additions of about 0·3% molybdenum are usually sufficient in this respect, and the resultant steels retain a high impact value, irrespective of the rate of cooling after tempering.

Among alloy steels the nickel–chrome–molybdenum steels possess the best all-round combination of properties, particularly where high tensile strength combined with good ductility is required in large components. These steels are relatively free from the mass effects of heat-treatment, the transformation rates of the nickel–chromium steels being still further reduced by the presence of molybdenum, which therefore contributes considerably to depth of hardening.

13.51. Molybdenum is also added to chromium steels to produce a general improvement in machinability and mechanical properties, whilst nickel–molybdenum steels are very suitable for case-hardening. Molybdenum has a very strong carbide-stabilising influence and raises the high-temperature strength and creep-resistance of high-temperature alloys. It enhances the corrosion resistance of stainless steels, particularly to chloride solutions. Molybdenum-bearing steels are listed in Table 13.4.

Steels Containing Vanadium

13.60. In 1801 Del Rio expressed the opinion that a Mexican ore which he analysed contained a new metal which he called "Erythronium"—from the Greek *erythros*, meaning "red"—because it produced red compounds. Later, N. G. Sefström discovered what he thought was a new mineral in some Swedish iron ore and named it "vanadium" after the Scandinavian goddess Vanadis. Almost immediately, in 1831, the famous chemist Wöhler identified Sefström's "vanadium" as being Del Rio's "erythronium", but it was not until 1867 that the pure metal was produced. To-day the United States is the leading producer of the metal, followed by Peru, South-West Africa and Zambia.

13.61. Plain vanadium steels are used to a very limited extent, but chromium–vanadium steels containing up to 0·2% vanadium are widely used for small and medium sections. The mechanical properties resemble those of nickel–chromium steels, but usually show an advantage in respect of the limit of proportionality and percentage reduction in area. Chromium–vanadium steels are also easier to forge, stamp and machine, but are more susceptible to mass effects of heat-treatment than the corresponding nickel–chromium steels.

13.62. Vanadium has a strong carbide-forming tendency, forming V_4C_3. It also stabilises martensite and troostite on heat-treatment and increases hardenability. Like nickel, it restrains grain growth of the austenite and, since it combines readily with oxygen and nitrogen, it is often used as a "cleanser" during deoxidation, to produce a gas-free ingot. One of the most important effects of vanadium is that it induces resistance to softening at high temperatures provided that the steel is first heat-treated to absorb some of the vanadium carbide into solid solution. Consequently vanadium steels are used for hot-forging dies, extrusion dies, die-casting dies and other tools operating at elevated temperatures. The vanadium content of some high-speed steels has been increased of recent years.

Some steels containing vanadium are shown in Table 13.5.

TABLE 13.4.—*Steels Containing Molybdenum*

Type of Steel	B.S. 970 designation	Proprietary trade names and codes	Composition (%)					Condition	Typical mechanical properties								Heat-treatment	Uses
			C	Mn	Ni	Cr	Mo		Yield point		Tensile strength		Elongation (%)	Reduction area (%)	Izod (J)	Brinell		
									N/mm²	tonf/in²	N/mm²	tonf/in²						
Mn–Mo steel	605M36	Firth Brown MMO1	0·35	1·6	—	—	0·27	28·5 mm diameter bar, oil-quenched and tempered at 600°C	1 000	65	1 130	73	18	57	69	—	Oil-quench from 830–850° C; temper at 550–650° C and cool in oil or air.	A substitute for more highly alloyed nickel-chromium steels.
Ni–Mn–Mo steel	795M19	Firth Brown NMM1	0·2	1·6	0·55	—	0·25	Oil-quenched and tempered	525	34	649	42	25	25	124	—	Oil-quench from 860–900° C; temper at 600–650° C; cool in air.	*Railway engineering:* connecting-rods, spring hanger and draw gears, slide-bar bolts, etc. *General engineering:* parts of harvesters, dredging machines, crushers, rods for well-boring. Has very good fatigue- and shock-resisting properties.
1% Cr–Mo steel	709M40	Firth Brown CRM1 Dunford Hadfield's "Hecla 150 and 157" Carr's P602	0·4	0·65	—	1·1	0·3	28·5 mm diameter bar, oil-quenched and tempered at 600°C	927	60	1 000	65	18	57	65	—	Oil-quench from 840–860° C; temper at 550–700° C, and cool in oil or air.	Suitable for crankshafts and connecting-rods, stub axles, etc. As a substitute for 3% nickel steel (it machines more readily and is free from temper brittleness). Zinc die-casting dies.
1¼% Mn–Ni–Cr–Mo steel	945M38	Firth Brown NMCM "Hecla 100"	0·4	1·35	0·75	0·45	0·2	28·5 mm diameter bar, oil-quenched from 850° C and tempered at 600° C	958	62	1 040	67	21	—	85	320	Oil-quench from 830–850° C; temper at 550–660° C; cool in air.	Automobile and general-engineering components requiring a tensile strength of 700–1 000 N/mm²

Type	Designation	Trade names						Condition									Heat treatment	Applications
1¾% Ni–Cr–Mo steel	817M40	Firth Brown NCM1 "Hecla 152" Carr's P553	0·4	0·55	1·5	1·1	0·3	Oil-quenched and tempered at 200°C	—	—	2 010	130	14	40	27	555	Oil-quench from 830–850°C; "light" temper 180–200°C; "full" temper 550–650°C; cool in oil or air.	Differential shafts, crankshafts and other highly-stressed parts where fatigue and shock-resistance are important. In the lightly tempered condition it is suitable for automobile and machine-tool gears. (It can be surface-hardened effectively in cyanide bath.)
								Oil-quenched and tempered at 600°C	988	64	1 080	70	22	61	69	—		
3% Ni–Cr–Mo steel	830M31	Firth Brown NCM6 "Hecla 116"	0·3	0·6	3·25	0·8	0·3	28·5 mm diameter bar, oil-quenched and tempered at 600°C	880	57	973	63	22	66	103	—	Oil-quench from 820–840°C; temper at 550–650°C; cool in oil or air.	Parts of thin section where maximum ductility and shock-resistance are required, e.g. connecting-rods, inlet valves, cylinder studs, valverockers.
4¼% Ni–Cr–Mo steel	835M30	Firth Brown ANCM "Hecla 67B" Carr's P552 Jessop-Saville G1 Special (SVL) T. Andrews' "Monarch NCG"	0·3	0·45	4·25	1·25	0·25	Air-hardened and tempered at 200°C	1 470	95	1 700	110	14	45	35	—	Air-harden from 820–840°C; temper at 150–200°C; and cool in air.	An air-hardening steel for aero-engine connecting-rods, valve mechanisms, gears, differential shafts and other highly-stressed parts. Can be further surface-hardened by cyanide or carburising.
½% Mo–boron steel	—	Br. Steel Corp. "Fortiweld"	0·10	0·6	—	B / 0·003	0·5	Normalised at 950°C, then stress-relieved for 3 hours at 600°C	432	28	571	37	16	—	—	—	Normalise at 930–980°C; stress-relieve at 590–610°C for 3 hours.	High creep stress up to 400°C; pressure vessels; heat exchangers; reactor vessels; gas-turbine and power-plant components.

TABLE 13.5.—*Steels Containing Vanadium*

Type of steel	Proprietary trade names and codes	Composition (%)						Condition	Hardness (V.P.N.)	Heat-treatment	Uses
		C	Si	Mn	Cr	Mo	V				
¼% V steel	Dunford Hadfield's "Hecla 28"	0·7	—	—	—	—	—	—	—	Water-quenched from 850° C. Temper as required.	Cold-drawing dies, etc.
	Balfour Darwin's DEMDS Carr's O6S	1·0	0·25	0·25	—	—	0·2	—	—		Coining and embossing dies (12.70-Part 2)
¾% Cr-V steel	Jessop-Saville F2	0·57	0·25	0·8	0·75	—	0·2	Hardened and tempered	575	Pre-heat to 700° C and then oil-quench from 850° C. Temper as required.	Die-casting dies for zinc-base alloys.
5% Cr-Mo-V steel	Edgar Allen AM3 Balfour Darwin's NAT	0·4	1·0	0·3	5·0	1·35	0·45	Hardened and— (a) Tempered at 550°C	600	Pre-heat to 850° C and then to 1 000° C; soak for 10–30 minutes and air-cool; temper at 550–650° C for 2 hours.	Hot-forging dies for steel and copper alloys where excessive temperatures are not reached. Extrusion dies and mandrels for aluminium alloys. Pressure and gravity dies for aluminium die-casting.
	"Hecla 174" Carr's 53S Balfour Dawrin's ADIC Thos. Andrews' "Monarch BLA" Jessop-Saville H50 Br. Steel Corp. CVM3	0·35	1·0	0·3	5·0	1·5	1·0	(b) Tempered at 650°C	375		
High C–high Cr steel	"Hecla 159" Jessop-Saville H42	1·6	0·2	0·3	13·0	1·0	0·5	Hardened and— (a) Tempered at 200 °C	800	Pre-heat to 850° C and then to 1 000° C; soak for 14–45 minutes and quench in oil or air; temper at 200–400 °C for 30–60 minutes.	Fine press tools. Deep-drawing dies and forming dies for sheet metal. Hobbing dies. Thread-rolling dies. Wire-drawing dies and wortle plates. Blanking dies, punches and shear blades for hard metals.
	Edgar Allen "Double Six" Carr's 14S Balfour Darwin's SC25	1·9	0·3	0·2	12·5	0·8	0·25	(b) Tempered at 400 °C	700	Pre-heat to 800° C and then to 1 030° C; air-cool; temper at 180–450° C for 2–3 hours.	
Vanadium tool steel	Thos. Andrews' "No. 4 Hardenite Van"	1·4	—	0·4	0·4	0·4	3·6	—	—	Water-quench from 770° C. Temper at 150–300° C.	Cold-heading dies.
5% Cr-5% V steel	Jessop-Saville H7	2·3	—	—	5·25	1·1	4·5	Hardened and tempered at 150° C	825	Oil- or air-quench from 940° C. Temper at 150–220° C.	Refractory moulds and other uses where resistance to abrasion is necessary.

Heat-resisting Steels

13.70. Steels required for service at elevated temperatures are used for the manufacture of such components as exhaust valves for aero-engines; racks for enamelling stoves; conveyor chains; furnace arch and floor plates; rotors of gas turbines; Siemens open-hearth furnace valves; annealing boxes; pyrometer sheaths, retorts, superheater-element supports, locomotive fire-doors, etc. The main requirements of such steels are:

(*a*) resistance to oxidation, and attack by vapours and gases existing in the working atmosphere;

(*b*) a sufficiently high strength at the working temperature.

13.71. In general, the resistance to oxidation is effected by the addition of chromium (and sometimes silicon), whilst the inclusion of nickel will toughen the alloy by restricting grain growth at high temperatures. The large amount of chromium present contributes to some extent to the necessary high-temperature strength, but this is developed still further by "stiffening" the alloy with additions of carbon, tungsten, titanium, molybdenum or aluminium (the latter forms an intermetallic compound, $NiAl_3$, with some of the nickel). In this way the limiting creep stress is raised.

13.74. Short-time creep tests are generally employed in this connection. The "time-yield" value is approximately equivalent to a rate of creep of one part in a million per hour. When subjected to the time-yield stress the specimen must not show an extension greater than 0·5% of the gauge length in the first twenty-four hours, and during the next forty-eight hours must not show any further extension, within a sensitivity of measurement of 0·002 5 mm on a 50 mm gauge length. For many purposes it is safe to work at half of the "time-yield" value as the safe working stress, though for other classes of work longer-time creep tests are necessary to determine the stress which will produce a given amount of strain, in a given time, at a particular temperature.

The compositions and properties of some typical heat-resisting steels are given in Table 13.6.

Manganese Steels

13.80. Although the U.S.S.R. is by far the largest producer of manganese, India is the world's leading exporter of the metal. South Africa and Ghana are also large producers, whilst some mining is carried out in North Africa, Latin America and Japan. The bulk of Britain's imports of manganese come from India, Ghana and South Africa.

TABLE 13.6.—Heat-resisting Steels

Relevant specifications	Proprietary trade names and codes	Composition (%)						Condition	Typical mechanical properties at various temperatures						Maximum working temperature (°C)	Typical uses
		C	Si	Mn	Cr	Ni	Other elements		Testing temperature (°C)	Yield stress N/mm²	Yield stress tonf/in²	Tensile strength N/mm²	Tensile strength tonf/in²	Elongation (%)		
B.S. 3100: 1398 Grade A	F. H. Lloyd's A14	0·2	0·3	0·8	—	—	Mo 0·5	As cast	20	278	18	463	30	18	450	Steam-turbine and other castings.
B.S. 3100: 1398 Grade D	F. H. Lloyd's A26	0·1	0·2	0·6	0·4	—	Mo 0·5 V 0·25	As cast	20	309	20	510	33	17	510	A creep-resisting steel for turbine castings working at high pressures.
AISI Series 311	Balfour Darwin's "Pireks 26/19"	0·15	—	—	20·0	25·0	—	Forged or rolled	—	—	—	—	—	—	820	Conveyor chairs and skids, heat-treatment boxes, recuperator tubes, exhaust valves and other furnace parts.
B.S. 970 (Pt. 4): 401S45	Firth Vickers' SLV	0·45	3·5	0·5	8·0	—	—	Oil-quenched 1 050° C; tempered 800° C.	20 / 200 / 400 / 600	*0·2% Proof stress* 801 / 708 / 585 / 185	52 / 46 / 38 / 12	1 001 / 909 / 816 / 339	65 / 59 / 53 / 22	23 / 20 / 20 / 45	850	A low-cost, heat-resisting steel used mainly for valves.
B.S. 970 (Pt. 4): 352S52	Firth Vickers' 21/42	0·5	0·2	9·0	21·0	4·0	Nb 2·5 N 0·45	Water-quenched 1 200° C, then air-cooled from 800° C. (Precipitation treatment)	20 / 200 / 400 / 600 / 800	585 / 447 / 385 / 323 / 216	38 / 29 / 25 / 21 / 14	1 001 / 847 / 755 / 616 / 385	65 / 55 / 49 / 40 / 25	16 / 20 / 23 / 24 / 28	850	A valve steel which can be precipitation-hardened.
AISI Series 302B	Br. Steel Corp., "Red Fox 34"	0·1	1·5	1·0	19·0	11·0	—	Air-cooled from 1 100° C.	20 / 200 / 400 / 600 / 800 / 900	*0·1% Proof stress* 347 / 262 / 248 / 217 / 139 / 85	22·5 / 17 / 16 / 14 / 9 / 5·5	698 / 543 / 527 / 465 / 248 / 155	45 / 35 / 34 / 30 / 16 / 10	55 / 43 / 40 / 37 / 45 / 50	1 000 (air) 950 (flue gases)	A fairly cheap grade of heat-resisting steel with a good combination of properties.

Specification	Trade name	C	Si	Mn	Cr	Ni	Co	Condition	Test temp (°C)	1·0% Proof stress		Tensile strength		Elongation %	Max. temp (°C)	Application
AISI Series 309	"Pireks 12/25"	0·3	1·2	0·8	23·0	12·5	—	As cast	800 / 1 000 / 1 050	— / — / —	— / — / —	48·3 / 15·4 / 11·0	3·1 / 1·0 / 0·7	— / — / —	1 050	Oil- and gas-fired installations—heat-treatment furnaces, nozzle burners, salt pots, super-heater supports. Used in the cast, forged or rolled form.
—	"Pireks 35/15"	0·45	1·2	0·8	15·0	35·0	—	As cast	800 / 1 000 / 1 050	— / — / —	— / — / —	69·0 / 20·7 / 15·9	4·5 / 1·3 / 1·0	— / — / —	1 050	A casting alloy. Not suitable for use when highly sulphurous gases are present. Useful for conditions involving cyclic thermal shock.
—	"Pireks 60/13"	0·35	2·0	0·8	13·0	60·0	—	As cast	800 / 1 000 / 1 050	— / — / —	— / — / —	45·5 / 15·4 / 11·7	2·9 / 1·0 / 0·7	— / — / —	1 050 / 1 100 (air)	A casting alloy suitable for carburising and quenching equipment.
B.S. 1501 (Pt. 3): 310S24 AISI Series 310	Firth Vickers' "Immaculate 5" "Pireks 20/25"	0·07	0·8	1·0	25·0	20·0	—	Quenched from 1 100° C to retain carbides in solution	20	270	17·5	543	35	40	1 100	A wrought steel. Hearth plates, trays and muffles.
—	"Pireks UMCo 50"	0·05	0·8	0·8	28·0	—	Co 50·0	As cast	—	—	—	—	—	—	1 200	Quenching baskets, etc. Good wear-resistance at high temperatures, hence used in cement-production equipment.

13.81. Nearly all steels contain some manganese residual from the deoxidation and desulphurisation processes. Such amounts are usually less than 1·0%, and it is only when the manganese content exceeds this amount that it is regarded as a deliberate alloying element. Of recent years, however, there has been a tendency to increase the manganese content of ordinary carbon steels at the expense of a little carbon, this

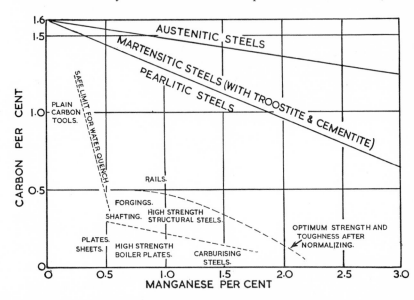

Fig. 13.10.—Diagram Illustrating the Effects of Manganese on the Structure, Properties and Uses of Low-manganese Steels.

giving improvements in respect of ductility and Izod values, in both the normalised and heat-treated conditions. The susceptibility of low-carbon steels to brittle fracture is reduced by raising the manganese/carbon ratio, using up to 1·3% manganese (7.63).

13.82. Like nickel, manganese stabilises austenite by raising the A_4 point and lowering the A_3 point; but, unlike nickel, manganese has a stabilising effect on the carbides, itself forming the carbide Mn_3C. Manganese also has a considerable strengthening effect on the ferrite, and also increases the depth of hardening to a useful degree. Despite these advantageous properties, the low-carbon, low-manganese steels are not widely used, though manganese has been employed to some extent of recent years to replace nickel in the low-alloy steels.

13.83. As recorded at the beginning of this chapter, the development of high-manganese steel in 1882 by Sir Robert Hadfield was one of the landmarks in the history of alloy steels. Hadfield's steel contained

13.2A.—High-manganese Steel (12·5% Manganese; 1·2% Carbon).

Water quenched from 1050° C. Austenite. ×100. Etched in 2% Nital.

Courtesy of Messrs. Hadfields Ltd., Sheffield.

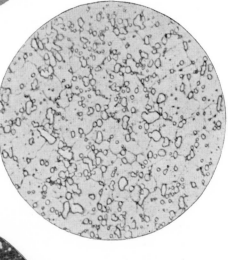

3.2B.—High-speed Steel (18% Tungsten; 4% Chromium; 1% Vanadium).

Air quenched from 1250° C. Carbide globules in an austenite/martensite matrix. ×500.

Courtesy of Messrs. Edgar Allen & Co. Ltd., Sheffield.

13.2C.—High-speed Steel (18% Tungsten; 4% Chromium; 1% Vanadium).

Air quenched from 1250° C and then "secondary hardened" at about 600° C. Austenite is transformed and the matrix etches darker. ×500.

Courtesy of Messrs. Edgar Allen & Co. Ltd., Sheffield.

PLATE 13.2

approximately 12·5% manganese and 1·2% carbon, and, both as far as composition and heat-treatment are concerned, this is basically the same steel as that used to-day. It is austenitic in type, though if the steel is allowed to cool slowly to room temperature some precipitation of carbides will take place. The steel is therefore water-quenched from 1 050° C in order to keep the carbon in solution.

Being austenitic, high-manganese steel is extremely tough and shock-resistant, and although relatively soft, it nevertheless wears extremely well. The reason for this apparently paradoxical situation is that mechanical disturbance causes the surface layers to become extremely hard so that the resultant component has a soft, though shock-resistant, core and a very hard, wear-resistant case. Abrasion of the surface will cause the Brinell figure to rise from 200 to 550. The reason for this phenomenon is still uncertain. Some claim that mechanical disturbance of the austenite causes martensite to form at the points of high stress concentration, whilst others have suggested that the cause is simply work-hardening of the austenite. There is little positive evidence of martensite formation and it seems most probable that strain-hardening occurs as a result of mechanical disturbance.

High-manganese steel is available as castings, forgings or hot-rolled sections, but is very difficult to machine because of its tendency to harden as soon as this is attempted. Despite this drawback, no really suitable substitute has been found for Hadfield steel, for applications where toughness combined with resistance to conditions of extreme abrasion are required, e.g. rock-crushing machinery, dredging equipment and track-work points and crossings. The austenitic core can be further toughened by the addition of carbide-forming elements, such as chromium and vanadium.

Some manganese steels are shown in Table 13.7.

Steels Containing Tungsten

13.90. Up to the middle of the eighteenth century the name "tungsten" (heavy stone) was allocated to the mineral scheelite, but in 1781 the renowned chemist K. W. Scheele demonstrated that scheelite contained a peculiar acid which he called tungstic acid. Metallic tungsten was isolated two years later by J. J. y Don Fausto d'Elhuyar. To-day tungsten is mined in some thirty countries, the leading producers being the United States, Bolivia, Brazil, Portugal, Spain, China, Korea, Burma, Thailand, Canada and Australia. The United Kingdom imports the bulk of her supplies from her ancient ally Portugal, from Australia and from Burma.

13.91. Tungsten raises the A_3 temperature and, like chromium,

TABLE 13.7.—*Manganese Steels*

B.S. 970 designation	Proprietary trade names and codes	Composition (%) C	Composition (%) Mn	Condition	Yield point N/mm²	Yield point tonf/in²	Tensile strength N/mm²	Tensile strength tonf/in²	Elongation (%)	Reduction area (%)	Izod (J)	Brinell	Heat-treatment	Uses
150M28	Firth Brown CMN1	0·25	1·5	Normalised	355	23	587	38	20	—	—	—	Oil-quench from 860° C (water-quench for sections over 38 mm dia.). Temper as required.	Automobile axles, crankshafts, connecting rods, etc., where a cheaper steel is required.
150M36	Firth Brown CMN2 Carr's "O"	0·35	1·5	28·5 mm bar, oil-quenched and tempered at 600° C	510	33	711	46	27	64	68	—	Oil-quench from 840–860° C, temper at 550–650° C and cool in oil or air.	Automobile and general engineering where a carbon steel is not completely satisfactory, but where the expense of a nickel-chromium steel is not justified.
—	Balfour Darwin's NSS3	1·0	2·0	Hardened and tempered	—	—	—	—	—	—	—	620	Oil-quench from 770° C; temper at 200–300° C).	Plug gauges; thread and precision gauges; blanking dies; press tools; reamers; taps; plastics moulds; cams; general purpose low-cost, oil-hardening steel.
—	"Hadfield" Austenitic steel F. H. Lloyds AMn (Castings only)	1·2	12·5	Water-quenched from 1050° C	—	—	849	55	40	—	—	Case-550 Core-200	Finish by quenching from 1050° C to keep the carbides in solution.	*Jaw-crushing machinery*; liner plates for ball- and rod-mills, parts for gyratory crushers. *Dredging equipment*; tumblers, buckets, bucket lips, heel plates. *Track work*; points and crossings. Digger teeth, crane track wheels, elevator chains, axle-bar liners, bottom plates for magnets and where non-magnetic properties are required.

Typical mechanical properties

TABLE 13.8.—Steels Containing Tungsten (Other than High-speed Steels)

Type of steel	B.S. 4659 designation	Proprietary trade names and codes	Composition (%)							Heat-treatment	Uses
			C	Si	Cr	Mo	V	W	Other elements		
Tool steel	—	Edgar Allen K9	1·0	—	0·75	—	—	0·4	Mn 0·85	Oil-quench from 790° C and temper 200–250° C	Dies; stay-taps; delicate broaches; milling cutters; plugs; gauges; circular cutters; fine press tools and master tools
	BO1	Balfour Darwin's TOH, Thos. Andrews' "Hardenite 7OH"	0·9	0·2	0·5	—	—	0·5	Mn 1·20		
Hot-working die steel	BH12	Edgar Allen AM1, Br. Steel Corp. CVM2, Dunford Hadfield's "Hecla 177", Carr's 58S, Balfour Darwin's CTU	0·35	1·0	5·0	1·5	0·4	1·35	—	Pre-heat to 800° C; then heat quickly to 1 020° C; soak and air-cool. Temper for 1½ hours at 540–620° C.	Extrusion dies, mandrels and noses for aluminium and copper alloys. Hot-forming, piercing, gripping and heading tools. Brass-forging and hot-pressing dies. Other hot-working operations.
Shock-resisting hot- and cold-working tool steel	—	Br. Steel Corp. DUX 4, Jessop-Saville "OK Crown"	0·5	0·6	1·1	—	—	2·25	—	Oil-quench from 880–920° C; temper at 200–300° C for cold-working tools; temper at 400–600° C for hot-working tools.	Hand and pneumatic chisels, caulking tools and general boilermaker's tools; punches; dies; blanking tools and shear blades. Coal-cutter picks. Cold-heading and nail-making dies. Engraving punches; hot-blanking tools. Hot-forging dies.
		Edgar Allen "Minerva", Balfour Darwin's "OO"	0·53	—	1·8	—	0·2	1·9	—		
Hot-working tool steel	—	Balfour Darwin's PDX	0·28	0·3	0·85	0·5	0·3	2·25	Ni 2·25	Pre-heat to 750° C and then to 1 000° C. Quench in oil. Temper at 650° C for 30 minutes.	Hot-extrusion mandrels for steels and non-ferrous metals. Hot-forging tools. Die holders. Zinc die-casting dies.
Hot-working die steel	BH19	Jessop-Saville J28, Balfour Darwin's PAC	0·4	—	4·25	0·4	2·25	4·25	Co 4·25	Pre-heat to 850° C and then to 1 150° C. Quench in oil. Temper at 550–680° C.	Extrusion dies. Hot-piercing punches. Brass-stamping dies. Die-casting dies.
Hot-working die steel	BH21	Br. Steel Corp. HSM/W9A, Edgar Allen "No. 5", Carr's 12S, Jessop-Saville "Special BB", Balfour Darwin's 227	0·3	0·2	2·85	—	0·35	10·0	—	Pre-heat to 850° C and then heat rapidly to 1 150–1 200° C; oil-quench (or air-cool small sections); temper for 2–3 hours at 600–700° C	Hot-forging dies and punches for making bolts, nuts, rivets and similar small components where tools reach high temperatures. Hot-forging dies for copper alloys. Extrusion dies, mandrels, pads and liners for the extrusion of copper alloys. Die-casting dies for copper alloys and for pressure die-casting of aluminium alloys.

forms a closed γ-loop. Unlike chromium, however, it inhibits grain growth. It combines readily with carbon, forming the extremely hard and very stable carbides W_2C and WC and, in steel, a double carbide, Fe_4W_2C. These carbides dissolve very slowly in steel when the latter is heated, and then only at elevated temperatures. Once in solution, the carbides can be made to precipitate only by cooling the steel very slowly from a high temperature. At temperatures up to 700° C precipitation is negligible. At the same time tungsten renders transformation very sluggish. For all of these reasons tungsten is an essential constitutent of high-speed steel (14.10), in which it develops high-temperature ("red") hardness following suitable heat-treatment.

13.92. Since tungsten carbide is so hard, it is commonly used in other tool steels and die steels, in amounts varying between 0·5 and 10·0% (Table 13.8). It is also used in heat-resisting steels, in which it raises the limiting creep stress at high temperatures.

Steels Containing Cobalt

13.100. Cobalt is chemically very similar to nickel and the ores of these metals are generally associated with each other. Until recently cobalt was used principally in permanent-magnet alloys (14.35) and in "super" high-speed steels (14.11). In the latter it gives a useful increase in red-hardness by promoting extreme sluggishness in transformation.

Currently, cobalt is a constituent of some "maraging" steels. A typical steel in this group contains 18% nickel, 8% cobalt, 4% molybdenum, up to 0·8% titanium and—because of the graphitising effect of the high nickel content—less than 0·03% carbon. If such a steel is solution-treated (17.55) at 870° C to absorb precipitated compounds and so give uniform austenite, and then air-cooled, a martensitic structure is produced. Unlike ordinary tetragonal martensite (12.22), this is of body-centred cubic form and consequently softer but tougher. However, if the steel is now precipitation-hardened (9.82) at 500° C for two hours, *coherent* precipitates of intermetallic compounds such as $TiNi_3$ are formed and tensile strengths of up to 2 100 N/mm² develop. The term "maraging" is of course derived from this "age-hardening"—or "ageing" (17.50)—of martensite. The main function of cobalt is to adjust the M_s line to a convenient temperature.

These steels are finding a wide variety of applications in aerospace machines and in tooling and industrial components. Possibly the most interesting use is in various types of rocket motor cases in the American space programme.

RECENT EXAMINATION QUESTIONS

1. Discuss the statement "Carbon is the most important alloying element in steels."
    (University of Aston in Birmingham, Inter B.Sc.(Eng.).)

2. Discuss the purpose and the effects produced by the addition of alloying elements to steel.   (Nottingham Regional Technical College.)

3. What metallurgical reasoning must be applied when alloy steels are being specified for machine parts?
    University of Aston in Birmingham, Inter B.Sc.(Eng.).)

4. (i) What effect does a substitutional element have upon the iron–iron carbide equilibrium diagram if (a) it is a γ stabiliser and (b) it is an α stabiliser? Relate these ideas to the ability of alloy steels to be heat treated.
    (ii) What effects do these substitutional elements have upon the other properties of steels?   (Aston Technical College, H.N.D. Course.)

5. Discuss the effects on the structure and properties of steels of the addition of alloying elements which (i) go into solid solution; (ii) form carbides.
    (West Bromwich Technical College.)

6. (a) Discuss the effects, upon the iron–carbon equilibrium diagram, of a substitutional alloying element that exhibits a "closed gamma loop". Sketch representative equilibrium diagrams at various fixed percentages of the substitutional element.
    (b) What effect will (a) have on whether or not the material is hardenable by heat-treatment?
    (c) What are the major problems associated with welding Fe–Cr–C alloys? How may these difficulties be overcome?
    (Aston Technical College.)

7. Most low-alloy steels contain less than 7% total alloy content. Detail the fundamental reasons for such additions when considering (i) hardenability and distortion during heat-treatment, and (ii) the strength, ductility and toughness of hardenable and non-hardenable steels.
    (Aston Technical College, H.N.D. Course.)

8. Discuss the uses of chromium as an alloying addition to carbon steels.
    (University of Glasgow, B.Sc.(Eng.).)

9. Discuss the use of nickel and chromium on the heat-treatment properties and microstructures of hardenable steels.
    (U.L.C.I., Paper No. C111–2–4.)

10. What is meant by the term "temper-brittleness"?
    Show how the mechanical properties of the steel are affected and indicate a way in which the effect may be minimised.
    (Derby and District College of Technology, H.N.D. Course.)

11. Name the three groups of stainless steels, indicating which are capable of being hardened by heat-treatment.
    Discuss the difficulties that may arise with welded stainless steels.
    (Nottingham Regional Technical College.)

12. Write an essay on "The Stainless Steels—the Relationship between their Composition, Classification and Applications". (S.A.N.C.A.D.)
13. Chromium and nickel, either singly or in combination, are added to steel in varying proportions to produce a range of now well-established alloys. Enumerate the most important of these alloys and discuss their properties and applications.
(Constantine College of Technology, Middlesbrough.)
14. Explain, with the aid of constitutional diagrams (where necessary), the effects of the following alloying elements when added singly to a plain carbon steel: (i) nickel; (ii) chromium; (iii) manganese.
Why are nickel and chromium often added in conjunction to many low-alloy steels? (West Bromwich Technical College.)
15. Weld-decay and temper-brittleness are defects associated with certain types of steels. Say in which materials these defects occur, under what conditions they arise and how they may be avoided.
(Rutherford College of Technology, Newcastle upon Tyne.)
16. Discuss the function of alloy additions to steels for the following applications: (a) good hardenability; (b) wear-resistance; (c) creep-resistance; (d) corrosion-resistance. (University of Strathclyde, B.Sc.(Eng.).)
17. A tool is intended for forging hot steel (forging temperature 1 250° C). What conditions will the tool have to withstand? Suggest a material suitable for this application and justify your choice.
(University of Aston in Birmingham, Inter B.Sc.(Eng.).)
18. Select THREE of the steels below, suggest a typical application for each and explain the role of the additions. Where appropriate, describe the heat-treatments required and the microstructures obtained.

    (a) 1·2% carbon.
    (b) 0·35% carbon; 1·5% manganese; 0·45% molybdenum.
    (c) 0·3% carbon; 12% chromium.
    (d) 0·03% carbon; 18% chromium; 8% nickel.
(Lanchester College of Technology, Coventry, H.N.D.)

19. Give the approximate compositions of alloy steels which would be suitable for any FOUR of the following:

    (i) a rustless kitchen sink unit;
    (ii) an aero-engine connecting-rod;
    (iii) the lip of a dredger bucket;
    (iv) a rustless fruit-knife blade;
    (v) an extrusion die for copper alloys.

Give reasons in each case for the suitability of the steel and outline the heat-treatment necessary for each steel you specify.
(Institution of Production Engineers Associate Membership Examination, Part II, Group A.)

CHAPTER XIV

COMPLEX FERROUS ALLOYS

High-speed Steels

14.10. The development of high-speed steel began in 1861 when Robert Mushet of Sheffield was investigating the value of manganese in the production of Bessemer steel. He found that one composition containing manganese hardened when cooled in air. Analysis showed approximately 6% tungsten to be present in addition, and this alloy subsequently found its way on to the market under the name of "air-hardening steel". It was discovered soon afterwards that cooling from 1 100° C in an air blast gave better results for tools. Some twenty years later Messrs. Maunsel White and Frederick Taylor, both of the Bethlehem Steel Company, replaced the manganese in Mushet's steel by chromium, and also increased the tungsten content. These changes in composition enabled the steel to be forged readily and produced a superior tool steel. Production engineers will already be familiar with the name of Taylor, who devoted a great deal of attention to time-study and machine-shop methods.

Modern high-speed steel was first introduced to the public at the Paris Exposition in 1900. The tools were exhibited cutting at a speed of about 0·3 m/s with their tips heated to redness. It was found later that maximum efficiency was obtained with a chromium content of 4% and about 18% tungsten, the carbon content having been reduced from 2·0%, in the original Mushet steel, to between 0·6 and 0·8%. In 1906 vanadium was first added to high-speed steel. Molybdenum was investigated as a substitute for some of the tungsten, but increases in the price of molybdenum in the 1920s suspended this development, particularly since molybdenum steels were regarded as being inferior in properties. Only in more recent years has molybdenum replaced much of the tungsten in many grades of high-speed steel. This is particularly true in the United States, where the greatest annual tonnage of high-speed steel is of the molybdenum type.

14.11. The main features of a high-speed steel are its great hardness in the heat-treated condition, and its ability to resist softening at relatively high working temperatures. Thus, high-speed steel tools can be used at cutting speeds far in excess of those possible with ordinary steel tools, since high-speed steel resists the tempering effect of the heat generated.

In high-speed steel ordinary iron carbide, Fe_3C, is replaced by double carbides such as Fe_4W_2C or $(MoW)_6C$, and the solid-solution phases based on these substances are far more temper-resistant than those based on simple iron carbide. Vanadium is present as the extremely abrasion-resistant carbide V_4C_3, which, because it dissolves but slowly in austenite during heat-treatment, also acts as a grain-refiner by impeding the grain growth of austenite. Since it is a ferrite stabiliser, vanadium tends to stabilise the δ-ferrite phase (11.20) at high temperatures, leading to carbide precipitation and to some difficulties in heat-treatment. Despite this it is now added advantageously in amounts up to 5·0% to modern "super" high-speed steels.

Up to 12·0% cobalt is also added to these "super" high-speed steels. Although its carbide-forming tendency is no stronger than that of iron, it promotes a useful increase in red-hardness when dissolved in the matrix of such a steel.

As mentioned above, high-speed steels containing greater proportions of molybdenum are now widely used. Generally these alloys require greater care during heat-treatment, being more susceptible to decarburisation than the tungsten varieties. With modern heat-treatment plant, however, this is not an unsurmountable difficulty. Molybdenum high-speed steels are considerably tougher than the corresponding tungsten types and are widely used for drills, taps and reamers.

14.12. Obviously, when we have a complex alloy containing up to six different elements, we cannot represent the system by a simple binary diagram (9.90). However, we can construct a two-dimensional diagram showing the relationship between microstructure, temperature and carbon content for a steel containing a *fixed* quantity of other elements (in this case 18% tungsten, 4% chromium and 1% vanadium). Such a diagram (Fig. 14.1) is generally called a "pseudo-binary" diagram and, whilst not being a true equilibrium diagram, serves as a useful "constitutional chart".

In the cast condition the structure of a high-speed steel is very like that of a cast iron. This cast structure is broken up by forging at temperatures between 900 and 1 150° C, followed by reheating to about 750° C to relieve working stresses. If required in the soft condition high-speed steel is usually annealed by "soaking" for about four hours at 850–900° C, followed by a very slow rate of cooling (10 to 20° C per hour) down to 600° C, when it can be cooled to room temperature in still air. At this stage the ferrite present contains relatively little dissolved alloying elements. These are present mainly as massive carbide particles which have precipitated during annealing.

14.13. As shown in the constitutional chart, austenite forms when the steel is heated to 840° C. Since the eutectoid point has been displaced so far to the left by the influence of tungsten and other added elements, this austenite contains little more than 0·2% dissolved carbon. If the

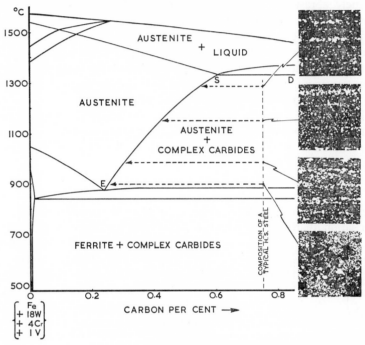

FIG. 14.1.—A "Pseudo-binary" Diagram Representing the Structures of 18-4-1 High-speed Steel.

Here the tungsten–chromium–vanadium content is kept constant and carbon made variable. The photomicrographs show the effect of the solution temperature upon the solubility of the complex carbide (light) in the matrix (dark) following quenching. The higher the quenching temperature, the greater the solubility of the carbide.

steel were quenched from 840° C, the martensite ultimately produced would temper easily and show little advantage over that in ordinary types of tool steel. High-speed steel is therefore heated further, so that more carbide is absorbed by the austenite in accordance with the increase in solubility of the carbide, as indicated by the sloping phase boundary *ES* in Fig. 14.1. Quenching from temperatures above 840° C will produce increasing red-hardness with rise in temperature, due to more carbon and alloying elements being taken into solution, and thus rendering the resultant martensite more sluggish to tempering influences. Even when the solidus temperature is reached at S (approximately 1 330° C), no more than 0·6% carbon has been absorbed and the remainder exists as isolated globules of complex carbides.

14.14. Therefore, in order to obtain the highest cutting efficiency in a high-speed steel, it must be hardened from a temperature little short

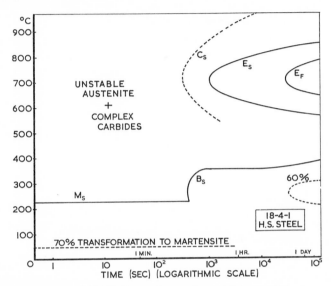

FIG. 14.2.—T.T.T. Curves for an 18–4–1 High-speed Steel.

The initial structure consists of saturated austenite together with some undissolved carbides. The line C_s indicates the commencement of precipitation of primary carbides (since the steel is not of eutectoid composition). E_s and E_f indicate the start and finish of eutectoid ("pearlite") formation. B_s shows the start of bainite formation whilst the associated broken line indicates the time required for 60% transformation to bainite. Note that at room temperature the transformation of austenite to martensite is incomplete (70% at 50° C) so that air- or oil-quenching will produce a mixture of martensite and retained austenite.

of the solidus SD, at which fusion commences. This temperature will vary with steels of different composition but is usually between 1 170° and 1 320° C. Heat-treatment at such temperatures has its attendant difficulties, and requires some skill and experience in management, as well as adequate plant. Grain growth and oxidation can be excessive at high temperatures unless the conditions under which the steel is heated are suitably controlled.

Oxidation is effectively prevented by heating the tool in a slightly carburising atmosphere; whilst grain growth is minimised by using a two-chamber furnace. The tool is first pre-heated in the lower-temperature chamber to about 850° C, and then transferred to the higher-temperature chamber, in which it is heated rapidly to the hardening

TABLE 14.1 — *High-speed Steels*

| Type of steel | B.S. 4659 designation | Proprietary trade names and codes | Composition (%) | | | | | | Heat-treatment | | Hardness (V.P.N.) | Characteristics and uses |
			C	Cr	W	V	Mo	Co	Quench in oil or air blast from:	Secondary-hardening treatment		
14% Tungsten	—	Carr's "Motor Special" Thos. Andrews "No. 1 Monarch"	0·65	4·0	14·0	0·25	—	—	1 250–1 300° C.	Double temper at 565° C for 1 hour	800–860	General shop practice—all-round work, moderate duties. Also for some cold-working tools (blanking tools and shear blades).
18% Tungsten (18–4–1)	BT1	Br. Steel Corp. VAP Edgar Allen "Stag Special", Carr's "Motor Maximum", Jessop-Saville "Ark Superior", Balfour Darwin's "Ultra Capital"	0·75	4·2	18·2	1·2	—	—	1 290–1 310° C.	" "	800–890	Lathe, planer and shaping tools; millers and gear cutters. Router bits; hobs; reamers; broaches; taps; dies; drills; slitting discs. Hacksaws; bandsaws; etc. Roller-bearings at high temperatures, e.g. gas turbines.
22% Tungsten	BT20	Carr's "Plutocrat" Balfour Darwin's "Ultra Capital 22"	0·78	4·5	21·5	1·4	0·5	—	1 280–1 320° C.	" "	830–950	Large hobs and milling cutters, automatic lathe tools, accurate form-tools, large broaches, etc. Holds its edge longer than the 18% tungsten type of steel.
6% Cobalt	BT4	Edgar Allen "Stag Extra Special" Carr's "Pluto Perfectum" Balfour Darwin's "Ultra Capital Plus One"	0·8	4·5	18·5	1·2	0·5	5·5	1 290–1 320° C.	" "	800–900	Lathe, planer and shaper tools on very hard materials and for extra heavy work, e.g. high-tensile steels, hard cast irons, new and brake-hardened railway tyres, etc. Austenitic steels which work-harden quickly.
12% Cobalt	BT6	Edgar Allen "Stag Major"	0·8	4·75	21·5	1·5	0·5	12·0	1 300–1 320° C.	" "	900–950	Lathe, planer and shaper tools, milling cutters, hobs, twist drills, etc., for exceptionally hard materials. Has maximum red-hardness and toughness. Suitable for severest machining duties—manganese steel, steels of over 1 200 N/mm². T.S., close-grained cast irons.
	—	Carr's "Pluto Paramount" Jessop-Saville TTQ	0·78	4·5	18·8	2·0	0·7	10·0				

Type	Code	Proprietary names	C	Cr	W	V	Mo	Co	Hardening temp.	Tempering	Hardness	Remarks
12% Tungsten–5% vanadium	BT15	Firth Brown "Supercut Vanleda", Edgar Allen "Stag V55", Jessop-Saville J36	1·5	4·5	12·5	5·0	—	5·0	1 230–1 300° C.	Double temper at 560° C for 1 hour	900–1 000	EXTREMELY hard due to high vanadium %, but also tough. Excellent red-hardness and edge retention. High accuracy and finish—form tools, turning, milling, shaping. Also cutting gas-turbine alloys.
18% Tungsten–5% vanadium	—	Edgar Allen "Stag Vanco"	1·5	5·0	17·5	5·0	0·5	8·75	1 250° C.	Double temper at 560° C for 2 hours	900–990	For very hard materials—gas-turbine alloys, etc. Tool bits, form tools, milling cutters, etc.
Molybdenum "562"	BM2	Br. Steel Corp. "Cyclone 56", Edgar Allen "Stag Mo 562", Jessop-Saville J34, Thos. Andrews' "Monarch 652", Carr's "Motor Magnus"	0·83	4·25	6·5	1·9	5·0	—	1 250° C.	Double temper at 565° C for 1 hour	850–900	Roughly equivalent to standard 18-4-1 high-speed steel, but tougher. Used for drills, reamers, taps, milling cutters, punches, threading dies, cold-forging dies.
9% molybdenum	BM1	Br. Steel Corp. "Cyclone 92", Balfour Darwin's "Capital 305"	0·8	3·75	1·6	1·25	9·0	—	1 200–1 230° C.	" "	830–900	A general-purpose Molybdenum-type high-speed steel—drills, taps, reamers, cutters. More susceptible to decarburisation during heat-treatment than ordinary 18-4-1.
9%-molybdenum–8% cobalt	BM42	Br. Steel Corp. "Cyclone Mc33", Carr's "Pluto Plus", Balfour Darwin's "Capital 405"	1·0	3·75	1·65	1·1	9·5	8·25	1 180–1 210° C.	Triple temper at 530° C for 1 hour	830–935	Similar uses to 12% Cobalt–21% Tungsten high-speed steel.

temperature. As soon as it reaches the hardening temperature it is oil-quenched or cooled in an air blast according to the composition of the steel (see Table 14.1). Pre-heating reduces the time during which the surface of the tool would otherwise be in contact with the high-temperature atmosphere.

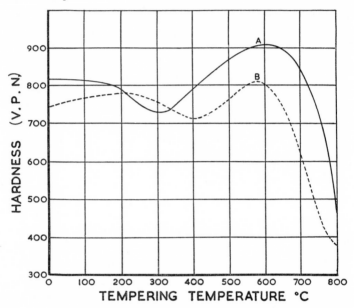

FIG. 14.3.—The Relationship between Hardness and Tempering Temperature for High-speed Steels.

(A) Represents the 12% cobalt type, and (B) the ordinary 18% tungsten type.

14.15. Due to the sluggishness produced in the austenite $\longrightarrow$ martensite transformation by the presence of alloying elements, the quenched high-speed steel will contain considerable amounts of retained austenite (Fig. 14.2). It is necessary to transform this to martensite, and this can be accomplished by tempering. Low-temperature tempering between 300 and 400° C reduces the hardness slightly but increases the toughness. On tempering between 400 and 600° C the hardness increases again, often to a figure higher than the original (Fig. 14.3). This phenomenon, known as "secondary hardening", is associated with the presence of vanadium. Secondary hardening is usually carried out by heating the tool for one-half to three hours at 550–600° C according to the composition of the steel. Some austenite may still be present following tempering, and for this and other reasons double tempering is now generally used. The retained austenite is "conditioned" during the first tempering process and some of it changes to martensite on cooling. During the second tempering process the transformation goes still further towards completion as more of the retained austenite changes to martensite. This second tempering process also relieves internal stresses which have been set up as austenite transformed to martensite during the first tempering process. In the case of high-speed steel the initial quench from a high temperature tends not to produce internal stresses, since an adequate "cushion" of soft austenite is retained. The first tempering process causes much of the retained austenite to transform to hard martensite and this may lead to the introduction of internal micro-stresses.

The final structure of a correctly heat-treated high-speed steel will consist of isolated globules of complex carbide, in a matrix of martensite which contains sufficient dissolved alloying elements to render it sluggish to tempering influences.

Cemented and Cast Cutting Materials

14.20. Mention has already been made (6.50) of the production of cutting tools by powder metallurgy, and though, strictly speaking, not ferrous alloys, these materials will be discussed here because of their close association with high-speed steels.

Cemented carbide tools are prepared by sintering at high temperature and pressure, in an atmosphere of hydrogen, a mixture of tungsten carbide powder with powdered cobalt. Sometimes tungsten carbide is replaced by powdered carbides of other "hard metals", tantalum, molybdenum, titanium and vanadium, but in each case incipient fusion of the cobalt occurs, and it forms a ductile bond between the carbide particles.

The resultant structure resembles that of a high-speed steel, in that spherical particles of carbide are held in a tough matrix, but whereas the properties of a high-speed steel are developed by heat-treatment, the structures of cemented carbide materials are produced by mechanical methods.

14.21. In "Cermets", which are sintered products similar in principle to the cemented carbides mentioned above, the object is to combine the high-temperature strength and hardness of a suitable metallic oxide (or carbide) with sufficient ductility provided by a metallic bonding element, thus giving a material with considerable resistance to thermal and mechanical shock and suitable for jet-engine or other high-temperature applications. Typical hard phase–ductile bond mixtures include: aluminium oxide–chromium; tungsten carbide–nickel; beryllium oxide–beryllium; and silicon carbide–silicon.

14.22. The "Stellites" are extremely hard alloys which, however, cannot be forged and must be cast to shape. They retain hardness, strength and resistance to oxidation at high temperatures. A typical stellite composition is 50% cobalt, 20% chromium, 18% molybdenum, 10% tungsten and 2% carbon. Its structure consists of the hard carbides of tungsten, molybdenum and chromium, bonded with a cobalt–chromium matrix. Similar compositions are used for plug-, screw- and tap-gauges; valves and valve seatings for aero engines; shear blades for thin, hard metal; and for turbo superchargers.

Magnet Steels and Alloys

14.30. The type of magnetism, exhibited by a group of the "transition metals",* is derived basically from the axial spin of electrons. This spin produces a magnetic field just as do electrons (i.e. an electric current) flowing in a coil of wire. The electrons in an atom are generally grouped in pairs, and since each electron in a pair spins in the opposite direction, the resultant magnetic field is zero. In the transition metals, however, unpaired electrons can exist, and since all spin in the same direction, a resultant field is produced. In iron, nickel and cobalt the lattice structures are such that atoms can become aligned, with their magnetic fields parallel, thus giving rise to a powerful resultant field.

Magnetic materials used in engineering can be divided into two groups:

(*a*) alloys which retain magnetism after the magnetising influence has been removed, i.e. magnetically "hard" materials used in permanent magnets;

(*b*) alloys which lose most of the induced magnetism when the magnetising field is removed, i.e. magnetically "soft" materials, such as are used in transformer cores and dynamo pole-pieces.

14.31. A permanent magnet is designed to provide a source of magnetism that does not involve the application of power derived from an external source. It must also retain its magnetism for, to all intents and purposes, an unlimited period, provided that it is not submitted to external demagnetising influences. The actual magnetic strength, or flux, in a permanent magnet is dependent upon its design and the steel or other alloy from which it is made. The essential characteristics of permanent-magnet materials are:

(*a*) coercive force;
(*b*) residual magnetism or "remanence";
(*c*) energy product value.

---

* This group includes a large number of the metallic elements, but only a few, the most important of which is iron, exhibit strong "ferromagnetism".

Coercive force refers to the ability of the magnet to withstand demagnetising influences. Remanence is the amount of magnetism remaining in the magnet after the magnetising field has been switched off. The energy product value is the quantity of magnetic energy possessed by the magnet, and is calculated from the demagnetising curve RU (Fig. 14.4). The magnetising force is expressed in "ampere-turns per metre", whilst the SI unit of induced flux density is the "tesla" (T), symbolised by H and B respectively.

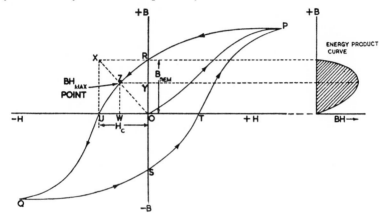

FIG. 14.4.—Magnetic Hysteresis Loop.

14.32. We will now consider what is known as the hysteresis loop. Fig. 14.4 shows a typical hysteresis loop ($PRUQST$) and magnetisation curve (OP) for a permanent magnet. When a piece of steel is placed in a solenoid with a current passing through it, magnetic flux is induced in the steel. The amount of this flux can be measured in webers (Wb) and the flux density in teslas (T), where Wb = Vs and T = Wb/m^2. In Fig. 14.4 magnetic flux values (B) are plotted vertically, whilst the magnetising force values (H), calculated from the current passing through the solenoid and the number of turns of wire on it, are plotted horizontally.

14.33. Beginning with a small value of H and gradually increasing it, corresponding values of B and H are plotted, and the curve OP produced. This is the magnetisation curve, which is continuous until it runs almost parallel to the horizontal axis at P. This state represents complete magnetic saturation of the specimen.

If we now gradually reduce the magnetising field to zero, the corresponding fall in induced magnetic flux is represented by the line PR. The intercept OR represents the value of the magnetic flux remaining in the steel when all external magnetising influence is removed. This is termed the remanent magnetism or remanence B_{rem}.

To demagnetise the steel, and at the same time to measure the amount of force required to achieve this, a current is once more passed through the solenoid but in the reverse direction, and gradually increased. This produces the curve RU, which is called the demagnetisation curve. The intercept OU corresponds to the amount of force required to completely demagnetise the steel, and is called, as already stated, the coercive force. To complete the loop the operations outlined are carried on to negative saturation and then repeated in the reverse sense.

The hysteresis loop, therefore, represents the lagging of the magnetic flux produced behind the magnetising force producing it. The area of the loop is proportional to the amount of energy lost when the specimen it represents is magnetised or demagnetised, and for permanent magnets the area of the loop must be as large as possible.

14.34. The ultimate criterion of a permanent magnet is the maximum product of B and H obtained from a point on the demagnetisation curve. In Fig. 14.4 it is shown that by completing the rectangle $RXUO$ and drawing the diagonal OX, the point Z at which the diagonal cuts the curve RU gives an approximation to the point at which the product of YZ and WZ is a maximum for the curve. This is called $BH_{max.}$, and corresponds to the maximum energy the magnet can provide in a circuit external to itself.

14.35. Fully hardened carbon steels containing about $1 \cdot 0\%$ carbon were originally used for the manufacture of permanent magnets. These were later replaced by alloy steels containing varying amounts of chromium, tungsten and cobalt. The cobalt steels are the only ones of any practical value to-day, and even these have been very largely replaced by alloys of the Alni–Alnico–Alcomax–Hycomax series. These have been developed from magnetic aluminium–nickel alloys discovered by Japanese workers in 1930.

14.36. The Alni–Alnico series have a lower remanence than cobalt steels but a much higher coercive force. Moreover, they retain magnetism much better under influences of shock and of heat (Fig. 14.5). The most modern alloys of the Alcomax–Hycomax–Columax series are superior also in remanence. They are *anisotropic* materials; that is, they have better magnetic properties along a preferred axis. This effect is obtained by heating the alloy to a high temperature and allowing it to cool in a magnetic field. Groups of atoms become orientated as the magnet cools, and this orientation is "frozen in".

14.37. Permanent-magnet alloys find many uses, but particular applications include the magnets for moving-coil loudspeakers, magnetic chucks, generators, magnetos, small electric motors, etc. The modern aluminium–nickel–cobalt type of alloy has enabled the size of a loudspeaker magnet to be reduced considerably over the last few years. Typical compositions and magnetic properties of the more important permanent-magnet alloys are given in Table 14.2.

14.38. The manufacture of such components as transformer cores and dynamo pole-pieces demands the use of a material which is magnetically "soft" but of high magnetic permeability. Permeability (μ)

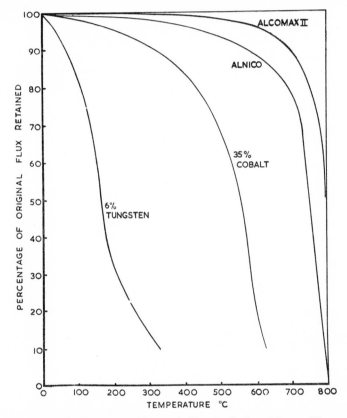

Fig. 14.5.—The Retention of Magnetic Flux by Various Magnet Alloys at Elevated Temperatures.

is denoted by the ratio B/H, and its value changes with the magnetising field, rising from a low figure to a maximum and then falling. Since a transformer core works on alternating current, the core is first magnetised, then demagnetised and then remagnetised with reverse polarity. This type of cycle produces the closed curve $PRUQST$ (Fig. 14.4), and the area which it encloses is a measure of the amount of energy wasted in overcoming remanent magnetism each time the current is reversed (which occurs many times per second). This energy is converted into

TABLE 14.2.—Permanent-magnet Alloys

Type of alloy	Composition % (balance iron)										Magnetic properties		
	C	Cr	W	Co	Al	Ni	Cu	Nb	Ti	Other elements	B_{rem} (T)	H_c (A/m)	BH_{max}
Carbon steel (hardened)	1·0	—	—	—	—	—	—	—	—	—	0·9	4 400	1 560
Tungsten steel (hardened)	0·7	0·6	6·0	—	—	—	—	—	—	—	1·1	5 250	2 620
3% Cobalt steel	1·0	9·0	—	3·0	—	—	—	—	—	Mo 1·5	0·72	10 300	2 730
15% Cobalt steel	1·0	9·0	—	15·0	—	—	—	—	—	Mo 1·5	0·82	14 300	4 830
35% Cobalt steel	0·9	6·0	5·0	35·0	—	—	—	—	—	—	0·90	20 000	7 800
"Alni"	—	—	—	—	12·5	26·0	4·0	—	—	—	0·56	46 100	9 750
"Alnico"	—	—	—	12·0	9·5	17·0	5·0	—	—	—	0·725	44 500	13 500
*"Alcomax II"	—	—	—	23·0	8·0	12·0	3·0	0·2	—	—	1·30	46 100	38 000
*"Alcomax III"	—	—	—	24·5	8·0	13·5	3·0	0·6	—	—	1·26	51 700	38 000
*"Alcomax IV"	—	—	—	24·5	8·0	13·5	3·0	2·0	—	—	1·15	59 700	31 700
*"Hyomax II"	—	—	—	29·0	7·0	15·0	4·0	2·0	4·0	—	0·85	95 500	28 100
*"Hycomax III"	—	—	—	34·0	7·0	15·0	4·0	—	5·0	—	0·88	115 400	35 200
*"Columax"	—	—	—	24·5	8·0	13·5	3·0	0·6	—	—	1·35	58 800	52 800
*"Feroba III" (Fired ceramic)	—	—	—	—	—	—	—	—	—	Barium 13·0 Oxygen 27·0	0·34	200 000	19 000

* Anisotropic alloys, properties measured along preferred axis.

heat, and it is therefore necessary to use materials, for transformer cores, which give very narrow hysteresis loops, and hence, a small "hysteresis loss". The difference in properties between a material suitable for a permanent magnet and one suitable for a transformer core is indicated by the different shapes of the hysteresis loops, as shown in Fig. 14.6.

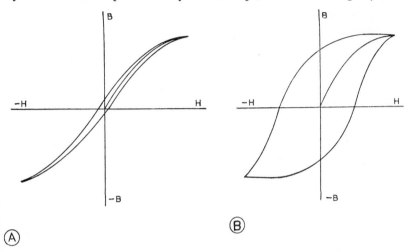

FIG. 14.6

(A) The Type of Hysteresis Loop Obtained from a Material Unsuitable for the Production of Permanent Magnets, but Suitable for the Manufacture of Transformer Cores.
(B) The Type of Hysteresis Loop for a Permanent-magnet Alloy.

14.39. Soft iron and iron–silicon alloys ("Stalloy") are most useful in respect of low hysteresis. The iron–silicon alloys contain up to $4\cdot5\%$ silicon and practically no carbon, and are usually supplied in the form of dead-soft sheet from which transformer-core laminations can be stamped. The use of laminations reduces losses from eddy currents, which would be more prevalent in a solid core.

Nickel–iron alloys are also of importance as low-hysteresis, high-permeability alloys. "Permalloy" contains 78% nickel and 22% iron, with carbon and silicon kept as low as possible. "Mumetal" contains about 74% nickel, 5% copper, up to $1\cdot0\%$ manganese and the balance iron. Both alloys are used in communications engineering, where the high permeability will utilise to the full the small currents involved. In addition to being used for transformer cores, these alloys are used as shields for submarine cables. Like the iron–silicon alloys, they are used in the dead-soft condition, the best treatment consisting of heating to $900°$ C, followed by slow cooling, and then heating to $600°$ C, followed by cooling in air.

RECENT EXAMINATION QUESTIONS

1. What considerations must be taken into account when choosing a tool material? How do carbon tool steels satisfy these requirements?
(University of Aston in Birmingham, Inter B.Sc.(Eng.).)

2. What are the basic requirements of a tool steel? Describe the heat-treatment of a high-speed tool steel, giving reasons for each step. Explain how the microstructure is related to the service requirements of the tool.   (S.A.N.C.A.D.)

3. (a) What do you understand by the term tool steel?
(b) Give typical approximate chemical compositions for tool steels suitable for thread rolls or other components that need great resistance to wear and abrasion.
(c) Describe how you would heat-treat such a tool, mentioning the precautions that would be necessary to produce a satisfactory article.
(Institution of Production Engineers Associate Membership Examinations, Part II, Group A.)

4. Why is tungsten so important as a constituent of high-speed steels? Outline the principles involved in the heat-treatment of such a steel.
(U.L.C.I., Paper No. 237–2–8).

5. (a) Discuss, with the aid of diagrams, how alloying elements affect the austenite→martensite reaction and show its importance in engineering design.
(b) "Nickel enters into solid solution with ferrite." Comment upon, and show the importance of, this statement with regard to case-hardening steels.
(c) Illustrate the effect of the addition of tungsten, chromium and vanadium upon the microstructure, heat-treatment and properties of a high-speed steel.   (Aston Technical College, H.N.D. Course.)

6. Draw T.T.T. curves for: (a) mild steel; (b) an oil-hardening steel; (c) high-speed steel.
With the aid of a T.T.T. diagram, describe the process of martempering.   (University of Aston in Birmingham, Inter B.Sc.(Eng.).)

7. Describe the changes in microstructure and properties which occur on increasing the tempering temperature applied to a hardened carbon tool-steel article.
What is secondary hardening?
(The Polytechnic, London W.1., Poly.Dip.)

8. Explain what is meant by *secondary hardening* and state the class of alloying element used to produce this phenomenon in steel. Detail the heat-treatment required for a component in which secondary hardening is the final stage.   (U.L.C.I., Paper No. B111–2–4.)

CHAPTER XV

CAST IRONS AND ALLOY CAST IRONS

15.10. Cast iron is very similar in composition and structure to the crude pig iron produced by the blast furnace. In fact, castings are occasionally made using pig iron direct from the blast furnace. It is usual, however, to remelt the pig iron in a cupola, any necessary adjustments in composition being made during this remelting process. Because production costs of pig iron are relatively low and no refining processes are necessary, cast iron is a cheap metallurgical material which is particularly useful where a casting requiring rigidity, resistance to wear or a high compressive strength is necessary. Other useful properties of cast iron include:

(a) its ease of machining when a suitable composition is selected;

(b) its fluidity and ability to make good casting impressions;

(c) its easily attainable melting temperatures (1 130–1 250° C) as compared with steels.

(d) the availability of high-strength cast irons when additional treatment is given to materials of suitable composition, e.g. spheroidal-graphite iron and pearlitic malleable iron.

The relationships between microstructure, composition and rate of cooling are similar, in principle, to those prevailing in the case of pig iron (2.50), but we shall deal with them more fully here.

The Effects of Composition on the Structure of Cast Iron

15.20. Ordinary cast iron is a complex alloy containing a total of up to 10% of the elements carbon, silicon, manganese, sulphur and phosphorus; the balance being iron. Alloy cast irons, which will be dealt with later in this chapter, contain also varying amounts of nickel, chromium, molybdenum, vanadium and copper.

15.21. Carbon can exist in two forms in cast iron, namely as free graphite or combined with some of the iron to form iron carbide (cementite). These two varieties are usually referred to as "graphitic carbon" and "combined carbon" respectively, and the total amount of both types in the specimen of iron as "total carbon". The form in which carbon is present depends largely upon the influence of silicon and, in alloy cast irons, those elements mentioned above. The effect of these elements on the stability of the cementite is generally the same as in steels (13.12).

Cementite is a hard, white, brittle compound, so that irons which contain much of it will present a white fracture when broken, and will have a low resistance to shock. At the same time they will possess a high resistance to wear. Such irons are called "white irons". The fractured surface of a cast iron containing graphite, however, will appear grey, and the iron will be termed a "grey iron".

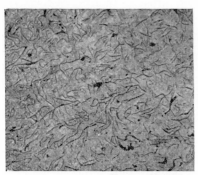

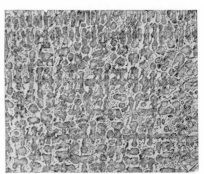

[*Courtesy of B.C.I.R.A.*   [*Courtesy of B.C.I.R.A.*

15.1A.—Grey Cast Iron.   15.1B.—White Cast Iron.

Graphite flakes in a matrix of pearl-   Primary cementite (white) and pearlite. × 60. Etched in 4% Picral.   ite. × 100. Etched in 4% Picral.

PLATE 15.1

The size and quantity of the graphite flakes will, to a large extent, govern the mechanical properties of a grey iron. If the reader refers to the complete iron–carbon diagram (Fig. 11.1) he will notice that a eutectic of austenite and cementite which contains 4·3% carbon is formed at 1 131° C. (This eutectic point, like the eutectoid composition in the case of steel at 0·83% carbon, may be moved farther to the left of the diagram by the elements present, so that a cast iron containing less than 4·3% carbon may be of eutectic composition.) In accordance with the equilibrium diagram, a cast iron containing more than the eutectic amount of carbon may be expected to begin solidification by depositing some cementite. Unless cooling is very rapid, however, graphite may precipitate instead, due to instability of the cementite caused by other elements (notably silicon) present. This primary graphite, which separates out from the melt during cooling, forms as large flakes, usually called "kish". The finer flakes of graphite occurring in grey cast iron are those which are formed by decomposition of the cementite after solidification Graphite flakes of this latter type exist in cast irons of compositions either above or below the eutectic, and produce rather better mechanical properties.

The presence of graphite gives a softer iron which machines well because of the effect of the graphite flakes in forming chip cracks in advance of the edge of the cutting tool (20.20).

15.22. Silicon dissolves in the ferrite of a cast iron and is the element which has the predominant effect on the relative amounts of graphite and cementite which are present. Silicon tends to make cementite unstable (13.12), so that it decomposes, producing graphite and hence, a grey iron. The higher the silicon content, the greater the degree of decomposition of the cementite, and the coarser the flakes of graphite produced.

Thus, whilst silicon actually strengthens the ferrite by dissolving in it, at the same time it produces softness by causing the cementite to break down to graphite. When, however, silicon is present in amounts in excess of that necessary to complete the decomposition of all the cementite, it will again cause hardness and brittleness to increase. Both the direct and indirect effects of silicon must therefore be considered.

The presence of silicon in a cast iron is beneficial in so far as it increases the fluidity of the molten iron, and so improves its casting properties.

15.23. Sulphur. The possible effect of the silicon present in an iron cannot be completely estimated without some reference to the sulphur content, since sulphur has the opposite effect, in that it tends to stabilise cementite. Sulphur, then, inhibits graphitisation and so helps to produce a hard, brittle, white iron. Moreover, its presence as the sulphide FeS in cast iron will also increase the tendency to brittleness.

15.24. Manganese. The effect of sulphur is governed, in turn, by the amount of manganese present. Manganese combines with sulphur to form manganese sulphide, MnS, which, unlike ferrous sulphide, is insoluble in the molten iron and floats to the top to join the slag. The indirect effect of manganese, therefore, is to promote graphitisation because of the reduction of the sulphur content which it causes. Manganese has a stabilising effect on carbides, however, so that this offsets the effect of sulphur reduction in promoting graphitisation. The more direct effects of manganese include the hardening of the iron, the refinement of grain and an increase in strength.

15.25. Phosphorus is present in cast iron as the phosphide Fe_3P, which forms a eutectic with the ferrite in grey irons, and with ferrite and cementite in white irons. These eutectics melt at about 950° C, so that high-phosphorus irons have great fluidity. Cast irons containing 1% phosphorus are, therefore, very suitable for the production of castings of thin section.

Phosphorus has a negligible effect on the stability of cementite, but its direct effect is to promote hardness and brittleness due to the large volume of phosphide eutectic which a comparatively small amount of phosphorus will produce. Phosphorus must therefore be kept low in castings where shock-resistance is important.

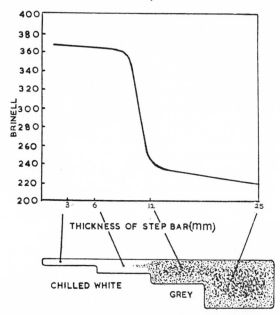

FIG. 15.1.—The Effect of Thickness of Cross-section on the Rate of Cooling, and Hence upon the Microstructure of a Grey Cast Iron.

The Effect of Rate of Cooling on the Structure of Cast Iron

15.30. A high rate of cooling during solidification tends to prevent the decomposition of cementite in an iron which, on slow cooling, would become graphitic. This effect is important in connection with the choice of a suitable iron for the production of castings of thin section. Supposing an iron which, when cooled slowly, had a fine grey structure containing small flakes of graphite were chosen for such a purpose. In thin sections it would cool so rapidly that decomposition of the cementite could not take place, and a thin section of completely white iron would result. Such a section would be brittle and useless.

15.31. This effect is illustrated by casting a "stepped bar" of iron of a suitable composition (Fig. 15.1). Here, the thin "steps" have cooled so quickly that no decomposition of the cementite has occurred, as indicated by the white fracture and the high Brinell figures. The thicker steps, having cooled more slowly, are graphitic and consequently

PLATE 15.2.—Illustrating the Effects of Thickness of Section and, Hence, Rate of Cooling on the Structure of a Grey Iron.

The thinnest part of the section has cooled quickly enough to produce a white iron structure, whilst the core of the thickest part has a grey iron structure. The relationships between sectional thickness and microstructure are similar to those indicated in Fig. 15.1 on the opposite page. Both micrographs × 300 and etched in 2% Nital. Macrosection × 3.

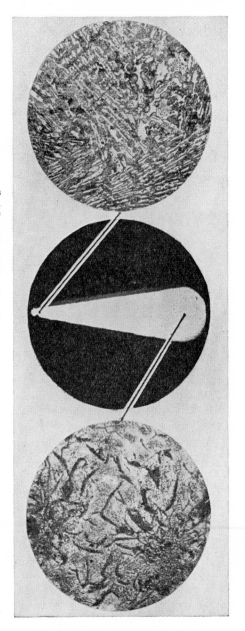

softer. Due to the chilling effect exerted by the mould, most casting have a hard white skin on the surface. This is often noticeable when taking the first "cut" in a machining operation.

In casting thin sections, then, it is necessary to choose an iron o rather coarser grey fracture than is required in the finished casting That is, the iron must have a higher silicon content than that used fo the production of castings of heavy section.

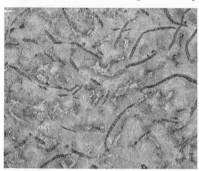
[Courtesy of B.C.I.R.A.
15.3A.—A Phosphoric Grey Cast Iron.
Flakes of graphite (dark) and patches of phosphide eutectic (light) in a matrix of pearlite. × 100. Etched in 4% Picral.

[Courtesy of B.C.I.R.A
15.3B.—An "Inoculated" Grey Cas Iron.
Note the refinement produced in th structure. × 60. Etched in 4% Picral.

PLATE 15.3

The Microstructure of Cast Iron

15.40. Neglecting patches of phosphide eutectic already mentioned the following structures are possible for cast irons of different composition and treatment:

(*a*) *Primary Cementite and Pearlite Only*. This type of structure is typical of the hard, white, low-silicon, high-sulphur irons, and i found also in other types of iron which have been chilled.

(*b*) *Primary Cementite, Graphite and Pearlite*. Mottled irons, in which some of the primary cementite has decomposed, forming graphite, are of this type.

(*c*) *Graphite and Pearlite*. This structure is typical of a grey high-duty iron in which all the primary cementite has transformed to graphite.

(*d*) *Graphite, Pearlite and Ferrite*. Coarser grey iron which will be weaker and softer.

(*e*) *Graphite and Ferrite*. Here all the pearlitic cementite, as well as the primary cementite, has broken down to graphite. This is usually due to a high silicon content. Such a cast iron will be very soft and easily machined. The ferrite present will contain dissolved silicon and manganese.

In addition to the phases enumerated above, iron phosphide, Fe_3P, as previously mentioned, may be present in the microstructure. In white or mottled iron it will be present as a ternary eutectic with cementite and ferrite, whilst in coarse grey irons of the type (*e*) above, a binary eutectic with ferrite will be formed. This binary eutectic contains only 10% phosphorus, so that an overall 1% phosphorus in the iron may produce sufficient phosphide eutectic to account for 10% by volume of the resulting structure. The embrittling effect of the phosphide eutectic network will then be evident.

The Growth of Cast Irons

15.50. The so-called "growth" of cast irons is caused by the breakdown of pearlitic cementite to ferrite and graphite when the iron is heated in the region of 700° C. This causes an increase in volume, which is further amplified as hot gases penetrate into the graphite cavities and oxidise the ferrite. Stresses are set up, and these lead to warping and the formation of cracks at the surface. Cast-iron moulds used for the casting of non-ferrous ingots develop surface cracks in this way, giving an effect reminiscent of crazy paving. Ultimately trouble will arise due to dressing oil collecting in these cracks.

Certain alloy cast irons have been developed to resist growth. "Silal" is relatively cheap and contains about 5·0% silicon with low carbon, so that its structure consists entirely of ferrite and graphite, and no cementite is present which can cause growth. Unfortunately "Silal" is rather brittle, so that, where the higher cost is justified, the alloy "Nicrosilal" can be used with advantage. This is an austenitic nickel–chromium cast iron (see Table 15.2).

Varieties and Uses of Ordinary Cast Iron

15.60. Foundry irons can be classified in the following groups according to their properties and uses:

15.61. Fluid Irons, which can be used where mechanical strength is of secondary importance, and in which high fluidity is obtained by means of high silicon (2·5–3·5%) and high phosphorus (up to 1·5%) contents. Both silicon and phosphorus improve fluidity, and a high silicon content will further ensure that thin sections will have a reasonably tough grey structure, since silicon prevents deep chilling. Such irons were used extensively at one time for the manufacture of lampposts, fireplaces and railings, but they have now been replaced by other materials for uses such as these.

15.62. Engineering Irons must have a reasonable mechanical strength, and the best type of microstructure is one containing small graphite flakes in a matrix of pearlite. An iron of this type will possess the best all-round mechanical properties and also good machinability. The silicon content will depend on the thickness of section to be cast, but in general it will not exceed 2·5% for castings of thin section, and may be as low as 1·2% for castings of heavy section.

The phosphorus content must be kept low where a shock-resistant casting is necessary, though up to 0·8% phosphorus may be present in some cases in the interests of fluidity. Sulphur must also be kept low (below 0·1%), as it leads to segregation, hard spots and general brittleness.

Irons of this type can have hard, wear-resistant, chilled surfaces introduced at various parts of the casting if desired. This is done by inserting chilling fillets at appropriate points in the sand mould. These fillets cause rapid cooling, and hence a layer of white iron on the surface of the casting at these points. At the same time the core of the casting remains grey and tough.

15.63. Heavy Castings do not require a high silicon content, as there is little danger of chilling. Usually, silicon is no higher than 1·5%, with up to 0·5% phosphorus for such castings as columns and large machine frames. In castings, such as ingot moulds, which are exposed to high temperatures a close-grained, non-phosphoric iron should be used.

High-strength Cast Irons

15.70. These are cast irons in which the properties of ordinary grey iron have been improved either by slight alterations in composition or by modifications in the casting technique.

15.71. Semi-steel is made by melting steel scrap along with pig iron in a cupola. This steel will dilute the quantities of silicon, manganese, sulphur and phosphorus in the resulting product. The final carbon content will not be greatly affected, however, since roughly the same amount (about 3·5% of the total charge) is always dissolved from the coke.

15.72. Inoculated Cast Iron. If an ordinary cast iron is superheated to, say, 1 500° C before casting, the formation of graphite is delayed and does not begin until well below the solidification range. This leads to the production of very small graphite flakes, which is a desirable feature. Unfortunately, much brittle cementite is also retained since graphitisation is incomplete under these conditions of cooling. Hence the super-heated iron is "inoculated" with some refractory material which will produce many small particles or nuclei, which, in turn, will promote the formation of fine graphite flakes (Plate 15.3B.) "Meehanite" (a registered proprietary name) is an iron which would normally solidify "white" but which has been inoculated with calcium silicide just before casting. Ferrosilicon, graphite and various commercial inoculants are used in other "trademarked" cast irons. In all of these products improved mechanical properties are obtained as a result of the small-flaked graphite induced by inoculation.

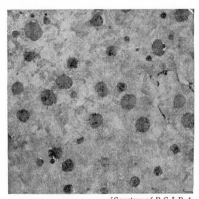

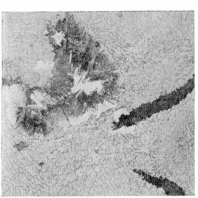

[Courtesy of B.C.I.R.A.   *[Courtesy of B.C.I.R.A.*

15.4A.—A Spheroidal-graphite Cast Iron.   15.4B.—An Acicular (or Martensitic) Iron
Here the graphite has been made to   Containing Molybdenum.
precipitate to nodular form by adding a   × 600. Etched in 4% Picral.
nickel–magnesium alloy. × 100. Etched
in 4% Picral.

PLATE 15.4

15.73. Spheroidal-graphite (S.G.) Cast Iron. (Also known as "nodular iron" or "ductile iron".) In a normal grey cast iron the graphite flakes are long and thin, and tend to be pointed at their ends. These long, thin flakes, having negligible tensile strength, act as discontinuities in the structure; whilst the sharp-pointed ends of the flakes introduce points of stress concentration. In S.G. cast iron the graphite flakes are replaced by spherical particles of graphite (Plate 15.4A), so that the metallic matrix is much less broken up, and the sharp "stress raisers" are eliminated.

The formation of this spheroidal graphite is effected by adding small amounts of cerium or magnesium to the molten iron just before casting. Since both of these elements have strong carbide-forming tendencies, the silicon content of the iron must be high enough (at least 2·5%) in order to prevent the formation, by chilling, of white iron in thin sections. Magnesium is the more widely used, and is usually added (as a nickel–magnesium alloy) in amounts sufficient to give a residual magnesium content of 0·1% in the iron. S.G. cast irons produced by the magnesium process have tensile strengths of up to 775 N/mm².

Recent patents claim the production of S.G. iron using the following substances instead of cerium or magnesium: calcium, calcium carbide, calcium fluoride, lithium, strontium, barium and the gas argon.

TABLE 15.1.—*Compositions and Uses of Some Typical Cast Irons*

Composition (%)					Uses
C	Si	Mn	S	P	
3·50	1·15	0·8	0·07	0·10	Ingot moulds and heat-resisting castings
3·30	1·90	0·65	0·08	0·15	Motor brake drums
3·25	2·25	0·65	0·10	0·15	Motor cylinders and pistons
3·25	2·25	0·50	0·10	0·35	Light machine castings
3·25	1·75	0·50	0·10	0·35	Medium machine castings
3·25	1·25	0·50	0·10	0·35	Heavy machine castings
3·60	1·75	0·50	0·10	0·80	Light and medium water pipes
3·40	1·40	0·50	0·10	0·80	Heavy water pipes
3·50	2·75	0·50	0·10	0·90	Ornamental castings requiring low strength

15.73. Lanz Perlit Iron is made by casting a low-silicon iron into a sand mould which has previously been pre-heated to 400° C. Thin sections would normally show a mottled structure if such an iron were cast into a cold mould, but the pre-heating reduces the rate of cooling, and so promotes the formation of fine graphite in these thin sections. At the same time the silicon content will regulate the growth of graphite in the heavier sections, so that the net result is that a casting of uneven section may have a remarkably uniform structure throughout. This method is also used for the permanent-mould casting of cylinder blocks and other high-quality castings.

Alloy Cast Irons

15.80. The microstructural effects which alloying elements have on a cast iron are, in most cases, similar to the effects these elements have on the structure of a steel. Alloying elements can therefore be used to improve mechanical properties, to refine grain, to increase the hardness by

TABLE 15·2.—*Compositions, Properties and Uses of Some Typical Alloy Cast Irons*

Type of iron	Composition (%)								Mechanical properties			Properties and uses
									Tensile strength		Brinell	
	C	Si	Mn	S	P	Ni	Cr	Other elements	N/mm²	tonf/in²		
"Chromidium"	3·2	2·1	0·8	0·05	0·17	—	0·32	—	278	18	230	Cylinder blocks, brake drums, clutch casings, etc.
Ni-Tensyl	2·8	1·4	—	—	—	1·5	—	—	355	23	230	An "inoculated" cast iron.
Wear-resistant	3·6	2·8	0·6	0·05	0·50	—	—	Vanadium 0·17	—	—	—	Piston rings for aero, automobile and diesel engines. Has wear-resistance and long life.
Ni—Cr-Mo iron	3·1	2·1	0·8	0·08	0·09	0·5	0·9	Molybdenum 0·9	371	24	300	Automobile crankshafts. Hard, strong and tough.
Heat-resistant	3·4	2·0	0·6	0·05	0·10	0·35	0·65	Copper 1·25	278	18	220	Good resistance to wear and heat-cracks. Used for brake drums and clutch plates.
Wear- and shock-resistant	2·9	2·1	0·7	0·05	0·10	1·75	0·10	Molybdenum 0·8 Copper 0·15	448	29	300	Crankshafts for diesel and petrol engines. Good strength, shock-resistance and vibration-damping capacity.
"Ni-hard"	3·3	1·1	0·5	—	—	4·5	1·5	—	—	—	600	A martensitic iron used to resist severe abrasion, e.g. chute plates in coke plant.
"Ni-resist"	2·8	1·3	—	—	—	21·0	2·0	—	170	11	140	Chemical plant handling sulphur or chloride solutions. Austenitic, non-magnetic and corrosion-resistant.
"Ni-resist"	2·9	2·1	1·0	0·05	0·10	15·0	2·0	Copper 6·0	216	14	130	Pump castings, handling concentrated brine; an austenitic, corrosion-resistant alloy.
"Nicrosilal"	2·0	5·0	1·0	—	—	18·0	5·0		263	17	330	An austenitic, corrosion-resistant alloy.
"No-mag"	2·8	1·5	6·0	—	—	11·0	—		232	15	210	An austenitic, corrosion-resistant alloy.
"Silal"	2·5	5·0	—	—	—	—	—	—	170	11	—	Resistant to high temperatures.

stabilising cementite and to stabilise, when desirable, martensitic and austenitic structures at room temperature.

15.81. Nickel is the most common alloying element used, and, as in steel, it tends to promote graphitisation. At the same time it has a grain-refining effect, so that whilst nickel will help to prevent chilling in thin sections, it will also prevent coarse grain in thick sections. Nickel also reduces the tendency of thin sections to crack.

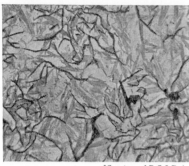

[*Courtesy of B.C.I.R.A.*]

15.5A.—A Martensitic Alloy Cast Iron.
Graphite in a matrix of martensite.
× 600. Etched in 4% Picral.

15.5B.—An Austenitic Alloy Cast Iron ("Ni-Resist").
Graphite and carbides in an austenite matrix. × 100. Etched in 4% Picral.

PLATE 15.5

15.82. Chromium is a strong carbide stabiliser, and so inhibits the formation of graphite. Moreover, the carbides formed by chromium are more stable and less likely to graphitise under the application of heat than is iron carbide. Irons containing chromium are therefore less susceptible to growth.

As in the low-alloy steels, the disadvantages attendant on the use of either nickel or chromium separately are overcome by using them in conjunction in the ratio of two to three parts of nickel to one part of chromium.

15.83. Molybdenum, when added in small amounts, dissolves in the ferrite, but in larger amounts forms double carbides. It increases the hardness of thick sections and promotes uniformity of microstructure. Impact values are also improved by amounts in the region of 0·5% molybdenum.

15.84. Vanadium promotes heat-resistance in cast irons, in so far as the stable carbides which it forms do not break down on heating. Strength and hardness are increased, particularly when vanadium is used in conjunction with other alloying elements.

15.85. Copper is only sparingly soluble in iron, and has a very slight graphitising effect. It has little influence on the mechanical properties, and its main value is in improving the resistance to atmospheric corrosion.

15.86. Martensitic irons, which are very useful for resisting abrasion, usually contain 4–6% nickel and about 1% chromium. Such an alloy is "Ni-hard", included in Table 15.2. It is martensitic in the cast state, whereas alloys containing rather less nickel and chromium would need to be oil-quenched in order to obtain a martensitic structure. Austenitic irons usually contain between 11 and 20% nickel and up to 5% chromium. These are corrosion-resistant, heat-resistant, non-magnetic alloys.

Malleable Cast Irons

15.90. These are irons which have been cast to shape in the ordinary way, but in which ductility and malleability have been increased from almost zero to a considerable amount by subsequent heat-treatment. The names of the two original malleabilising processes, the Blackheart and the Whiteheart, refer to the colour of a fractured section after heat-treatment has been completed. A more recent process for the manufacture of pearlitic malleable iron aims at the production of castings which, when suitably treated, will have tensile strengths of up to 775 N/mm². In all three processes the original casting is of white iron, which will be very brittle before heat-treatment. This white structure is achieved by keeping the silicon content low, usually not much more than 1·0%, whilst in the Whiteheart process some sulphur, too, is permissible, thus increasing still further the stability of the cementite.

15.91. Blackheart Malleable Iron castings are manufactured from white iron of which the following composition is typical:

Carbon	2·5%
Silicon	1·0%
Manganese	0·4%
Sulphur	0·08%
Phosphorus	0·1%

In the original process the castings were fettled and placed in white-iron containers along with some non-reactive material, such as gravel or cinders, to act as a mechanical support. The lids were luted on to exclude air and the containers loaded into an annealing furnace which was fired by coal, gas, fuel oil or pulverised fuel. Treatment temperatures varied between 850 and 950° C and were dependent upon the desired quality of product and the analysis of the original casting. The duration of heat-treatment was between 50 and 170 hours, again depending upon the type of casting and the analysis of the iron. The development of modern annealing furnaces in which castings need no longer be packed in an insulating material has resulted in a reduction in heating and cooling-off times, so that malleabilising can now be effected in 48 hours or less. Such furnaces are of the continuous type in which a moving hearth carries the castings slowly through a long furnace of small cross-section and in which a controlled non-oxidising atmosphere is circulated.

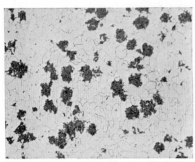

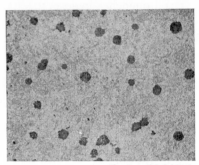

15.6A.—Blackheart Malleable Cast Iron.

15.6B.—Whiteheart Malleable Cast Iron.

"Rosettes" of temper carbon in a matrix of ferrite. × 60. Etched in 4% Picral.

Micrograph taken from the *core* of the material and showing rosettes of temper carbon in a matrix of pearlite. × 60. Etched in 4% Picral.

PLATE 15.6

The effect of this prolonged annealing is to cause the cementite to break down, but instead of forming coarse graphite flakes, the carbon is precipitated in the form of small "rosettes" of "temper carbon". A fractured section will thus be black, hence the term "blackheart". The final structure, which consists entirely of ferrite and finely divided temper carbon, is soft, readily machinable and almost as ductile as cast steel. Blackheart malleable castings find particular application in the automobile industries because of the combination of castability, shock-resistance and ease of machining they afford. Typical uses include rear-axle housings, wheel hubs, brake shoes, pedals, levers and door hinges.

15.92. Whiteheart Malleable Iron is manufactured from white iron of which the following composition is typical:

Carbon .	3·3%
Silicon .	0·6%
Manganese	0·5%
Sulphur	0·25%
Phosphorus	0·1%

In this process the castings are heated in contact with some oxidising material, such as hematite ore, for between 70 and 100 hours at a temperature of about 1 000° C.

During annealing, the carbon at the surface of the casting is oxidised by contact with the hematite ore, and lost as carbon dioxide. This causes more carbon to diffuse outwards from the core, and, in turn, this is lost by oxidation. Thus, after treatment a thin section may be completely ferritic, and on fracture present a steely white appearance; hence the name "whiteheart". Heavier sections will not be completely decarburised, and so, whilst the outer layer is ferritic, this will merge into an inner zone containing some pearlite, and at the extreme core some nodules of temper carbon. The surface layer may exhibit some oxide penetration.

Thin-sectioned components requiring high ductility are often made in the form of whiteheart malleable castings. Examples include fittings for gas, water, air and steam pipes; bicycle- and motorcycle-frame fittings; parts for agricultural machinery; switchgear equipment; and parts for textile machinery.

15.93. Pearlitic Malleable Iron is produced from raw material similar in composition to that of blackheart malleable iron, though in some cases alloying elements may be used to stabilise the required pearlitic matrix in the final structure. When an unalloyed iron is used it is first malleabilised either fully or partially at about 950° C to cause adequate breakdown of the primary cementite. It is then reheated to 900° C so that carbon will dissolve in the austenite present at that temperature. Subsequent treatment consists either of air-cooling or of oil-quenching and tempering to produce a pearlitic type of matrix.

If an alloyed iron is used it is often given a normal malleabilising treatment, the presence of carbide-stabilising elements causing retention of the pearlitic matrix during this process.

The final structure will consist of rosettes of temper carbon in a pearlitic matrix. Tensile strengths in the region of 775 N/mm^2 are possible and in many respects the product can compete with cast steel, despite the extra costs of the heat-treatment processes.

15.94. The choice between ordinary grey iron, S.G. iron and malleable iron for a specific application is, as with most materials problems, dictated by both economics and technical requirements. Grey iron is, of course, lowest in cost and also the easiest in which to produce sound castings. With regard to the relative merits of the two high-strength competitors, S.G. iron and malleable iron, the choice may be less easy. Malleable iron is the more difficult to cast since at this stage it will have a "white" structure. This also involves limitations as to cross-section as compared with S.G. iron. Moreover, the cost of the malleabilising process makes the product a little more expensive than S.G. iron. However, it is sometimes necessary to anneal S.G. iron in order to improve its machinability, and in this case malleable iron may prove more attractive cost-wise.

RECENT EXAMINATION QUESTIONS

1. Discuss the role of microstructure in determining the general properties of cast irons.
                    (University of Aston in Birmingham, Inter B.Sc.(Eng.).)

2. Discuss the importance of control of graphite formation in cast irons.
                    (University of Glasgow, B.Sc.(Eng.).)

3. What factors control the structure of cast iron? Explain how these principles are used in the foundry to control the strength of grey-iron castings.   (University of Aston in Birmingham, Inter B.Sc.(Eng.).)

4. Discuss the influence of the following elements on the structure and properties of cast iron: (i) silicon; (ii) manganese; (iii) sulphur; (iv) phosphorus.   (West Bromwich Technical College.)

5. Describe briefly the methods available to control the size and shape of graphite in cast iron, and explain how the mode of occurrence of the graphite affects the properties.   (U.L.C.I., Paper No. C111–2–4.)

6. Comment on the differences in structure, properties and uses of grey cast iron and white cast iron.
       Describe one method by which the properties of cast iron may be improved.   (Derby and District College of Technology.)

7. Explain fully what is meant by the term *growth* as applied to cast irons. Discuss methods which are used to overcome this phenomenon.
                    (U.L.C.I., Paper No. B111–2–4.)

8. Describe the effects of composition and heat-treatment on the structure of cast irons and show how this affects mechanical properties.
                    (Nottingham Regional College of Technology.)

9. Cast iron is a cheap, easily produced engineering material, but ordinary grey cast iron has relatively poor mechanical properties. Discuss, in general terms, the methods employed to overcome these limitations.
                    (S.A.N.C.A.D.)

10. Discuss the effect of chemical composition and cooling rate on the structure and properties of cast irons. Briefly describe one method for producing (*a*) malleable iron and (*b*) nodular iron.
                    (Rutherford College of Technology, Newcastle upon Tyne.)

11. Discuss briefly the main factors which govern the state of carbon and the mode of graphite precipitation in cast irons. Proceeding from these factors, and taking into account the general effects of alloying elements, explain the origin of the microstructure in the following materials:

(a) spheroidal-graphite Ni-resist cast iron;
(b) whiteheart and blackheart malleable irons;
(c) martensitic Ni-hard cast iron.

(Lanchester College of Technology, Coventry, H.N.D.)

12. Give an account of the production, microstructure, properties and uses of each of the following:

(a) grey cast iron;
(b) inoculated high-duty cast iron;
(c) whiteheart malleable cast iron.

(Carlisle Technical College.)

13. Account for the structural differences between ordinary grey cast iron and the following "special" irons:

(i) spheroidal-graphite iron;
(ii) acicular cast iron;
(iii) blackheart malleable cast iron.

What are the characteristic properties associated with each of the above special irons?   (S.A.N.C.A.D.)

14. Differentiate between white cast iron and whiteheart malleable cast iron. Describe the process by which one may be made into the other.

(Derby and District College of Technology, H.N.D. Course.)

15. How are malleable castings produced? What are the most important engineering properties of these materials?

(University of Aston in Birmingham, Inter B.Sc.(Eng.).)

16. Distinguish between blackheart and whiteheart malleable cast iron. Sketch and fully label a representative microstructure for each iron.

(Derby and District College of Technology.)

17. (a) Explain briefly the difference between malleable and spheroidal-graphite cast irons.

(b) Choose an application for which either material might be suitable and outline the factors which would be considered in making a decision which to use.   (C. & G., Paper No. 293/23.)

18. "Most cast irons consist of a dispersion of graphite in a steel-like matrix." Illustrate this by sketches of microstructures and discuss the variations in properties that can arise, by reference to: (a) pearlitic grey cast iron; (b) blackheart malleable iron; (c) S.G. cast iron.

Indicate briefly the treatment necessary to produce blackheart malleable iron and S.G. cast iron.

(Constantine College of Technology, H.N.D. Course.)

CHAPTER XVI

COPPER AND THE COPPER-BASE ALLOYS

16.10. Copper was undoubtedly the first metal to be used by Man. In many countries it is found in small quantities in the metallic state and, being soft, it was readily shaped into ornaments and utensils. Moreover, many of the ores of copper can easily be reduced to the metal, and since these ores often contain other minerals, it is very probable that copper alloys were produced as the direct result of smelting. It is thought that bronze was accidentally produced in Cornwall by smelting ores containing both tin- and copper-bearing minerals in the camp fires of ancient Britons.

In ancient times, before Man understood the role played by carbon when present in iron, bronze was the best material for making knives and other cutting implements. Although the "Bronze Age" is generally assumed to have followed the "Stone Age", some archaeologists believe that a "Copper Age" separated the two. It is fairly certain that the Egyptians used copper compounds for colouring glazes some 15 000 years ago, and they may have been the first to extract the metal. By 400 B.C. the knowledge accumulated by early Egyptian metallurgists had spread to Europe.

16.11. At the end of the eighteenth century practically all the world's requirement of copper was smelted in Swansea, the bulk of the ore coming from Cornwall, Wales or Spain. Deposits of copper ore were then discovered in the Americas and Australia, and subsequently imported for smelting in Swansea. Later it was found more economical to set up smelting plants near to the mines, and Britain ceased to be the centre of the copper industry.

Climbers and ramblers who ascend Snowdon by the popular "Pyg" track will pass the ruins of an old copper-mine, perched on the mountain side below the Crib Goch ridge. This small mine was one of many located in the mountainous areas of Britain, but all have long since ceased production. Their output was insignificant compared with that of a modern mine like that at Chuquicamata, in Chile.

To-day the United States is by far the greatest producer of copper, as of many other metals, but increased production in Chile, Canada, Zambia and Katanga has taken from the United States her former control of the world's markets. In Europe the U.S.S.R., Jugoslavia and

Spain are leading producers. The bulk of Britain's supply of copper comes from Zambia and Canada.

Properties and Uses of Copper

16.20. A very large part of the world's production of metallic copper is used in the unalloyed form, mainly in the electrical industries. Copper has a very high specific conductivity, and is, in this respect, second only to silver, to which it is but little inferior. When relative costs are considered, copper is naturally the metal used for industrial purposes demanding high electrical conductivity.

16.21. The Electrical Conductivity of Copper. The International Standard of Electrical Resistance for annealed copper was adopted in 1913, and copper which reached this standard was said to be 100% conductive. The fact that, working to the 1913 standard, figures of 101 and 102% are frequently quoted is explained by reason of improvements that have taken place in the production of copper since 1913.

As indicated in Fig. 16.1, the presence of impurities reduces electrical conductivity. To a less degree, cold-work has the same effect. Reduction in conductivity caused by the presence of some elements in small amounts is not great, so that up to 1% cadmium, for example, is added to telephone wires in order to strengthen them. Such an alloy, when hard-drawn, has a tensile strength of some 460 N/mm² as compared with 340 N/mm² for hard-drawn, pure copper, whilst the electrical conductivity is still over 90% of that for soft pure copper. Other elements have pronounced effects on conductivity; 0·04% phosphorus, for example, will reduce the electrical conductivity to about 75% of that for pure copper.

16.22. The Commercial Grades of Copper include both furnace-refined and electrolytically refined metal. High-conductivity copper (usually referred to as O.F.H.C. or "oxygen-free high conductivity") is of the highest purity, and contains at least 99·9% copper. It is used where the highest electrical and thermal conductivities are required, and is copper which has been refined electrolytically.

Fire-refined grades of copper can be either "tough pitch" or "de-oxidised" according to their subsequent application. The former contains small amounts of oxygen (present as cuprous oxide, Cu_2O) absorbed during the manufacturing process. It is usually present in amounts of the order of 0·04–0·05% oxygen (equivalent to 0·45–0·55% cuprous oxide). The cuprous oxide is present as tiny globules, which are part of a copper–cuprous oxide eutectic. These globules have a negligible effect as far as electrical conductivity, and most of the other properties, are concerned. The presence of cuprous oxide is, in fact, often advantageous, since harmful impurities, such as bismuth, appear to collect as oxides associated with the cuprous oxide globules, instead of occurring as brittle intercrystalline films, as they would otherwise do.

For processes such as welding and tube-making, however, the existence of these globules is extremely deleterious, since reducing atmospheres containing hydrogen cause "gassing" of the metal. Hydrogen is soluble in solid copper, so that it comes into contact with the subcutaneous globules of cuprous oxide, reducing them thus:

$$Cu_2O + H_2 \rightleftharpoons 2Cu + H_2O$$

Although this is what is called a "reversible reaction", the high concentration of hydrogen relative to cuprous oxide makes the reaction proceed to the right, and the water formed is present as steam, which is almost insoluble in solid copper. It is therefore precipitated at the crystal boundaries, thus, in effect, pushing the crystals apart and reducing the ductility by as much as 85% and the tensile strength by 30–40%. Under the microscope "gassed" tough-pitch copper is recognised by the thick grain boundaries, which are really minute fissures, and by the absence of cuprous oxide globules.

For such purposes as welding, therefore, copper is deoxidised before being cast by the addition of phosphorus, which acts in the same way as the manganese used in deoxidising steels. A small excess of phosphorus, of the order of 0·04%, dissolves in the copper after deoxidation, and this small amount is sufficient to reduce the electrical conductivity by as much as 25%. So, whilst copper which is destined for welding or thermal treatment in hydrogen-rich atmospheres should be of this type, copper deoxidised by phosphorus would be unsuitable for electrical purposes, where either electrolytic copper or good-quality tough-pitch copper must be used.

16.23. Up to 0·5% arsenic was added to much of the copper used in the construction of locomotives. This addition considerably increased the strength at elevated temperatures by raising the softening temperature from about 190° C for the pure metal to 550° C for arsenical copper. This made arsenical copper useful in the manufacture of steam locomotive fire-boxes, boiler tubes and rivets, since the alloy, whilst being stronger at high temperatures, still had a high thermal conductivity.

The addition of 0·5% lead or tellurium imparts free-cutting properties to copper (20·24) and provides a material which can be machined to close tolerances whilst still retaining 95% of the conductivity of pure copper.

16.24. The Effects of Impurities on the electrical properties of copper have already been mentioned. Quite small amounts of some impurities will also cause serious reductions in the mechanical properties.

Bismuth is possibly the worst offender, and even as little as 0·002% will sometimes cause trouble, since bismuth is insoluble in amounts in excess of this figure, and, like ferrous sulphide in steel, collects as brittle films at the crystal boundaries. Antimony produces similar effects and, in particular, impairs the cold-working properties.

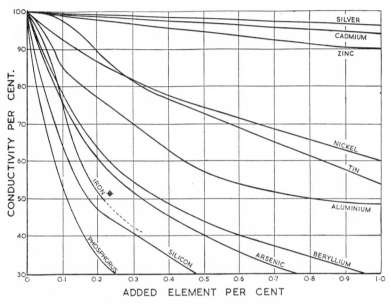

FIG. 16.1.—The Effect of Impurities on the Electrical Conductivity of Copper.

Selenium and tellurium make welding difficult, in addition to reducing the conductivity and cold-working properties; whilst lead causes hot-shortness, since it is insoluble in copper and is actually molten at the hot-working temperatures.

The Copper-base Alloys

The copper-base alloys include the brasses and bronzes, the latter being copper-rich alloys containing either tin, aluminium, silicon or beryllium; though the tin bronzes are possibly the best known.

The Brasses

16.30. The brasses comprise the useful alloys of copper and zinc containing up to 45% zinc, and constitute one of the most important groups of non-ferrous engineering alloys.

As shown by the equilibrium diagram (Fig. 16.2), copper will dissolve up to 32·5% zinc at the solidus temperature of 902° C, the proportion increasing to 39·0% at 454° C. With extremely slow rates of cooling, which allow the alloy to reach structural equilibrium, the solubility of zinc in copper will again decrease to 35·2% at 250° C. Diffusion is very sluggish, however, at temperatures below 450° C, and with ordinary industrial rates of cooling the amount of zinc which can remain in solid solution in copper at room temperature is about 39%. The solid solution so formed is represented by the symbol α. Since this solid solution is of the disordered type, it is prone to the phenomenon of coring, though this is not extensive, indicated by the narrow range between liquidus and solidus.

If the amount of zinc is increased beyond 39% another phase, β', will appear in the microstructure of the slowly cooled brass. This phase is hard, but quite tough at room temperature, and plastic when it changes to its modification β above 454° C. Further increases in the zinc content beyond 50% cause the appearance of the phase γ in the structure. This is extremely brittle, rendering alloys which contain it unsuitable for engineering purposes.

Due to coring effects, an alloy which is nominally α-phase in structure may contain some β'-phase in the as-cast condition. This will depend upon the rate of cooling and the nearness of the composition of the alloy to the α/α + β' phase boundary. Such β'-phase will usually be absorbed fairly quickly on subsequent mechanical working whilst hot. The α + β' alloys usually have a Widmanstätten structure (11.53) on cooling, due to the manner in which the particles of the α-phase precipitate as the alloy cools from out of the β-phase area.

16.31. As mentioned above, the α-phase is quite soft and ductile at room temperatures, and for this reason the completely α-phase brasses are excellent cold-working alloys. The presence of the β'-phase, however, makes them rather hard and with a low capacity for cold-work; but since the β-phase is plastic at red heat, the α + β' brasses are best shaped by hot-working processes, such as forging or extrusion. The α-phase tends to be rather hot-short within the region of 30% zinc and between the temperatures of 300 and 750° C, and is therefore much less suitable as a hot-working alloy unless temperatures and working conditions are strictly controlled. The α-phase would also introduce difficulties during the extrusion of the α + β' alloys were it not for the fact that it is absorbed into the β-phase when the 60–40 composition (one of the most popular alloys of this group) is heated to a point above the α + β/β phase boundary in the region of 750° C, thus producing a uniform plastic structure of β-phase only. The α-phase is usually in

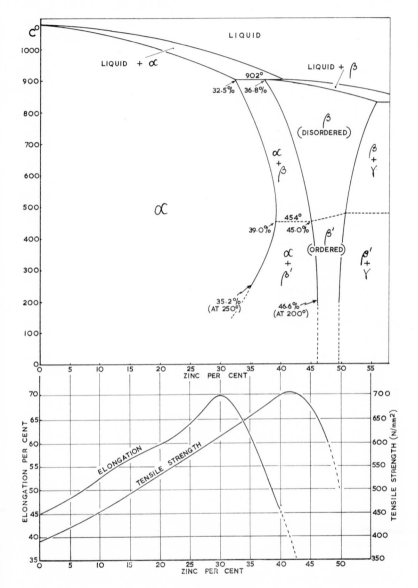

FIG. 16.2.—The Copper–Zinc Thermal-equilibrium Diagram.

The lower diagram indicates the relationship between composition and mechanical properties.

process of being precipitated whilst hot-working is taking place, so that, instead of the Widmanstätten structure being formed again as the temperature falls, it is replaced by a refined granular α + β′ structure which possesses superior mechanical properties to those of the directional Widmanstätten structure. The needle-shaped crystals of the α-phase are prevented from forming by the mechanical disturbances which accompany the working process.

Thus the brasses can conveniently be classified according to whether they are hot-working or cold-working alloys.

16.32. The Cold-working α-brasses. These are generally completely α-phase in structure, though a limited amount of cold-work may also be applied to those α + β′ alloys which contain only small amounts of the β′-phase. The α + β′ alloys proper are, however, shaped by hot-working processes in the initial stages, and such cold-work as is applied is merely for finishing to size or to produce the correct degree of work-hardening for subsequent use.

The α-brasses are useful mainly because of their high ductility, which reaches a maximum at 30% zinc, as shown in Fig. 16.2. Such alloys need to be of very high purity, since the inclusion of even small amounts of impurity will lead to a big loss in ductility. The α-brasses are also rather sensitive to annealing temperatures and, since grain growth is rapid at elevated temperatures, it is easy to "burn" the alloy. (This trade term should not be confused with oxidation of the metal. It is widely used industrially to signify overheating.) α-Brasses should be annealed at about 600° C. If overheated to 750° C, grain growth is so rapid that on subsequent pressing an "orange peel" effect is apparent on the surface. This is due to coarse grain being large enough to be visible on the surface.

16.33. Hard-drawn brasses are subject to "season cracking". In such brasses any corrosion which takes place is usually intercrystalline, so that cracks will ultimately arise as internal stresses are relieved by fracture at the relatively weak grain boundaries. This effect can often be seen in old brass electric-light switch-covers, particularly when they have been subjected to corrosion in a damp atmosphere. Season cracking can be avoided by giving the component a low-temperature, stress-relief anneal at about 250° C, after fabrication.

16.34. Other elements may be added in small amounts to the α-brasses in order to improve either corrosion-resistance or mechanical properties. *Tin* is added in amounts up to 1·0% in order to improve corrosion-resistance, particularly in naval brass and "Admiralty" brass for condenser tubes. Such small quantities of tin are retained in solid solution. Alternatively, small amounts of *arsenic* (0·01–0·05%) may be

16.1A.—70/30 Brass, as Cast.

Cored crystals of α solid solution. ×35. Etched in ammonia—hydrogen peroxide.

16.1B.—70/30 Brass, in the Cold-worked Condition.

Distorted α crystals showing strain bands. ×100. Etched in ammonia—hydrogen peroxide.

16.1C.—70/30 Brass, Cold-worked and then Annealed.

Twinned α crystals. All coring removed. ×100. Etched in ammonia—hydrogen peroxide.

PLATE 16.1

added to 70–30 brass destined for the manufacture of condenser tubes, as it is said to improve corrosion-resistance and inhibit dezincification. *Lead*, in amounts of the order of 2·0%, is added to improve machinability. It is insoluble in brass, and exists as small globules, which cause local fracture during machining (20.24). *Aluminium* is sometimes added, in amounts up to 2·0%, to brass for the manufacture of naval condenser tubes, since it imparts excellent corrosion-resistance to the alloy. *Nickel* is retained in solid solution, and small amounts may be added to brass to improve corrosion-resistance. The now obsolete twelve-sided threepenny pieces were made from a brass containing 20% zinc, 1% nickel and the balance copper.

16.35. The Hot-working $\alpha + \beta'$ brasses. The hot-working alloys usually contain not more than 60% copper, though some of the α-brasses are often hot-worked in the initial "breaking-down" stages. The $\alpha + \beta'$ alloys, however, are shaped almost entirely by hot-working.

The only important "straight" brass in this group is 60–40 brass, or Muntz metal. As already mentioned, the α-phase is entirely absorbed into the β-phase when the alloy is heated to about 750° C, so that the best hot-working temperature range is while cooling between 750 and 650° C, during which range the α-phase is being deposited. The mechanical-working process breaks down the α-phase into small particles as it is deposited, and prevents the reintroduction of the coarse Widmanstätten structure.

16.36. "Free-cutting" Brasses of the 60–40 type contain about 2·0% lead, whilst a similar alloy of higher purity is used extensively for hot-forging where a machining operation is to follow.

16.37. High-tensile Brasses are misleadingly called "Manganese Bronze", possibly because the manganese they often contain produces an oxidised-bronze effect on the surface of extruded rod. These brasses contain 54–62% copper, up to 7·0% "other elements" and the balance zinc. In addition to being hot-working alloys, they are also used in the cast form for such applications as marine propellers, water-turbine runners, rudders, gun mountings and sights, and locomotive axle boxes. The wrought sections are used for pump rods, and for stampings and pressings for automobile fittings and switch gear. The tensile strength is increased, by the addition of these "other elements", to as much as 700 N/mm^2 in the chill-cast or forged condition. The additions, usually in amounts up to 2·0% each, include manganese, iron and aluminium; whilst up to 1·0% tin may also be included to improve corrosion-resistance.

Some of the more important brasses are included in Table 16.1.

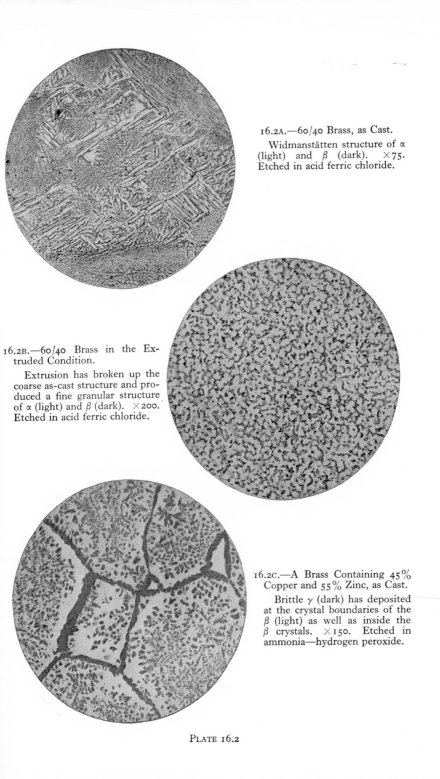

16.2A.—60/40 Brass, as Cast.

Widmanstätten structure of α (light) and β (dark). ×75. Etched in acid ferric chloride.

16.2B.—60/40 Brass in the Extruded Condition.

Extrusion has broken up the coarse as-cast structure and produced a fine granular structure of α (light) and β (dark). ×200. Etched in acid ferric chloride.

16.2C.—A Brass Containing 45% Copper and 55% Zinc, as Cast.

Brittle γ (dark) has deposited at the crystal boundaries of the β (light) as well as inside the β crystals. ×150. Etched in ammonia—hydrogen peroxide.

PLATE 16.2

TABLE 16.1.—*Typical Commercial Brasses*

B.S. 2870/5 designation	Composition (%)					Condition	Typical mechanical properties							Characteristics and uses
	Cu	Zn	Pb	Sn	Other elements		0·1% Proof stress		Tensile strength		Elongation (%)	Hardness (V.P.N.)		
							N/mm²	tonf/in²	N/mm²	tonf/in²				
CZ101	90	10	—	—	—	Annealed / Hard	77·2 / 463	5 / 30	278 / 510	18 / 33	55 / 4	60 / 150		*Gilding Metal.* Used for architectural metal-work, imitation jewellery, etc., on account of its gold-like colour and its ability to be brazed and enamelled.
CZ106	70	30	—	—	—	Annealed / Hard	77·2 / 510	5 / 33	324 / 695	21 / 45	70 / 5	65 / 185		*Cartridge Brass.* Deep-drawing brass having the maximum ductility of the copper-zinc alloys. Used particularly for the manufacture of cartridge and shell cases.
CZ107	65	35	—	—	—	Annealed / Hard	92·7 / 510	6 / 33	324 / 695	21 / 45	65 / 4	65 / 185		"*Standard*" *Brass.* A good general-purpose cold-working alloy useful where the high ductility of the 70-30 quality is not necessary. Used for press-work and limited deep-drawing.
CZ108	63	37	—	—	—	Annealed / Hard	92·7 / 541	6 / 35	340 / 726	22 / 47	55 / 4	65 / 185		"*Common*" *Brass.* A general-purpose alloy suitable for simple forming operations by cold-work.
CZ109	60	40	—	—	—	Hot-rolled	108	7	371	24	40	75		*Yellow or Muntz Metal.* Hot-rolled plate used for tube plates of condensers and similar purposes. Can be cold-worked to a limited extent. Also as extruded rods and tubes.
CZ120	59	39	2·0	—	—	Annealed / Hard	92·7 / 463	6 / 30	371 / 618	24 / 40	45 / 5	75 / 190		*Clock Brass.* Used for the plates and wheels in clock and instrument manufacture. Lead imparts free-cutting properties.
CZ121	58	39	3·0	—	—	Extruded rod	139	9	448	29	30	100		*Free-cutting Brass.* Most suitable material for high-speed machining, but can be only slightly deformed by bending, etc. A 61% Cu alloy has greater impact strength.
CZ111	70	29	—	1·0	0·01-0·05 As	Annealed / Hard	77·2 / 432	5 / 28	340 / 587	22 / 38	70 / 10	65 / 175		*Admiralty Brass.* A standard composition for condenser tubes. Tin gives improved corrosion-resistance over plain 70-30 brass. Arsenic inhibits dezincification.
CZ112	62	37	—	1·0	—	Extruded	154	10	417	27	35	100		*Naval Brass.* For structural applications and forgings. Tin reduces corrosion, especially in sea-water.
CZ110	76	22	—	—	0·02-0·06 As 2·0 Al	Annealed / Hard	108 / 463	7 / 30	371 / 618	24 / 40	70 / 8	65 / 175		*Aluminium Brass.* Possesses excellent corrosion-resistance, and is a popular alloy for condenser tubes.
CZ114	58	Rem.	—	—	Total 7·0 max.	Grade A / Grade B	232 min. / 278 min.	15 min. / 18 min.	463 min. / 541 min.	30 min. / 35 min.	20 min. / 15 min.	—		*High-tensile Brass.* Some of the alloys are lightly cold-worked to give the necessary properties and hence, if heated, subsequently show a loss of tensile strength.

The Tin Bronzes

16.40. Bronzes containing approximately 10% tin were probably the first alloys to be used by Man. In Britain bronze articles almost four thousand years old have been found, and during the Roman occupation of Britain the copper-mines of Cumberland and Wales were in a state of rapidly increasing production. Centuries before the Roman invasion, however, Phoenician traders from Tyre and Sidon brought their ships to Cornwall in search of tin.

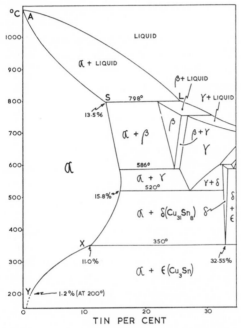

FIG. 16.3.—The Copper–Tin Equilibrium Diagram.

One of the significant factors in the early Roman conquests was undoubtedly the bronze sword, and it is thought that in even earlier times metal-workers realised that a high tin content, in the region of 10%, produced a hard bronze whilst less tin gave a softer alloy.

16.41. The relationship between the equilibrium diagram and the actual microstructure produced for a given alloy is rather more complex in the case of tin bronzes than with the brasses. The rate of diffusion of copper and tin into each other is much lower than it is with copper and zinc. This is indicated by the wide range of composition at any temperature, between the liquidus *AL* and the solidus *AS* (Fig. 16.3), and leading to a high degree of coring during the actual solidification process.

Moreover, structural changes below approximately 400° C take place in copper–tin alloys with extreme sluggishness. Both of these factors generally lead to the establishment of some microstructure other than that indicated by the equilibrium diagram for a cast bronze cooled to room temperature under normal industrial conditions.

In short, whilst the peritectic reaction ($\alpha \longrightarrow \beta$) at 789° C and the eutectoid transformations ($\beta \longrightarrow \alpha + \gamma$) at 586° C and ($\gamma \longrightarrow \alpha + \delta$) at 520° C will take place as indicated by the diagram during ordinary rates of cooling, the eutectoid transformation ($\delta \longrightarrow \alpha + \varepsilon$) at 350° C would occur only under conditions of extremely slow cooling, such as would never be encountered industrially. Hence the phase ε (Cu_3Sn) is never seen in the structure of a cast bronze containing more than 11·0% tin. Further, due to the slow rate of diffusion of copper and tin atoms below 350° C, the precipitation of ε from α in alloys containing less than 11·0% tin, in accordance with the phase boundary XY, will not occur. For practical purposes the reader can therefore ignore that part of the equilibrium diagram below 400° C and assume that whatever structure has been attained at 400° C will persist to room temperature under normal industrial rates of cooling.

16.42. As with brasses, the α-phase, being a solid solution, is tough and ductile, so that α-phase alloys can be cold-worked successfully. The δ-phase, however, is an intermetallic compound of composition equivalent to $Cu_{31}Sn_8$, and is a hard, brittle, blue substance, whose presence renders the $\alpha + \delta$ bronzes rather brittle. The δ-phase must, therefore, be absent from alloys destined for any degree of cold-work.

Due to heavy coring arising from slow rates of diffusion, cast alloys with as little as 6·0% tin will show particles of δ at the boundaries of the cored α crystals. The skeletons of the α-phase crystals will be much richer in copper than the nominal 94%, thus making the outer fringes correspondingly richer in tin to such an extent that the δ-phase is formed. In order to make such an alloy amenable for cold-work, the δ-phase can be absorbed by prolonged annealing (say six hours at 700° C), which will promote diffusion so that equilibrium is attained in accordance with the equilibrium diagram and a *uniform* α-phase structure produced. Subsequent air-cooling—or even furnace-cooling at the usual industrial rates—will be too rapid to permit precipitation of any of the ε-phase when the phase boundary XY is reached; and so the uniform α structure will be retained at room temperature. By using such initial heat-treatment to produce a uniform α structure it is possible to cold-work bronzes containing as much as 14% tin, though in general industrial practice only alloys with up to 7% tin are produced in wrought form.

The tin bronzes can be classified as follows:

16.43. Plain Tin Bronzes. These comprise both wrought and cast alloys, the former usually containing up to 7% tin and the latter as much as 18% tin. The wrought alloys contain none of the δ-phase, the absence of which makes them amenable to shaping by cold-working operations. These alloys are usually supplied as rolled sheet, drawn rod or drawn turbine blading.

The cast alloys are used mainly for bearings, since the structure fulfils the requirements for that duty, namely, hard particles of the δ-phase, which will resist wear, embedded in a matrix of the α-phase, which will resist shock.

16.44. Phosphor-bronzes. Most of the tin bronzes mentioned above contain small amounts of phosphorus (up to 0·05%) residual from de-oxidation which is carried out before casting. They are often incorrectly called phosphor-bronzes. True phosphor-bronze contains phosphorus as a deliberate alloying addition, generally present in amounts between 0·1 and 1·0%.

Wrought phosphor-bronzes contain up to 8·0% tin and up to 0·3% phosphorus, and, like the plain tin bronzes, are supplied in the form of rod, wire and turbine blading. The phosphorus not only increases the tensile strength but also, it is claimed, improves the corrosion-resistance.

Cast phosphor-bronzes contain up to 13·0% tin and up to 1·0% phosphorus, and are used mainly for bearings and other components where a low friction coefficient is desirable, coupled with high strength and toughness. The phosphorus is usually present in these alloys as copper phosphide, Cu_3P, a hard compound which forms a ternary eutectoid with the α- and δ-phases.

16.45. Bronzes Containing Zinc. These, again, comprise both wrought and cast alloys. The wrought alloys are used chiefly for the manufacture of coinage and contain up to 3·0% tin and up to 2·5% zinc. In recent years the tin content of British "copper" coinage has been reduced from the pre-war figure of 3·0% to as low as 0·5%. This began as a war-time measure when the Japanese occupation of Malaya cut off our main supplies of tin; and has continued since because of the high prices which have prevailed for the metal. The replacement of tin by some zinc cheapens the alloy, zinc being only about a tenth the price of tin. A subsidiary function of zinc is that, like phosphorus, it acts as a deoxidiser and forms zinc oxide, ZnO, which floats to the top of the melt. The zinc-bearing bronzes are all α-phase alloys with similar structures to those of straight tin bronzes of like compositions.

The best known of the cast alloys is "Admiralty gunmetal", or

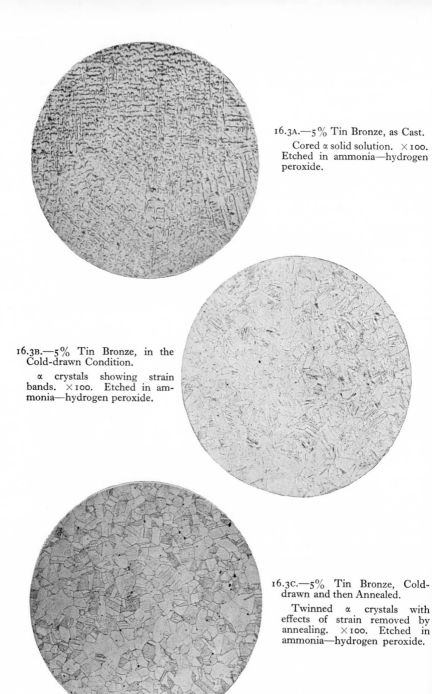

16.3A.—5% Tin Bronze, as Cast.

Cored α solid solution. ×100. Etched in ammonia—hydrogen peroxide.

16.3B.—5% Tin Bronze, in the Cold-drawn Condition.

α crystals showing strain bands. ×100. Etched in ammonia—hydrogen peroxide.

16.3C.—5% Tin Bronze, Cold-drawn and then Annealed.

Twinned α crystals with effects of strain removed by annealing. ×100. Etched in ammonia—hydrogen peroxide.

PLATE 16.3

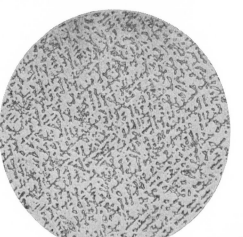

16.4A.—Phosphor Bronze Containing 10% Tin and 0·5% Phosphorus, as Cast.

Cored α (dark) with an infilling of α, δ and Cu₃P eutectoid. ×100. Etched in ammonia—hydrogen peroxide.

16.4B.—The Same Bronze at Higher Magnification Showing the Nature of the Eutectoid.

×1100. Etched in ammonia —hydrogen peroxide.

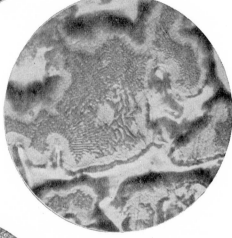

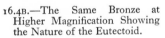

16.4C.—90/10 Aluminium Bronze, Cast and Annealed.

Primary α (light) and α + γ₂ eutectoid. ×100. Etched in acid ferric chloride.

PLATE 16.4

"88–10–2 gunmetal", containing 10% tin and 2% zinc. It is no longer used in naval ordnance, but is still widely employed where a strong, corrosion-resistant casting is required. Its structure is similar to that of a straight tin bronze containing 11% tin, so that, due to considerable coring, a lot of $\alpha + \delta$ eutectoid will be present. The zinc not only cheapens the alloy and acts as a deoxidiser but also improves casting fluidity.

16.46. Bronzes Containing Lead. Up to 2·0% lead is sometimes added to bronzes as well as to brasses in order to improve machinability. Larger amounts are added for some special bearings; and such bronzes permit 20% higher loading than do lead-base or tin-base white metals. The thermal conductivity of these bronzes is also higher, so that they can be used at higher speeds, since heat is dissipated more quickly. They also have a higher resistance to wear, and are used widely for aero and automobile crankshaft bearings. Representative alloys include D.T.D. 229A, which contains 2·0% tin and 24·0% lead; and "red brass" (5·0% tin, 5·0% zinc and 5·0% lead), which is occasionally used as a bearing metal, but more often for pressure-tight castings.

Aluminium Bronze

16.50. Like the brasses, the aluminium bronzes can be divided into two main groups, the hot-working and cold-working alloys respectively. The equilibrium diagram (Fig. 16.4) indicates that a solid solution (α), containing up to 9·4% aluminium at room temperature, is formed. Like the other α solid solutions based on copper, it is quite ductile. With more than 9·4% aluminium the phase γ_2 is formed. This is an intermetallic compound of the formula Cu_9Al_4 and, in common with compounds of this type, is very hard and brittle, resulting in an overall brittleness of alloys containing the γ_2-phase.

16.51. Further inspection of the equilibrium diagram reveals similarities between it and the iron–carbon diagram. The two α-phases are analogous; the β-phase solid solution of the copper–aluminium diagram corresponds to the γ (austenite) phase of the iron–carbon diagram; and the $\alpha + \gamma_2$ eutectoid is similar to the ferrite + cementite eutectoid (pearlite) of the steels. As a result of these similarities in structural transformation, a 10% aluminium bronze can be heat-treated in a manner parallel to that of steel.

Consider a 10% aluminium bronze; this will consist entirely of the phases α and γ_2 if allowed to cool slowly to room temperature. If it is reheated the $\alpha + \gamma_2$ eutectoid is transformed to the solid solution β when the eutectoid temperature (565° C) is reached, and as the temperature rises further, the α-phase is absorbed until at about 900° C

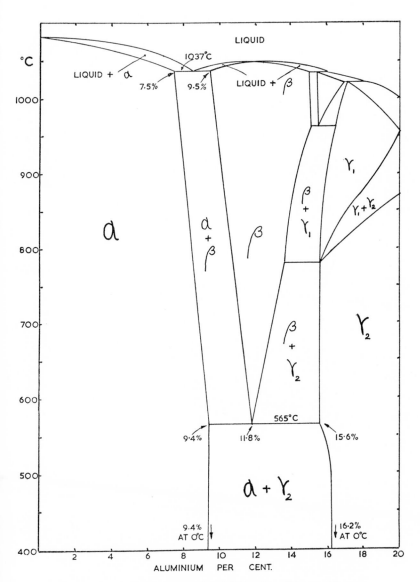

°C

LIQUID

1037°C

LIQUID + α

7.5%   9.5%

LIQUID + β

γ₁

1000

β
+
γ₁

γ₁ + γ₂

900

α

α
+
β

β

β
+
γ₁

800

γ₂

700

β
+
γ₂

γ₂

600

565°C

9.4%   11·8%   15·6%

500

α + γ₂

400

9.4%
AT 0°C

16·2%
AT 0°C

2   4   6   8   10   12   14   16   18   20

ALUMINIUM PER CENT.

FIG. 16.4.—The Copper–Aluminium Thermal-equilibrium Diagram.

TABLE 16.2.—*Tin Bronzes, Phosphor-bronzes and Gunmetal*

Relevant specifications	Composition (%)					Condition	Typical mechanical properties							Characteristics and uses
	Cu	Sn	P	Zn	Other elements		0.1% Proof stress		Tensile strength		Elongation (%)	Hardness (V.P.N.)		
							N/mm²	tonf/in²	N/mm²	tonf/in²				
—	95.5	3	—	1.5	—	Annealed Hard	92.7 571	6 37	324 726	21 47	65 5	60 200		*Coinage Bronze.* Used for British "copper" coinage.
B.S. 2870: PB101	96	3.75	0.1	—	—	Annealed Hard	108 618	7 40	340 741	22 48	65 5	60 210		*Low-tin Bronze.* Used where good elastic properties combined with resistance to corrosion and corrosion fatigue are necessary. Widely used as springs and instrument parts.
B.S. 2870/4: PB102	94	5.5	0.1	—	—	Annealed Hard	123 618	8 40	355 695	23 45	65 15	65 180		*Drawn Phosphor-bronze.* Generally used in the work-hardened condition. Useful for engineering components subjected to friction. Also for steam-turbine blading and other corrosion-resisting applications.
B.S. 1400: CT1/C	Rem.	10	0.1	—	—	Sand-cast	123	8	278	18	15	90		*Cast Phosphor-bronze.* A suitable alloy for general sand-castings. With phosphorus raised to 0.5% (B.S.S. 1400/PB1/C) the alloy is a standard phosphor-bronze for bearings. It is often supplied as cast sticks for turning small bearings, etc.
—	Rem.	18	Up to 1.0	—	—	Sand-cast	154	10	170	11	2	—		*High-tin Bronze.* Used for bearings subjected to heavy compression loads—bridge and turntable bearings, etc.
B.S. 1400: G1/C	88	10	—	2	—	Sand-cast	123	8	293	19	16	85		*Admiralty Gunmetal.* Widely used for pumps, valves and miscellaneous castings, particularly for marine purposes because of its corrosion-resistance. Also used for statuary because of good casting properties.
B.S. 1400: LG2/C	85	5	—	5	Pb 5	Sand-cast	92.7	6	216	14	13	65		85-5-5-5 *Leaded Gunmetal* or "*Red Brass*". Often used as a substitute for Admiralty gunmetal and also where pressure tightness is required.
DTD 229A	74	Up to 2.0	—	—	Pb 24	Cast	—		139	9	—	30		*Copper-Lead Bearing Alloy.* Can be bonded to steel shells for bearing purposes. Alloys containing up to 45% lead are in use.

TABLE 16.3.—*Aluminium Bronzes*

Relevant specifications	Composition (%)				Condition	Typical mechanical properties							Characteristics and uses
	Cu	Al	Fe	Other elements		0·1% Proof stress		Tensile strength		Elonga-tion (%)	Hard-ness (V.P.N.)		
						N/mm²	tonf/in²	N/mm²	tonf/in²				
B.S. 2870/5: CA101	Rem.	5	—	MN and/or Ni up to 4·0	Annealed / Hard	123 / 587	8 / 38	386 / 772	25 / 50	70 / 4	80 / 220	Cold-worked for decorative purposes, imitation jewellery, etc. Also useful in various engineering applications, especially in tube form. Excellent resistance to corrosion and to oxidation on heating.
B.S. 2870/5: CA102	Rem.	7	—	Fe, Mn and Ni up to 2·0	Hot-worked	154	10	432	28	45	100	Suitable for chemical-engineering applications, especially at moderately elevated temperatures.
B.S. 2872: CA103	Rem.	9·5	Fe + Ni = 4·0		Forged	247	16	556	36	30	180	These alloys are cold-worked to a limited extent only and are shaped by hot-working processes, including forging. The properties quoted are for the hot-worked condition. The materials can be heat-treated by quenching and tempering (16.51).
B.S. 2872: CA104	80	10	5	Ni 5	Forged	463	30	726	47	20	215	
B.S. 1400: AB1/C	Rem.	9·5	2·5	Ni and Mn up ot 1·0 each optional	Cast	185	12	525	34	30	115	The best-known aluminium bronze for both die-casting and sand-casting.
—	Rem.	12	—	5–8 Fe, Mn and Ni	Cast	479	31	556	36	3	250	A hard alloy of service where heavy compressive loads are involved. Good resistance to wear.

the structure consists entirely of the solid solution β. Water-quenching from this temperature produces a structure consisting of the phase β′. This is not shown in the equilibrium diagram, since, like martensite in steels, it is not an equilibrium phase. The β′-phase is hard and brittle like martensite, and is in fact very similar in microstructural appearance. Tempering this β′-phase at 500° C causes the precipitation of a fine agglomerate of the phases α and γ_2, closely resembling sorbite in steels.

16.52. In spite of these heat-treatment phenomena, the main industrial uses of aluminium bronzes depend upon other features such as:

(*a*) the ability to retain strength at elevated temperatures, particularly when certain other elements are present;

(*b*) the high resistance to oxidation at elevated temperatures;

(*c*) good corrosion-resistance at ordinary temperatures;

(*d*) good wearing properties;

(*e*) the pleasing colour which makes some of these alloys useful for decorative purposes, particularly as a substitute for gold in imitation jewellery.

16.53. The α-phase Cold-working Alloys contain between 4·0 and 7·0% aluminium and occasionally up to 4·0% nickel, which improves corrosion-resistance still further. This latter type of alloy is particularly useful in the manufacture of condenser tubes, where high strength and corrosion-resistance are required. Since the composition of these α-phase alloys can be adjusted to give a colour similar to that of 18-carat gold, rolled sheet is used in the manufacture of imitation-gold cigarette-cases and for other decorative articles of this type.

16.54. The $\alpha + \gamma_2$ Hot-working and Casting Alloys, with between 7·0 and 12·0% aluminium, may also contain other elements, such as iron, nickel or manganese. The hot-working alloys contain from 7·0 to 10·0% aluminium and are shaped by forging or by hot-rolling according to their subsequent application. These alloys may also contain up to 5·0% each of iron and nickel, the iron acting as a grain-refiner. Such alloys are used for chemical-engineering purposes (especially for components exposed to high temperatures); also for a variety of purposes where a corrosion-resistant forging is required.

The alloys used for sand- and die-casting contain between 9·5 and 12·0% aluminium with varying amounts of iron and nickel up to 5·0% each, and manganese up to 1·5%. These alloys are used widely in marine engineering, e.g. for pump rods, valve fittings, propellers, propeller shafts and bolts. They are also used for valve seats and sparking-plug bodies in internal-combustion engines and for brush-holders in generators; for bearings to work at heavy duty; for gear wheels; for

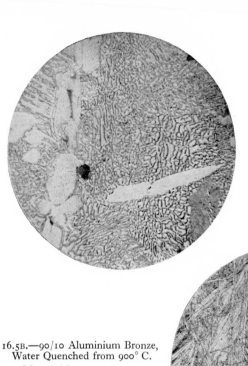

16.5A.—90/10 Aluminium Bronze, Cast and Annealed.

Crystals of primary α (light) in a matrix of $\alpha + \gamma_2$ eutectoid. $\times 1100$. Etched in acid ferric chloride.

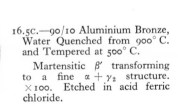

16.5B.—90/10 Aluminium Bronze, Water Quenched from 900° C.

Martensitic type of structure (β'). $\times 100$. Etched in acid ferric chloride.

16.5C.—90/10 Aluminium Bronze, Water Quenched from 900° C. and Tempered at 500° C.

Martensitic β' transforming to a fine $\alpha + \gamma_2$ structure. $\times 100$. Etched in acid ferric chloride.

PLATE 16.5

pinions and worm wheels; and in the manufacture of non-sparking tools in the gas industry, though they are softer and, hence, inferior to beryllium bronze in this direction.

Copper–Nickel Alloys

16.60. As the thermal-equilibrium diagram (Fig. 9.5) shows, the metals copper and nickel are soluble in each other in all proportions, forming a continuous series of solid solutions. In the cast condition cored crystals will be present, but, though coring may be extensive, it can never lead to the precipitation of a brittle secondary phase, as may happen with the other copper alloys dealt with. On annealing, all copper–nickel alloys consist of uniform solid solutions.

16.61. Being completely solid solution in structure, the copper–nickel alloys are all ductile. They are also very accommodating with regard to the methods of manufacture, and can be hot- or cold-worked by rolling, forging, stamping, pressing, drawing and spinning. The absence of any second phase in the microstructure eliminates the possibility of electrolytic corrosion (22.21) in the presence of an electrolyte, and this contributes to the high corrosion-resistance of copper–nickel alloys. This corrosion-resistance reaches a maximum with Monel Metal. Some of the more important copper–nickel alloys are given in Table 16.4.

16.62. Nickel Silvers are alloys containing from 10 to 30% nickel, 55 to 63% copper and the balance zinc. They are all completely solid solution in structure, and comparable with similar brasses as far as mechanical properties are concerned. They are, however, white in colour, which makes them admirably suitable for the manufacture of spoons, forks and other table-ware. Being ductile, the alloys can be cold-formed for these applications and are usually silver-plated (the stamp "E.P.N.S." means "electro-plated nickel silver"). The addition of 2·0% lead makes them easy to engrave, and such alloys are useful for the manufacture of Yale-type keys, where the presence of lead increases the ease with which the blank can be cut to shape.

Heat-treatable Copper Alloys

16.70. It is often assumed that precipitation hardening (9.80) is a phenomenon confined to a study of the light alloys. Some of the copper alloys, however, can be treated in order to utilise similar structural changes.

16.71. Beryllium Bronze is one of the most outstanding of these alloys. It contains 1·75–2·5% beryllium and up to 0·5% cobalt, the

TABLE 16.4.—Copper–Nickel Alloys

Relevant specifications	Composition (%)			Condition	Typical mechanical properties						Characteristics and uses
	Cu	Ni	Other elements		0·1% Proof stress		Tensile strength		Elonga-tion (%)	Hard-ness (V.P.N.)	
					N/mm²	tonf/in²	N/mm²	tonf/in²			
B.S. 2870/5: CN101	95	5	Fe 1·2 Mn 0·5	Annealed Hard	— —	— —	263 463	17 30	50 5	65 130	Has slightly better mechanical properties and corrosion-resistance than pure copper. Can be regarded as toughened copper.
B.S. 2870/5: CN103	85	15	Mn 0·25	Annealed Hard	— —	— —	324 494	21 32	45 5	70 145	The alloy of lowest nickel content which has a more or less white colour.
B.S. 2870/5: CN104	80	20	Mn 0·25	Annealed Hard	108 463	7 30	340 541	22 35	45 5	75 165	Used for manufacture of bullet envelopes because of its high ductility and corrosion-resistance. Will withstand severe cold-working.
B.S. 2870/5: CN105	75	25	Mn 0·25	Annealed Hard	— —	— —	355 602	23 39	45 5	80 170	Used mainly for coinage, e.g. the present British "silver" coinage.
B.S. 2870/5: CN106	70	30	Mn 0·4	Annealed Hard	108 541	7 35	355 649	23 42	45 5	80 175	Used for condenser and cooler tubes where good resistance to corrosion is required.
—	60	40	—	Annealed Hard	— —	— —	386 649	25 42	45 5	90 190	"Constantan." Largely used as a resistance material and for thermocouples. Has a high specific resistance and a very low temperature coefficient. "Eureka" is a similar alloy.
B.S. 3073/6: NA13	29	68	Fe 1·25 Mn 1·25	Annealed Hard	216 571	14 37	541 726	35 47	45 20	120 220	Monel Metal. Combines good mechanical properties with excellent corrosion-resistance. Used in chemical-engineering plant, etc.
B.S. 3073/6: NA18	29	66	Al 2·75 Fe 0·9 Mn 0·4 Ti 0·6	Annealed Hard Heat-treated	340 571 788	22 37 51	680 757 1 070	44 49 69	40 25 22	175 225 310	"K" Monel. A heat-treatable modification of monel metal. Used for motor-boat propeller shafts.

balance being copper. It is solution-treated at 800° C (Fig. 16.5), quenched, cold-worked (if necessary) and then precipitation-hardened at 275° C for one hour. After such treatment the alloy attains the remarkable tensile strength of 1 300–1 400 N/mm² due to the formation of the coherent precipitate β'. It is particularly suitable for the manufacture of tools where non-sparking properties are required (chisels and

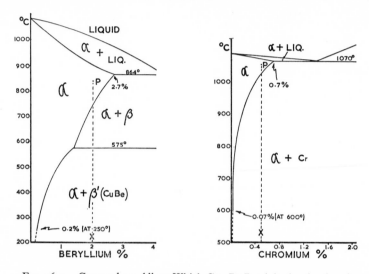

FIG. 16.5.—Copper-base Alloys Which Can Be Precipitation-hardened.

In each case the commercial alloy is of approximate composition X and is solution-treated at P, at which temperature the structure is completely unsaturated solid solution α.

hacksaw blades in gas-works, "dangerous" mines or where explosives are being manufactured), since it is both hard and tough. It also has a high fatigue limit and is therefore used for the manufacture of springs and for diaphragms and tubes of pressure-recording instruments.

The metal beryllium (20.20—Part II) is lighter than aluminium, but its melting point (1 280° C) is much higher. It would undoubtedly find greater application than it does at present were it not for its scarcity and, consequently, its high price. Beryllium occurs in the beautiful grass-green and transparent precious stone emerald, and in the semi-precious gemstone aquamarine, which, as its name implies, is a pale bluish-green in colour, rather like sea-water. Both stones are a form of the mineral beryl, which constitutes the only important source of beryllium. Beryl is mined in Argentina, Brazil, the United States, India and South Africa.

16.72. Chromium Copper, containing 0·4–0·6% chromium (Fig. 16.5), is another heat-treatable alloy which finds use in the electrical industries, since it combines a reasonable strength of 540 N/mm² with a conductivity of 80% in the hardened state.

16.73. "K" Monel. The mechanical properties of ordinary monel metal (16.61) can be further improved by the addition of 2·0–4·0% aluminium to give an alloy designated "K" Monel. This alloy can be solution treated, cold-worked and then precipitation-hardened to give a tensile strength as high as 1 160 N/mm².

Some brasses and bronzes which contain nickel (6·0%) and aluminium (1·5%) can be precipitation-hardened. Such brasses are used for machine components, such as gear wheels, instrument pinions and pressings where strength and hardness are most important.

RECENT EXAMINATION QUESTIONS

1. Give an appraisal of copper and its alloys as an engineering material.
        (University of Aston in Birmingham, Inter B.Sc.(Eng.).)
2. Describe the effects of alloying on the following properties with respect to copper: (*a*) electrical conductivity; (*b*) machinability; (*c*) formability; (*d*) corrosion-resistance; (*e*) mechanical strength.
        (University of Aston in Birmingham, Inter B.Sc.(Eng.).)
3. Give an account of EITHER:
      (*a*) the effect of small amounts of oxygen and other elements on the properties and uses of pure copper;
   OR
      (*b*) the structure and uses of copper–zinc base alloys.
        (University of Aston in Birmingham, Inter B.Sc.(Eng.).)
4. Discuss the chief properties of copper and its alloys.
        (Nottingham Regional College of Technology.)
5. The copper–zinc alloy system forms the basis for a large number of useful engineering materials.
   Discuss the properties and applications of useful alloys in this system, for zinc contents within the range 10–40%.
   What elements would you add to give: (*a*) an improvement in resistance to corrosion; (*b*) an improvement in machinability?
        (The Polytechnic, London W.1., Poly.Dip.)
6. By reference to the appropriate thermal-equilibrium diagram, show how the composition and structure of a commercial brass dictates the method by which it will be shaped and the subsequent uses to which it will be put.
        (U.L.C.I., Paper No. C111–1–4.)
7. With reference to the equilibrium diagram for copper–tin alloys, give an account of the structure, properties and uses of tin bronzes. Discuss the reasons for adding: (*a*) phosphorus; (*b*) zinc; and (*c*) lead to these alloys.   (Rutherford College of Technology, Newcastle upon Tyne.)

8. A number of alloys of copper are called "bronzes". Give the standard composition, properties and a suitable application for each of four "bronzes" containing different principal alloying elements other than tin.

(Derby and District College of Technology.)

9. Why is zinc or phosphorus commonly added in the making of tin bronzes? Give the standard composition and a typical use for a tin bronze containing (a) lead and (b) phosphorus.

(Derby and District College of Technology, H.N.D. Course.)

10. Discuss the constitution, structure, properties, heat-treatment and uses of the industrially important aluminium bronzes.

(Rutherford College of Technology, Newcastle upon Tyne.)

11. (a) Discuss the relationship that exists between the phases and subsequent applications of the Cu–Zn, Cu–Sn and Cu–Al series of binary alloys.

(b) What are the underlying requirements for a bearing alloy? Show how closely these are realised in the Cu–Sn alloys, referring specifically to the phases present. (Aston Technical College.)

12. Compare the industrial brasses and aluminium bronzes as engineering materials. (S.A.N.C.A.D.)

13. Show the effect which additions of up to 13% aluminium to copper have on the microstructure and properties of the resulting alloys.

Give the standard compositions of two alloys, stating two uses for each alloy. (Derby and District College of Technology.)

14. Discuss EITHER:

    (a) solidification of 70–30 and 60–40 brasses under equilibrium and non-equilibrium conditions;

OR

    (b) the solidification of 90–10 aluminium bronze and the subsequent heat-treatment to obtain the best mechanical properties.

(Nottingham Regional College of Technology.)

15. Compare the Cu–Zn and Cu–Ni ranges of alloys from the following aspects where applicable:

    (a) suitability for hot- or cold-working;
    (b) effect of additional alloying elements;
    (c) susceptibility to, and effect of, heat-treatment;
    (d) typical compositions and uses.

Sketch the relevant parts of both thermal-equilibrium diagrams to amplify your discussion. (Aston Technical College.)

16. Sketch the copper–nickel thermal-equilibrium diagram. Give reasons why the two metals form an equilibrium system of this type.

Name THREE copper–nickel alloys of engineering importance, giving composition and essential properties in each case.

(West Bromwich Technical College.)

17. The various alloys which contain copper and nickel as principal elements have a very wide range of compositions and also of properties.

Discuss the various commercial copper–nickel alloys in relation to their composition and properties. (The Polytechnic, London W.1.)

18. Write notes on the properties of ANY FOUR of the following: (a) oxygen-free high-conductivity copper; (b) 70–30 brass; (c) 60–40 brass; (d) 95–5 aluminium bronze; (e) gunmetals.

(Lanchester College of Technology, Coventry, H.N.D.)

19. (a) "The basic requirements of a precipitation-hardening alloy system is that the solid solubility limit should decrease with decreasing temperature."

Discuss this statement with reference to non-ferrous alloys. Show the changes in mechanical properties that occur after ageing for increasing times at various fixed temperatures.

(b) How can further strengthening be obtained with Cu–Be alloys? At what stage should this further hardening be applied?

(Aston Technical College.)

20. Write an account of the ways in which the properties of copper may be made suitable for particular industrial applications by the addition of small amounts of an alloying element.

Indicate the composition, condition and an appropriate application for three of these types of copper.

(Derby and District College of Technology.)

21. Pure Copper is weak. How has it been found possible to increase the strength of the metal and still retain a relatively high conductivity?

(University of Aston in Birmingham, Inter B.Sc.(Eng.).)

22. Discuss the reasons why copper-base alloys have tended to be replaced by other materials during the past twenty years or so.

(University of Aston in Birmingham, Inter B.Sc.(Eng.).)

CHAPTER XVII

ALUMINIUM AND ITS ALLOYS

17.10. The properties of aluminium which chiefly affect its use as a metallurgical material are its low relative density and its high affinity for oxygen. Since its relative density is only about one-third that of steel, or of a copper-base alloy, it is used, when alloyed with small amounts of other elements, for castings in aero, automobile and constructional engineering. Moreover, alloying and heat-treatment can produce alloys of aluminium which are, weight for weight, stronger than steel, so that the use of aluminium alloys is further extended to stress-bearing members in both aero-engine and air-frame.

17.11. The high affinity of aluminium for oxygen is both disadvantageous and useful. It is disadvantageous in so far as it increases the cost of extraction of the metal by making necessary a relatively expensive electrolytic extraction process. Usually a metal is extracted by heating its oxide ore with a cheap reducing agent, such as carbon (in the form of coke), and the resulting crude metal is refined by allowing the bulk of the impurities present in it to be oxidised by the action of air. This is the basis of the production of pig iron and its subsequent conversion to steel. Aluminium, however, has a greater affinity for oxygen than has carbon, so aluminium can be separated from oxygen economically only by electrolytic means, since such reducing agents as possess greater affinities for oxygen than aluminium itself are far too expensive.

It was using an expensive reducing agent in the form of metallic potassium which enabled the Danish physicist and chemist H. C. Oersted to produce the first samples of aluminium in 1825. Consequently aluminium was worth about £250 per kg in those days, and, even later, it is said that the more illustrious foreign guests to the Court of Napoleon III were privileged to use forks and spoons made from aluminium, whilst the French nobility had to be content with mere gold plate and silver cutlery. In the "Gay Nineties" aluminium was still regarded as being in the nature of a curious precious metal, though it was in 1886 that C. M. Hall, a twenty-two-year-old student, had discovered a relatively cheap method for producing aluminium by electrolysing a fused mixture of aluminium oxide and the mineral cryolite.

Aluminium cannot be purified by blowing air through it, as in the case of iron. This treatment would oxidise the aluminium and leave behind the impurities. Therefore the ore must be purified before being electrolysed, and this involves an expensive chemical process. The ore of aluminium is bauxite, which is named after Les Baux, in the south of France, where it was discovered in 1821. France is still an important producer of bauxite, whilst the United States, Jamaica, Malaya, Ghana, India, Russia, Hungary, Guyana, Jugoslavia and Italy all mine the ore. Recently large deposits have been discovered in Queensland. Britain is wholly dependent upon imported ore to maintain production at the Kinlochleven and Lochaber works, both of which are sited within easy access of water transport and relatively cheap hydro-electric power in the heart of the Highlands of Scotland.

17.12. Although aluminium has a great affinity for oxygen, its corrosion-resistance is relatively high. This is due to the dense impervious film of oxide which forms on the surface of the metal and protects it from further oxidation. The corrosion-resistance can be further improved by anodising (22.61), a treatment which artificially thickens the natural oxide film. Since aluminium oxide is extremely hard, wear-resistance is also increased by the oxide layer; and the slightly porous nature of the surface of the film allows it to be coloured with either organic or inorganic dyes. In this respect high oxygen-affinity is an asset.

The high affinity of aluminium for oxygen also makes it useful as a deoxidant in steels and also in the Thermit process of welding (21.54).

17.13. The fact that aluminium has over 50% of the specific conductivity of copper means that, weight for weight, it is a better conductor of electricity than is copper. Hence it is now widely used, generally twisted round a steel core for strength, as a current carrier in the electric "grid" system.* Pure aluminium is relatively soft and weak (it has a tensile strength of about 90 N/mm² in the annealed condition), so that for most other engineering purposes it is used in the alloyed condition.

Alloys of Aluminium

17.20. The addition of alloying elements is made principally to improve mechanical properties, such as tensile strength, hardness, rigidity and machinability, and sometimes to improve fluidity and other casting properties.

17.21. One of the chief defects to which aluminium alloys are prone is porosity due to gases dissolved during the melting process. Molten aluminium will dissolve considerable amounts of hydrogen if, for any reason, this is present in the furnace atmosphere. When the metal is cast and begins to solidify the solubility of hydrogen diminishes almost

* The current high price of copper has resulted in an even greater use of aluminium in the electrical industries, though aluminium wire is sometimes copper-coated in order to increase its electrical conductivity.

TABLE 17.1.—*Wrought Aluminium Alloys Which Are Not Heat-treated*

Relevant specifications	Proprietary British trade names	Composition (%)†						Condition	Typical mechanical properties					Uses
		Cu	Mg	Si	Fe	Mn	Other elements		0.1% Proof stress		Tensile strength		Elongation (%)	
									N/mm²	tonf/in²	N/mm²	tonf/in²		
B.S. 1470/S1C; B.S. 1477/P1C; B.S. 1476/E1C; etc., B.S. 2L54; 3L34; 3L36; 4L16; 4L17; etc	"Al.3", ALCAN GB-2S, BA.99, BMB.2C, Hiduminium 1c, Impalco.102	0.1	—	0.5	0.7	0.1	—	Soft Hard	30.9 139	2 9	84.9 154	5.5 10	35 5	Panelling and moulding; lightly stressed and decorative assemblies especially in architecture; holloware; electrical conductors; equipment for chemical, food and brewing industries; packaging.
B.S. 1470/NS3, B.S. 1472/NP3, B.S. 1477/NP3; etc., B.S. 2L59; 2L60; 2L61; etc.	ALCAN GB-3S, BA.60, BMB-3, Hiduminium 11, Impalco.190, Manganal	0.1	—	0.6	0.7	1.0–1.5	Ti 0.2*	Soft Hard	46.3 170	3 11	108 201	7 13	34 4	Metal boxes; bottle caps; domestic and commercial food containers and cooking utensils; roofing sheets; panelling of land-transport vehicles.
B.S. 1470/NS4, B.S. 1476/NE4, B.S. 1477/NP4; etc., B.S. 3L44; L80; L81	ALCAN GB-M57S, BA.21, Birmabright 2, Hiduminium 22, Impalco.510, MG.2	0.1	1.7–2.8	0.6	0.5	0.5	Ti 0.15* Cr 0.25	Soft ½ hard	77.2 216	5 14	185 263	12 17	24 4	Marine superstructures; life-boats; panelling exposed to marine atmospheres; fencing wire; chemical plant; panelling for road and rail vehicles.
B.S. 1470/NS5, B.S. 1476/NE5, B.S. L82	Alcan Gb-54S, BA.27, Birmabright 3, Hiduminium 33, Impalco.520, MG.3	0.1	2.8–4.0	0.6	0.5	0.6	Ti 0.15* Cr 0.25	Soft ¼ hard	92.7 170	6 11	216 293	14 19	18 8	Shipbuilding; deep-pressing for car bodies.
B.S. 1470/NS6, B.S. 1471/NT6, B.S. 1474/NV6, B.S. 1476/NE6, B.S. 1477/NP6, B.S. 2L58; L82	Alcan GB-A56S, BA.28, Birmabright 5, Hiduminium 05, Impalco.560, MG.5	0.1	4.5–5.5	0.6	0.5	1.0	Ti 0.15* Cr 0.25	Soft ¼ hard	123 216	8 14	263 293	17 19	18 8	Shipbuilding and applications requiring high strength and corrosion-resistance; rivets.
B.S. 1472/NF7, D.T.D.18aB, D.T. D186B, D.T. D.297A	Alcan GB-58S, BA.227, Birmabright 7, Hiduminium 07	0.1	6.5–7.5	0.6	0.5	0.5	Ti 0.15* Cr 0.25	Extruded	185	12	340	22	35	Roofing supports in mines; other applications requiring strength and corrosion-resistance.

| B.S. 1475/NG21 | Alcan GB-33S AWCO.45 Impalco.460 | | 0·1 | 0·25 | 4·5–6·0 | 0·6 | — | Extruded | 46·3 | 3 | 123 | 8 | 4 | Wire and welding rods. |
| B.S. 1475/NG2 | AWCO.40 Impalco.450 | | 0·1 | 0·25 | 10·0–13·0 | 0·6 | — | Extruded | — | — | — | — | — | Wire for welding (this alloy is also used in the sheet form for panelling and light marine construction). |

† Single values are maxima unless otherwise stated.   * Optional.

TABLE 17.2.—Cast Aluminium Alloys Which Are Not Heat-treated

Relevant specifications	Proprietary trade names	Composition (%)					Condition	Typical mechanical properties					Uses
		Si	Cu	Mg	Mn	Zn		0·1% Proof stress		Tensile strength		Elongation (%)	
								N/mm²	tonf/in²	N/mm²	tonf/in²		
B.S. 1490/LM1	LM.1.* Alloy "Z.3"	3	7	—	—	3	Sand-cast Chill-cast	84·9 108	5·5 7	131 170	8·5 11	1 1·5	Gravity die-casting. General-purpose alloy for lightly stressed parts not subjected to shock. Little used now.
B.S. 1490/LM2	LM.2.* Hiduminium 00	10	2	—	—	—	Chill-cast Pressure-cast	84·9 92·7	5·5 6	185 247	12 16	2 1·5	Pressure die-castings. General purposes. Will withstand moderate stresses.
B.S. 1490/LM4 B.S. L79	LM.4.* Hiduminium 20 Alcan GB-117	5	3	—	0·5	—	Sand-cast Chill-cast	69·5 77·2	4·5 5	147 170	9·5 11	2 3	Sand-castings; gravity and pressure die-castings. Good foundry characteristics and inexpensive. General-purpose alloy where mechanical properties are of secondary importance. Can be heat-treated.
B.S. 1490/LM5 B.S. L74	LM.5.* Birmabright Hiduminium 30 Alcan GB-B320	—	—	5	0·5	—	Sand-cast Chill-cast	77·2 92·7	5 6	170 185	11 12	5 9	Sand-castings and gravity die-castings. Suitable for moderately stressed parts. Good corrosion-resistance for marine work. Takes an excellent polish.
B.S. 1490/LM6 B.S. 4L-33	LM.6.* Alpax Hiduminium 10 Alcan GB-160	11·5	—	—	—	—	Sand-cast Chill-cast Pressure-cast	54·1 61·8 84·9	3·5 4 5·5	170 201 216	11 13 14	7 9 4	Sand-castings; gravity and pressure die-castings. Excellent foundry characteristics. Large castings for general and marine engineering. Radiators; sumps; gear-boxes; etc. One of the most widely used aluminium alloys.

* The current tendency is not to use proprietary trade names for the casting alloys. Many leading companies, such as Messrs. Birmingham Aluminium Casting (1903) Co., Ltd., William Mills Ltd., International Alloys Ltd., use the "LM" number to specify their alloys.

to zero, so that tiny bubbles of gas are formed in the partly solid metal. These cannot escape, and give rise to "pinhole" porosity. The defect is eliminated by treating the molten metal before casting with a suitable flux, or by bubbling nitrogen or chlorine through the melt. As a substitute for chlorine, which is a poisonous gas, tablets of the organic compound hexachlorethane can be used.

17.22. Aluminium alloys are used in both the cast and wrought conditions. Whilst the mechanical properties of many of them, both cast and wrought, can be improved by the process known as "precipitation-hardening", a number are used without any such treatment being applied. It is convenient, therefore, to classify the somewhat bewildering multitude of aluminium alloys into the following four main groups, according to the condition in which they are employed.

Wrought Alloys Which Are Not Heat-treated

17.30. The main requirements of alloys in this group are sufficient strength and rigidity in the work-hardened state, coupled with good corrosion-resistance. These properties are typical of the alloys shown in Table 17.1. As will be seen, these alloys are widely used in the manufacture of panels for land-transport vehicles. Here the high corrosion-resistance of the aluminium–magnesium alloys is utilised, those with a higher magnesium value having an excellent resistance to sea-water and marine atmospheres, so that they are used extensively for marine superstructures. The desired mechanical properties are produced by the degree of cold-work applied in the final cold-working operation, and these alloys are commonly supplied as "soft", "quarter hard", "half hard", "three-quarter hard" or "full hard". The main disadvantage is that, once the material has been finished to size, no further variation can be made in mechanical properties (other than softening by annealing), whereas, with the precipitation-hardening alloys, the properties can be varied, within limits, by heat-treatment.

17.31. As would be expected, most of these alloys (the main exception being those of aluminium and silicon) have structures consisting entirely of solid solutions. This is a contributing factor to their high ductility and high corrosion-resistance. Some of the alloys listed are occasionally used in the cast condition.

Cast Alloys Which Are Not Heat-treated

17.40. This group of alloys contains those which are widely used as general-purpose materials as sand-castings and as die-castings.

These alloys are mainly used when rigidity, fluidity in casting and good corrosion-resistance are more important than strength.

17.41. Undoubtedly the most widely used alloys in this class are those containing between 9·0 and 13·0% silicon, with occasionally small amounts of copper. These alloys are of approximately eutectic composition (Fig. 17.1), a fact which makes them eminently suitable as die-

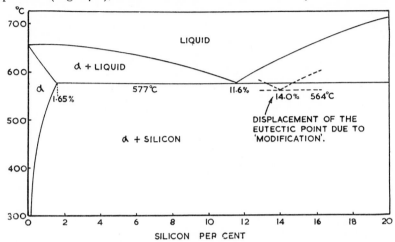

FIG. 17.1.—The Aluminium–Silicon Thermal-equilibrium Diagram, Showing the Effect of "Modification".

casting alloys, since their freezing range will be small. The rather coarse eutectic structure can be refined by a process known as "modification". This consists of adding small amounts of sodium (about 0·01% by weight of the charge) to the melt just before casting. The effect is to delay the precipitation of silicon when the normal eutectic temperature is reached, and also cause a shift of the eutectic composition towards the right of the equilibrium diagram. Therefore, as much as 14% silicon may be present in a modified alloy without any *primary* silicon crystals forming in the structure (Plates 17.1B and c). It is thought that sodium acts in much the same way as "inoculants" in cast iron, namely, to cover non-metallic inclusions with a film which will prevent them from acting as centres of crystallisation. Thus, supercooling occurs and is responsible for the relatively fine-grained structure which ultimately forms, since crystallisation will begin spontaneously from a large number of nuclei. Remelting tends to restore the original structure due to a loss of sodium by oxidation. The modification process raises the tensile strength from 120 to 200 N/mm² and the elongation from 5·0 to over 15·0%. The relatively high ductility of this cast eutectic alloy is due to the fact that the α-solid-solution phase in the eutectic constitutes nearly 90% of the total structure. As will be seen (Plates 17.1B and c), it is therefore continuous in the microstructure and

17.1A.—A Wrought Aluminium-Manganese Alloy Containing $1\frac{1}{4}\%$ Mn in the Annealed Condition, Showing Uniform Crystals of Solid Solution.

Etched electrolytically and photographed under polarised light which emphasises grain-contrast. ×100.

Courtesy of Aluminium Laboratories Ltd.

17.1B.—Aluminium–Silicon Alloy Containing 12% Silicon, in the Cast Condition (Unmodified).

Eutectic of silicon (dark) and α solid solution. The structure is coarse due to lack of "modification". ×100. Unetched.

Courtesy of Aluminium Laboratories Ltd.

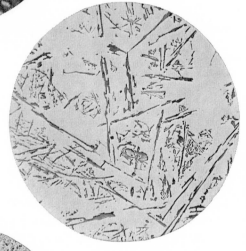

17.1C.—The Same Alloy Modified with Sodium.

The eutectic structure is considerably refined and there are now dendrites of primary α (light) due to the displacement of the eutectic point to the right by modification. ×100. Unetched.

Courtesy of Aluminium Laboratories Ltd.

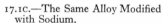

PLATE 17.1

acts as a "cushion" against much of the brittleness arising from the hard silicon phase.

These aluminium–silicon alloys can therefore be produced in the wrought form, though they are better known as materials for die-casting and other types of casting where intricate sections are required, since high fluidity and low shrinkage, as well as the narrow freezing range mentioned above, make them very suitable for such purposes. Their high corrosion-resistance makes these alloys useful for marine work, and the fact that they are somewhat lighter than the aluminium–copper alloys makes them suitable for aero and automobile construction.

17.42. The aluminium–copper alloys contain considerable amounts of a eutectic composed of the solid solution α and an intermetallic compound of the formula $CuAl_2$ (see Fig. 17.2), and are useful where rigidity and good casting properties are required in components not subjected to shock. These alloys also machine well, but their corrosion-resistance is inferior to that of the aluminium–silicon alloys. Moreover, they cannot be successfully anodised and are best protected by painting. Aluminium–copper alloys were at one time widely used for gear-boxes, sumps and other automobile castings, for cases for switch gear, for household and industrial fittings, etc., but of recent years they have become almost obsolete.

17.43. The main feature of the aluminium–magnesium–manganese alloys is good corrosion-resistance, which enables them to receive a high polish. These include the well-known "Birmabright" series of alloys, whose combination of corrosion-resistance, rigidity and resistance to shock makes them admirably suitable for moderately stressed parts working under marine conditions.

Wrought Alloys Which Are Heat-treated

17.50. Without doubt the most significant feature of the aluminium alloys is the ability of some of them to undergo a change in properties under suitable heat-treatment. Although this phenomenon is not confined to aluminium alloys, but is possible with any alloy in a series in which a change in solubility of some constituent in the parent metal takes place (9.80), it is more widely used in the aluminium-base alloys than in any others.

The phenomenon was first observed by a German research metallurgist, Dr. Alfred Wilm, in 1906, who noticed that an aluminium alloy containing small amounts of magnesium, silicon and copper, when quenched from a temperature in the region of 500° C, subsequently hardened unassisted, if allowed to remain at room temperature over a period of days. The strength increased in this way, reaching a maximum value in just under a week, and the effect was subsequently known as "age-hardening". About four years later Wilm transferred the sole rights of his patent to the Dürener Metal Works at Duren in Western

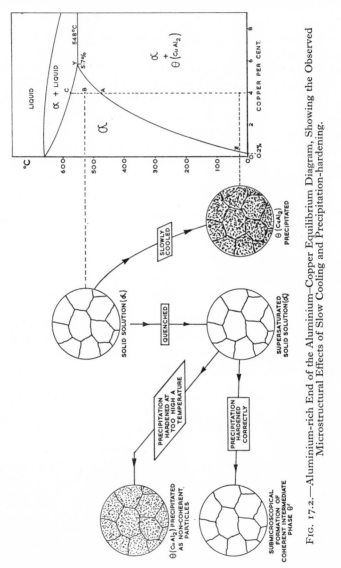

FIG. 17.2.—Aluminium-rich End of the Aluminium–Copper Equilibrium Diagram, Showing the Observed Microstructural Effects of Slow Cooling and Precipitation-hardening.

Germany, and the alloy produced was named "duralumin". Although this name is often used to describe any wrought aluminium alloy of this type, it is, strictly speaking, the proprietary name given to a series of alloys produced by certain companies both here and abroad. The first significant use of duralumin was during the First World War, when it found application in the structural members of the airships bearing the name of Graf von Zeppelin.

17.51. The age-hardening process can be accelerated by heating the quenched alloy at temperatures up to 180° C. Such treatment was referred to originally as "artificial age-hardening", but both of these slightly misleading titles are now covered in modern metallurgical nomenclature by the general term "precipitation-hardening". Much speculation has taken place over the years as to the fundamental nature of this phenomenon. Metallurgists now generally agree that the increases in strength and hardness produced are directly connected with the formation of coherent precipitates (9.82) within the lattice of the parent solid solution. We will consider the application of this principle to the precipitation-hardening of aluminium–copper alloys.

17.52. Let us assume that we have an aluminium–copper alloy containing 4·0% copper. At temperatures above 500° C this will consist entirely of α solid solution as indicated by the equilibrium diagram (Fig. 17.2). If we now allow the alloy to cool *very slowly* to room temperature, equilibrium will be reached at each stage and particles of the intermetallic compound $CuAl_2$ (θ) will form as a non-coherent precipitate. This precipitation will commence at A and continue until at room temperature only 0·2% (X) copper remains in solution in the aluminium. The resulting structure will lack strength because only 0·2% copper is left in solution, and it will be brittle because of the presence of coarse particles of $CuAl_2$.

If the alloy is now slowly reheated, the particles of $CuAl_2$ will be gradually absorbed, until at A we once more have a complete solid solution α (in industrial practice a slightly higher temperature, B, will be used to ensure complete solution of the $CuAl_2$). On quenching the alloy we retain the copper in solution, and, in fact, produce a supersaturated solution of copper in aluminium at room temperature. In this condition the alloy is somewhat stronger and harder because there is more copper actually in solid solution in the aluminium, and it is also much more ductile, because the brittle particles of $CuAl_2$ are now absent. So far these phenomena permit of a straightforward explanation, forthcoming from a simple study of the microstructure. What happens subsequently is of a sub-microscopical nature, that is, its observation is beyond the range of an ordinary optical microscope.

17.53. If the quenched alloy is allowed to remain at room temperature, it will be found that strength and hardness gradually increase (with a corresponding reduction in ductility) and reach a maximum in about

TABLE 17.3.—*Wrought Aluminium*

Relevant specifications	Proprietary trade names	Composition (%)†					
		Cu	Mg	Si	Mn	Fe	Other elements
B.S. 1471/HT9 B.S. 1472/HF9 B.S. 1474/HV9 B.S. 1475/HG9 B.S. 1476/HE9 D.T.D. 372B	Alcan GB-50S; AWCO.24 BA.24; Birmetal 055 Durcilium W Hiduminium 46 Impalco.910; Simgal	0·1	0·4– 0·9	0·3– 0·7	0·3	0·5	Ti 0·2*
B.S. 1470/HS30 B.S. 1471/HT30 B.S. 1474/HV30 B.S. 1475/HG30 B.S. 1476/HE30 B.S. 1477/HP30 D.T.D. 346A; 5080	Alcan GB-B51S; BA.25 Duralumin H Hiduminium 44 Impalco.920; AWCO.25 Birmetal 071 Durcilium S	0·1	0·4– 1·4	0·6– 1·3	0·4– 1·0	0·5	Ti 0·2*
B.S. 1472/HF11 B.S. 1474/HV11 B.S. 1476/HE11; L84; L85 D.T.D. 450A; 460A	Alcan GB-62S; BA.306 Birmetal 161; Duralumin E Durcilium M; Impalco.690 Hiduminium 03	1·0– 2·0	0·5– 1·2	0·8– 1·3	1·0	0·7	Ti 0·3*
B.S. 1472/HF12; L83 D.T.D. 246B	BA.307; Impalco.830 Hiduminium RR56	1·8– 2·8	0·6– 1·2	0·5– 1·3	0·5	0·6– 1·2	Ni 0·6– 1·4 Ti 0·3*
B.S. L86	Alcan GB-16S; AWCO.304 Birmetal 230; Duralumin M	1·5– 3·0	0·2– 0·5	0·7	0·5	0·7	Ti 0·3*
B.S. 1470/HS14; 2L70 B.S. 1471/HT14 B.S. 1474/HV14 B.S. 1476/HE14; 2L64 B.S. 1477/HP14	Alcan GB-17S; BA.301 Birmetal 477 Duralumin B; AWCO 301 Durcilium H Hiduminium 01	3·5– 4·7	0·4– 1·2	0·2– 0·7	0·4– 1·0	0·7	Ti} Cr} 0·3*
B.S. 1470/HS15; 2L70 B.S. 1471/HT15; 2L62; 2L63 B.S. 1472/HF15; L76; L77 B.S. 1473/HB15 B.S. 1474/HV15 B.S. 1475/HG15 B.S. 1476/HE15; 2L64; 2L65; L87 B.S. 1477/HP15 D.T.D. 5010; D.T.D. 5020A	Alcan GB-B26S BA.305 Birmetal 478 Duralumin S Hiduminium 66 AWCO.305 Durcilium K Impalco.660	3·8– 4·8	0·2– 0·8	0·5– 0·9	0·3– 1·2	0·7	Ti} Cr} 0·3*
D.T.D. 324A	Alcan GB-38S Hiduminium 08	0·7– 1·3	0·8– 1·5	10·5– 13·0	0·2	0·6	Ni 0·7– 1·3 Ti 0·2*
B.S. 1470/HS20 B.S. 1471/HT20 B.S. 1474/HV20 B.S. 1475/HG20 B.S. 1476/HE20 B.S. 1477/HP20	Alcan GB-65S; BA.22 Birmetal 016 Duralumin F Impalco.940; AWCO.22 Durcilium Q Hiduminium 43	0·15– 0·4	0·8– 1·2	0·4– 0·8	0·2– 0·8 (or Cr 0·15– 0·35)	0·7	Ti 0·2
D.T.D. 5044 D.T.D. 5024	Alcan GB-C75S; BA.705 Birmetal Z/52 Duralumin L Hiduminium RR.77 Impalco.740	0·3– 0·7	2·2– 3·2	0·5	0·3– 0·7	0·5	Zn 5·2– 6·2 Ti 0·3
D.T.D. 5074	Alcan GB-C77S; BA.704 Duralumin K (and KC) Durcilium P Birmetal Z/12 Hiduminium 89 Impalco.770	1·0– 2·2	2·0– 3·0	0·5	0·3	0·5	Zn 5·0– 7·5 Ti 0·3

Single values are maxima unless otherwise stated.   * Optional.

Heat-treatment	Typical mechanical properties					Characteristics and uses
	0·1% Proof stress		Tensile strength		Elongation (%)	
	N/mm²	tonf/in²	N/mm²	tonf/in²		
Solution-treated at 520° C; quenched; and precipitation-hardened at 170°C for 10 hours.	185	12	247	16	18	Good corrosion-resistance. Glazing bars and window sections; windscreen and sliding-roof sections for the automobile industry. Good surface-finish obtainable with this alloy.
Solution-treated at 510° C; quenched; and precipitation-hardened at 175° C for 10 hours.	152	10	247	16	20	Structural members for road, rail and sea transport vehicles; general architectural work; ladders and scaffold tubes. High electrical conductivity—hence used for overhead lines, etc.
Solution-treated at 525° C; quenched; and precipitation-hardened at 170° C for 10 hours.	402	26	432	28	12	Structural members for aircraft and road vehicles; tubular furniture.
Solution-treated at 530° C; quenched; and precipitation-hardened at 170° C for 10–20 hours.	324	21	417	27	10	Air-screw forgings; sheets; tubes.
Solution-treated at 495° C; quenched; and aged at room temperature for 5 days.	154	10	293	19	20	Rivets for built-up structures.
Solution-treated at 480° C; quenched; and aged at room temperature for 4 days.	278	18	402	26	10	General purposes; stressed parts in aircraft and other structures. The original "Duralumin".
Solution-treated at 510° C; quenched; and precipitation-hardened at 170° C for 10 hours.	463	30	510	33	10	Highly stressed components in aircraft—e.g. stressed-skin construction. Engine parts such as connecting-rods. Also supplied in the "Alcad" form—i.e. covered with a thin layer of high-purity aluminium which protects the alloy from corrosive influences.
Solution-treated at 490° C for 12 hours; quenched; and precipitation-hardened at 130° C for 6 hours.	293	19	371	24	4	Components operating at high temperatures—pistons, cylinder-heads, etc.
Solution-treated at 520° C; quenched; and precipitation-hardened at 170° C for 10 hours.	263	17	309	20	13	Plates; bars and sections for shipbuilding. Body panels for cars and rail vehicles. Containers.
Solution-treated at 465° C; quenched; and precipitation-hardened at 120° C for 24 hours.	541	35	587	38	11	Skinning and framework of aircraft. Other applications where a high strength/weight ratio is essential. Also supplied in the "Alclad" form.
Solution-treated at 465° C; quenched; and precipitation-hardened at 120° C for 24 hours.	587	38	649	42	11	Highly stressed aircraft components such as booms. Other military equipment requiring the highest strength/weight ratio. The strongest aluminium alloy produced commercially.

six days. After this time has elapsed no further appreciable changes occur in the properties. The completely α-phase structure obtained by quenching is not the equilibrium structure at room temperature. It is in fact super-saturated with copper so that there is a strong urge for copper to be rejected from the solid solution α as particles of the non-coherent precipitate CuAl₂ (θ). This stage is never actually reached at room temperature because of the sluggishness of diffusion of the copper atoms within the aluminium lattice. However, some movement does occur and the copper atoms take up positions within the aluminium lattice so that nuclei of the intermediate phase (θ') are formed. These nuclei are present as a coherent precipitate *continuous* with the original α lattice, and in this form, cause distortion within the α lattice. This effectively hinders the movement of dislocations (9.82) and so the yield strength is increased. The sub-microscopical change within the structure can be represented so:

$$\text{Cu} + 2\text{Al} \longrightarrow \theta'$$

Cu	2Al	θ′
[Al lattice]	[Al lattice]	[Intermediate coherent precipitate]

An improvement in properties over those obtained by ordinary "natural age-hardening" can be attained by "tempering" the quenched alloy at temperatures up to nearly 200° C for short periods. This treatment increases the amount of the intermediate coherent precipitate θ' by accelerating the rate of diffusion, and so strength and hardness of the alloy rise still further. If the alloy is heated to a higher temperature, a stage is reached where the structure begins to revert rapidly to one of equilibrium and the coherent intermediate phase θ' precipitates fully as non-coherent particles of θ (CuAl₂):

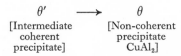

$$\theta' \longrightarrow \theta$$

θ′	θ
[Intermediate coherent precipitate]	[Non-coherent precipitate CuAl₂]

When this occurs both strength and hardness begin to fall. Further increases in temperature will cause the θ particles to grow to a size making them easily visible with an ordinary optical microscope, and this will be accompanied by a progressive deterioration in mechanical properties. Fig. 17.3 illustrates the general effects of variations in time and temperature during post-quenching treatment for a typical alloy of the precipitation-hardening variety. At room temperature (20° C) the tensile strength increases slowly and reaches a maximum of about 390 N/mm² (25 tonf/in²) after approximately 100 hours. Precipitation-treatment at temperatures above 100° C will result in a much higher maximum tensile strength being reached. Optimum strength is obtained by treatment at 165° C for about ten hours, after which, if the treatment time is prolonged, rapid precipitation of non-coherent particles of θ (CuAl$_2$) will cause a deterioration in tensile strength and hardness as shown by curve C. Treatment at 200° C, as represented by curve D, will give poor results because the rejection from solution of non-coherent θ is very rapid such that precipitation will "overtake" any increase in tensile

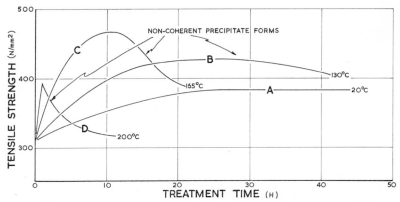

FIG. 17.3—The Effects of Time and Temperature of Precipitation-treatment on the Structure and Tensile Strength of a Suitable Alloy.

strength. This process of deterioration in structure and properties due to faulty heat-treatment is generally termed "reversion". Time and temperature of precipitation-treatment differ with the composition of the alloy, and must always be controlled accurately to give optimum results.

17.54. Precipitation-hardening is not confined to the aluminium-base alloys, but can take place with any alloy in a system where a sloping solubility boundary such as XY (Fig. 17.2) exists. As far as the aluminium alloys are concerned, those containing copper and those containing magnesium and silicon (which cause hardening due to the formation of Mg$_2$Si) are the most important. In addition, precipitation-hardening is utilised in some of the copper-base alloys, notably beryllium bronze and chromium–copper (16.70).

17.55. There are a number of alloy compositions sold under the general trade name of "Duralumin", but all of them rely on the presence of approximately 4·0% copper to effect hardening. In addition, both magnesium and silicon are usually present, so that hardening will to some extent be assisted by the formation of Mg$_2$Si. Thus, one standard grade of duralumin contains 4·0% copper, 0·6% manganese, 0·5% magnesium and 0·4% silicon, with small amounts of iron. The main function of manganese is to refine the grain.

TABLE 17.4.—*Cast Aluminium Alloys*

Relevant specifications	Proprietary trade names	Composition (%)				
		Cu	Si	Mg	Mn	Others
B.S. 1490/LM4.WP	*LM.4(WP) Hiduminium 20 Alcan GB-117	3·0	5·0	—	0·5	Fe 0·8 Ti 0·2
B.S. 1490/LM8 D.T.D. 716A, 722A, 727A, 735A	*LM.8; Alcan GB-B.116 Birmidal; Hiduminium 40	—	4·5	0·5	0·5	Ti 0·15
B.S. 1490/LM9 B.S. L75	*LM.9 Alpax β; Alpax γ	—	11·5	0·4	0·5	—
B.S. 1490/LM10 B.S. 2L53	*LM.10 Hiduminium 90 Alcan GB-350	—	—	10·5	—	Ti 0·2
B.S. 1490/LM11 D.T.D. 298B; 304B; 361B	*LM.11 Hiduminium 80 Alcan GB-226	4·5	—	—	—	Ti 0·15
B.S. 1490/LM13	*LM.13 Alcan GB-162	0·9	12·0	1·2	—	Ni 2·5
B.S. 1490/LM14 B.S. 3L35	*LM.14; "Y" Alloy Hiduminium Y Alcan GB-218	4·0	0·3	1·5	—	Ni 2·0 Ti 0·2
B.S. 1490/LM16 B.S. L78	*LM.16 Alcan GB-125	1·2	5·0	0·5	—	Ni 0·25
B.S. 1490/LM23 B.S. 2L51	*LM.23 Hiduminium RR50 Ceralumin B	1·8	2·5	0·2	—	Ti 0·15 Fe 1·0 Ni 1·2
—	Birmal MB.7; Alcan GB-730	1·0	0·6	1·0	—	Ni 1·6 Sn 6·5
—	Birmal L4 Bohn L4 (American)	3·0	9·5	1·0	—	Ni 1·0 Fe 1·0 Ti 0·1
—	Birmal D.12	6·5	5·5	0·4	—	Ti 0·1

* The current tendency is not to use proprietary trade names for the casting alloys. Many leading companies, such as Messrs. Birmingham Aluminium Casting (1903) Co., Ltd., William Mills Ltd., International Alloys Ltd., use the "LM" number to specify their alloys.

The initial hot-working of duralumin, either by hot-rolling or extrusion, is done between 400 and 450° C. This treatment breaks up the

Heat-treatment	Typical mechanical properties					Characteristics and uses
	0·1% Proof stress		Tensile strength		Elongation (%)	
	N/mm²	tonf/in²	N/mm²	tonf/in²		
Solution-treated at 520° C for 6 hours; quenched in hot water or oil; precipitation-hardened at 170° C for 12 hours.	278	18	324	21	1	General purposes (sand, gravity and pressure die-casting). Withstands moderate stresses, shock and hydraulic pressure. Also used in the non-heat-treated condition.
Solution-treated at 540° C for 4–12 hours; quenched in oil; precipitation-hardened at 165° C for 8–12 hours.	—	—	278	18	2	Good casting properties and corrosion-resistance. Mechanical properties can be varied by using different heat-treatments.
Solution-treated at 530° C for 2–4 hours; quenched in warm water; precipitation-hardened at 150–170° C for 16 hours.	247	16	293	19	2	Suitable for intricate castings, stressed but not subjected to shock. Good corrosion-resistance like the plain high-silicon alloy.
Solution-treated at 525° C for 8 hours; cooled to 390° C and then quenched in oil at 160° C or in boiling water.	201	13	324	21	18	High combination of proof stress, tensile strength and percentage elongation. High corrosion-resistance. Used in marine work and aircraft (sea-planes).
Solution-treated at 525° C for 12 hours; oil-quenched; precipitation-hardened at 120–170° C for 6–18 hours.	185	12	293	19	5	Good strength, ductility and shock-resistance. Casting necessitates special technique.
Solution-treated at 520° C for 4–12 hours; oil-quenched; precipitation-hardened at 165–185° C for 6–12 hours.	293	19	309	20	Nil	Notable for low thermal expansion. Used for pistons for high-performance engines.
Solution-treated at 510° C; precipitation-hardened in boiling water for 2 hours or aged at room temperature for 5 days.	216	14	278	18	—	Pistons and cylinder-heads for liquid- and air-cooled engines. General purposes.
Solution-treated at 520° C for 12 hours; water-quenched; precipitation-hardened at 150–160° C for 8–10 hours.	201	13	263	17	1	Useful for intricate shapes and pressure-tightness. Cylinder-heads; valve bodies; water jackets.
No Solution-treatment required; precipitation-hardened at 155–170° C for 8–16 hours.	123	8	201	13	3	A good general-purpose alloy for sand-casting and gravity die-casting. High rigidity and moderate shock-resistance.
Special heat-treatment: heat at 160–165° C for 12–24 hours.	0·2% Proof stress in compression— 170 N/mm² (11 tonf/in²)					Developed for the manufacture of die-cast bearings.
Precipitation-treated at 180–195° C for 8–12 hours.	—	—	232	15	Nil	Piston castings generally, particularly the autothermic piston.
Solution-treated at 490° C for 4–8 hours; quenched in hot water; precipitation-hardened at 175° C for 6–12 hours.	—	—	340	22	Nil	Pistons for moderate duty. Retains rigidity well at temperatures up to 250° C.

coarse eutectic to some extent, and the alloy can then be cold-worked. It is subsequently solution-treated at 480–500° C for one to three hours

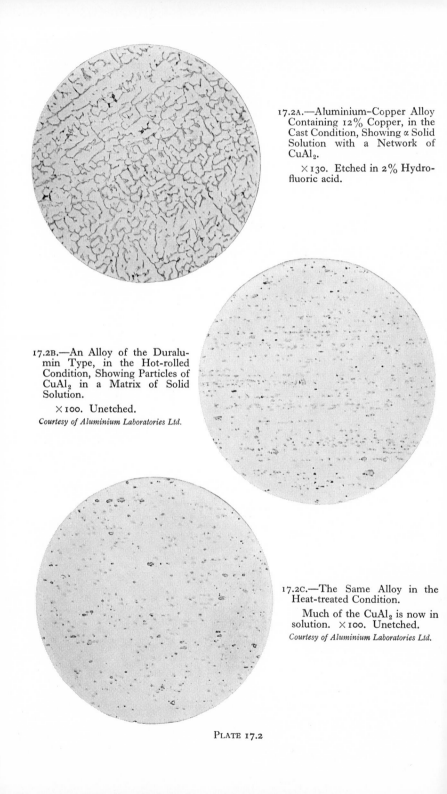

17.2A.—Aluminium–Copper Alloy Containing 12% Copper, in the Cast Condition, Showing α Solid Solution with a Network of $CuAl_2$.

×130. Etched in 2% Hydrofluoric acid.

17.2B.—An Alloy of the Duralumin Type, in the Hot-rolled Condition, Showing Particles of $CuAl_2$ in a Matrix of Solid Solution.

×100. Unetched.

Courtesy of Aluminium Laboratories Ltd.

17.2C.—The Same Alloy in the Heat-treated Condition.

Much of the $CuAl_2$ is now in solution. ×100. Unetched.

Courtesy of Aluminium Laboratories Ltd.

PLATE 17.2

in order to absorb the $CuAl_2$ and Mg_2Si and is then water-quenched. Accurate temperature control is essential, as the temperature of treat-

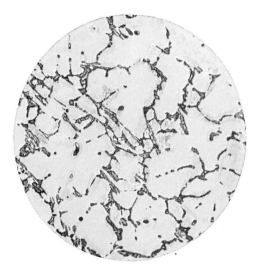

17.3A.—"Y Alloy" in the "As-cast" Condition, Showing Large Amounts of Inter-metallic Compounds in the Structure.

× 100. Etched in ½% Hydro-fluoric acid.

Courtesy of Aluminium Laboratories Ltd.

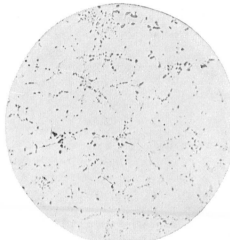

17.3B.—The Same Alloy in the Solution-treated and Pre-cipitation-hardened Condition.

Much of the intermetallic compounds are now in solution. × 100. Etched in ½% hydrofluoric acid.

Courtesy of Aluminium Laboratories Ltd.

PLATE 17.3

ment must be maintained between A and C (Fig. 17.2). After quenching, cold-working operations can be carried out on the resulting solid-solution structure if desired. The alloy is subsequently hardened, either by allowing it to remain at room temperature for about five days or by heating it to some temperature between 100 and 160° C for up to ten hours, according to the properties required. Heating at above 160° C may cause visible particles of $CuAl_2$ to appear in the microstructure, with consequent losses in strength and hardness.

17.56. Just as "tempering" will accelerate precipitation-hardening, so will refrigeration impede the process. This fact was utilised extensively in aircraft production during the Second World War. If a duralumin rivet is solution-treated it is in a soft condition, but if sufficient time is allowed to elapse before using the solution-treated rivet it will begin to harden and will possibly split whilst being driven. It was found that if, immediately after quenching, the duralumin rivets were stored in a refrigerator at about $-20°$ C, precipitation-hardening was considerably slowed down, and the rivets could therefore be stored at the sub-zero temperature until they were required.

17.57. Although the addition of copper forms the basis of many of the precipitation-hardening aluminium alloys, copper is absent from a number of them which rely instead on the presence of magnesium and silicon. Such alloys have a high electrical conductivity approaching that of pure aluminium, so that they can be used for the manufacture of overhead conductors of electricity. Most commercial grades of aluminium contain iron as an impurity, but in some of these alloys it is utilised in greater amounts in order to increase strength by promoting the formation of $FeAl_3$, which assists in precipitation-hardening. Titanium finds application as a grain refiner, whilst other alloys containing zinc and chromium produce tensile strengths in excess of 620 N/mm² in the heat-treated condition. A selection of typical wrought heat-treatable alloys of aluminium is given in Table 17.3.

Cast Alloys Which Are Heat-treated

17.60. Many of these alloys are of the 4·0% copper type, whilst others contain an addition of about 2·0% nickel. Precipitation-hardening is then due to the combined effects of the intermediate phases based on the intermetallic compounds $CuAl_2$ and $NiAl_3$. Possibly the best known of these alloys is "Y Alloy" ("Y" was the series letter used to identify the alloy during its experimental development at the National Physical Laboratory in the early days of the First World War). It contains approximately 4·0% copper, 2·0% nickel and 1·5% magnesium, and whilst fundamentally an alloy for casting, it can also be used in the wrought condition. As a cast alloy it is used where high strength is required at elevated temperatures, as, for example, in high-duty piston and cylinder heads.

17.61. There are other alloys similar in composition to Y alloy, but containing rather less copper. One of the interesting features of this type of alloy is that some of them (e.g. RR.50) can be precipitation-hardened by heating at 155–170° C without the preliminary solution-treatment at a high temperature. The 5·0 and 13·0% types of aluminium–silicon alloy are also brought into this group by the addition of small amounts of either copper and nickel or magnesium and manganese, which render the alloys amenable to precipitation-hardening.

A selection of the alloys representative of this group is shown in Table 17.4.

RECENT EXAMINATION QUESTIONS

1. What characteristics of aluminium and its alloys account for its versatility in uses? Suggest suitable competitors for two possible applications.
(University of Aston in Birmingham, Inter B.Sc.(Eng.).)

2. Write an essay on the methods available to improve the properties of aluminium alloys.   (Lanchester College of Technology, Coventry.)

3. Write an account of the structures and properties of the principal commercial aluminium–silicon alloys, dealing in particular with the process known as "modification".   (U.L.C.I., Paper No. B111–1–4.)

4. What is understood by the terms *modification* and *age-hardening* as applied to the alloys of aluminium? What is the structural basis of these procedures and why are they carried out?   (S.A.N.C.A.D.)

5. What is meant by the terms *solution-treatment* and *precipitation-treatment*? Illustrate your answer by reference to a suitable aluminium-base alloy.   (U.L.C.I., Paper No. C111–1–4.)

6. Describe the changes which occur during the usual treatment of an alloy containing 96% Al and 4% Cu.
(Rutherford College of Technology, Newcastle upon Tyne.)

7. What are the properties of aluminium and its alloys which make it suitable for many engineering applications?

Describe the heat-treatment which would be given to an alloy containing 4% copper, 0·5% magnesium, 0·5% silicon, 0·5% iron and 0·5% manganese to obtain: (*a*) a soft condition, suitable for further working; (*b*) the maximum strength.

Explain briefly the meaning of the term "modification" as applied to aluminium alloys containing 12% silicon, and state the use of these alloys.   (Constantine College of Technology, H.N.D. Course.)

8. Give an account of the theoretical basis of precipitation-hardening treatment.

Describe the application of this type of treatment to an aluminium-base alloy, stating the composition of the alloy and the mechanical properties before and after treatment.
(Derby and District College of Technology, H.N.D. Course.)

9. Write an essay on "The Heat-treatment of Aluminium Alloys". The essay should include a full account of the mechanism and theories of ageing.   (Aston Technical College.)

10. What are the essentials of precipitation-hardening treatments? Give three distinct examples of alloys in which precipitation-hardening is found.   (University of Aston in Birmingham, Inter B.Sc.(Eng.).)

11. Copper and aluminium form two useful commercial alloy series—the copper–aluminium alloys and the aluminium–copper alloys. Basing your answer on the appropriate thermal-equilibrium diagrams, discuss what you consider to be the most significant metallurgical aspect of each of the two alloy systems. Which of the two is put to greater practical use? Give reasons.

    Your diagrams may be approximate, sufficient only to show the principles involved.   (Aston Technical College, H.N.D. Course.)

12. High strength in aluminium alloys may be obtained in either the as-cast or heat-treated condition. Explain how the improvement in properties is obtained in each condition, giving examples of the structures developed and the alloying elements with the approximate concentrations used.

    (Lanchester College of Technology, Coventry, H.N.D. Course.)

13. Which types of alloy can be hardened (*a*) by cold-working; (*b*) by precipitation-hardening; (*c*) by a combination of (*a*) and (*b*)?

    Why is it necessary to exercise close control of heat-treatment variables in precipitation-hardening heat-treatment?

    (University of Aston in Birmingham, Inter B.Sc.(Eng.).)

14. Describe TWO methods of hardening aluminium alloys. Give the principles underlying each method.

    (University of Aston in Birmingham, Inter B.Sc.(Eng.).)

CHAPTER XVIII

OTHER NON-FERROUS METALS AND ALLOYS

Magnesium-base Alloys

18.10. It is perhaps surprising that magnesium-base alloys should be suitable for engineering purposes, since in composition they are similar to the incendiary bombs used by the R.A.F. and the Luftwaffe during the Second World War. Incendiary-bomb cases contained about 7% aluminium (which was added to introduce better casting properties) and small amounts of manganese; the balance being magnesium. Fortunately, combustion of such an alloy takes place only at a relatively high temperature, and this was generated by filling the bomb case with thermit mixture (21.54). Ordinarily, magnesium alloys can be melted and cast without mishap, provided that the molten alloy is not overheated.

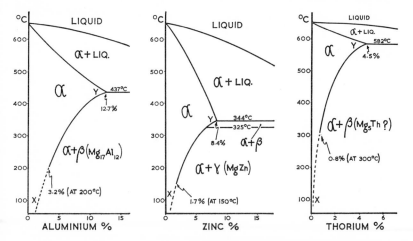

FIG. 18.1.—Changes in Solid Solubility Along XY Make Precipitation-hardening Possible in Each Case.

18.11. Although pure magnesium is a relatively weak metal, its alloys containing suitable amounts of aluminium, zinc or thorium can be strengthened considerably by precipitation-hardening. The thermal-equilibrium diagrams (Fig. 18.1) indicate the changes in solid solubility of these three metals in the respective α-phases; changes which are necessary if precipitation-hardening is to be effected (9.82).

TABLE 18.1.—*Cast Magnesium-base Alloys*

Composition (%)						Condition	Typical mechanical properties						Full heat-treatment
Al	Mn	Zn	Zr	Th	Rare earths		0.1% Proof stress		Tensile strength		Elongation (%)		
							N/mm²	tonf/in²	N/mm²	tonf/in²			
8.0	0.3	0.7	—	—	—	As cast Fully treated	77.2 97.3	5.0 6.3	139 210	9.0 13.6	2.0 2.0		Solution-treated for 8 hours at 380° C and then 16 hours at 410° C. Precipitation-hardened at 190° C for 10–12 hours.
10.0	0.3	0.7	—	—	—	Chill-cast	116	7.5	201	13.0	2.0		Solution-treated for 8 hours at 390° C and then 16 hours at 410° C. Precipitation-hardened at 200° C for 8 hours.
—	—	4.5	0.7	—	—	As cast Fully treated	121 151	7.8 9.8	232 203	15.0 17.0	10.0 7.0		Can be used in the as-cast condition or precipitation-hardened at 170° C for 12 hours without any preliminary solution-treatment.
—	—	4.2	0.7	—	1.2	As cast Fully treated	92.7 131	6.0 8.5	170 216	11.0 14.0	5.0 4.0		Solution-treated for 2 hours at 320° C. Precipitation-hardened at 170° C for 10 hours.
—	—	3.5	0.7	—	3.2	Stabilised	84.9	5.5	167	10.8	5.0		Castings are "stabilised" by heating at 170° C for 12 hours. They retain their strength up to 200° C.
—	—	—	0.7	3.2	—	Heat-treated	100	6.5	208	13.5	8.0		Solution-treated for 2 hours at 565° C and precipitation-hardened for 16 hours at 205° C.

TABLE 18.2.—*Wrought Magnesium-base Alloys*

Composition (%)					Condition	Typical mechanical properties					
Al	Mn	Zn	Zr	Th		0.1% Proof stress		Tensile strength		Elongation (%)	
						N/mm²	tonf/in²	N/mm²	tonf/in²		
—	1.5	—	—	—	Rolled Extruded	92.7 123	6 8	201 232	13 15	5.0 4.0	
3.0	0.3	1.0	—	—	Annealed Rolled	108 154	7 10	247 278	16 18	12.0 8.0	
6.0	0.3	1.0	—	—	Forged Extruded	154 139	10 9	278 216	18 14	8.0 8.0	
—	—	1.0	0.7	—	Rolled	154	10	247	16	8.0	
—	—	3.0	0.7	—	Rolled Extruded	170 216	11 14	263 309	17 20	8.0 8.0	
—	1.0	—	—	3.0	Rolled	216	14	278	18	10.0	

18.12. Small amounts of manganese, zirconium and the "rare earth" metals are also added to some magnesium alloys. Manganese improves corrosion-resistance, whilst zirconium is an effective grain refiner. Thorium and the rare-earth metals (in particular, cerium) give further increase in strength whilst allowing the alloy to retain a somewhat higher ductility, but their most important function is to give improved creep-resistance at high working temperatures.

The greater strength obtained by alloying, coupled with their low relative density of about 1·8, makes these alloys particularly useful where weight is a limiting factor. Such uses include various castings and forgings used in the aircraft industry—landing-wheels, petrol tanks, oil tanks, crankcases and airscrews, besides many engine parts in both piston and jet engines. These alloys, which are manufactured under such trade names as "Elektron" and "Magnuminium", include both cast and wrought alloys, examples of each being given in Tables 18.1 and 18.2.

18.13. The melting and casting of these alloys present some difficulty in view of the ease with which molten magnesium takes fire, particularly when overheated. They are generally melted under a flux containing the fluorides of calcium and sodium and the chlorides of potassium and magnesium. During the casting operation flowers of sulphur is often dusted on to the stream of molten alloy. The sulphur takes fire in preference to the magnesium, and thus protects the latter with a surrounding atmosphere of sulphur dioxide gas.

Zinc-base Die-casting Alloys

18.20. These are zinc–aluminium–copper alloys, typical compositions of which are given in Table 18.3. They are used for the die-casting of a

TABLE 18.3.—*Zinc-base Die-casting Alloys*

Composition (%)								Shrinkage (after 5 weeks normal ageing) (mm/metre)
Al	Cu	Mg max.	Fe max.	Pb max.	Cd max.	Sn max.	Zn	
4·1	2·7	0·03	0·07	0·003	0·003	0·001	Balance	−1·12
4·1	1·0	0·03	0·07	0·003	0·003	0·001	Balance	−0·83
4·1	—	0·04	0·07	0·003	0·003	0·001	Balance	−0·65

very wide range of components, both in the engineering industry and for domestic appliances under the trade name of "Mazak". These die-castings include parts for such diverse articles as motor-cars, washing-machines, accurately scaled children's toys, radio sets, alarm clocks and electric fires.

18.21. Difficulty was experienced in the development of these zinc-base alloys owing to swelling of the casting during use, coupled with intercrystalline embrittlement. These faults were found to be due to intercrystalline corrosion caused by small amounts of cadmium, tin and lead. For this reason high-grade zinc of "four nines" quality (i.e. 99·99% pure) is used for the manufacture of these alloys, and the impurities thus limited to the small quantities indicated in Table 18.3.

18.22. Normally, after casting, these zinc-base alloys undergo a slight shrinkage which is complete in about five weeks. When close dimensional limits are necessary a stabilising anneal at 150° C for three hours should be given before machining, in order to accelerate any volume change that is likely to take place.

In order to test these alloys for "growth" and intercrystalline corrosion, the "steam test" is applied. This accelerates the formation of these defects, and consists in suspending the test piece above water at 95° C in a closed vessel for ten days. The test piece is then examined for expansion, and mechanically tested to reveal any intercrystalline corrosion which may have occurred.

18.23. The corrosion-resistance of zinc-base alloys can be increased by a process called "chromate passivation". The components are degreased and then immersed for up to one minute in a solution containing sulphuric acid and sodium bichromate. They are then washed and dried. The resulting passive film varies in colour from iridescent pinkish-green to yellowish-brown, and produces considerable improvements in corrosion-resistance, particularly in moist atmospheres.

Nickel–Chromium High-temperature Alloys

18.30. The main feature of the nickel–chromium alloys is their ability to resist oxidation at elevated temperatures. If further suitable alloy additions are made the strength is increased under conditions of stress at high temperatures. Alloys rich in nickel and chromium have a high specific resistance to electricity, which makes them admirable materials for the manufacture of resistance wires, and (because of their low rate of oxidation at high temperatures) heater elements of many kinds capable of working at temperatures up to bright red heat. Such alloys (the "Brightray" series) are included in Table 18.4.

18.31. A further group of high-temperature, nickel–chromium-base alloys are those containing iron and, sometimes, molybdenum and tungsten. Of these "Inconel" is probably the best known and contains 80% nickel, 14% chromium and 6% iron. It is used for a number of purposes, including food-processing equipment, tubular electric-heating elements for cookers, exhaust manifolds, etc. "Hastelloy C"

(55% nickel, 17% molybdenum, 15% chromium, 5% tungsten and 5% iron) is resistant to corrosion by oxidising acids such as nitric and sulphuric, as well as to oxidising and reducing atmospheres up to 1 000 C; whilst "Hastelloy X" (47% nickel, 22% chromium, 18% iron, 9% molybdenum and 1% tungsten) maintains reasonable strength and good resistance to oxidation up to 1 200° C. It is used for furnace equipment as well as in jet-engine construction.

TABLE 18.4.—*High-temperature Resistance Alloys*

Composition (%)				Tensile strength of annealed rod		Resistivity $(10^{-8}\,\Omega\,m)$	Maximum working temp. (° C)	Uses
Ni	Cr	Fe	Cu	N/mm²	tonf/in²			
80	20	—	—	911	59	103	1150	Heaters for electric furnaces, cookers, kettles, immersion heaters, hair dryers, toasters, etc.
65	15	20	—	726	47	106	950	Similar uses to above, but for goods of lower quality—also for soldering-irons, tubular heaters, towel rails, laundry irons, etc.
34	4	62	—	664	43	91	700	Cheaper-quality heaters working at low temperatures, but mainly as a resistance wire for motor-starter resistances, etc.
45	—	—	55	402	26	48	300	Limited use for low-temperature heaters such as bed-warmers, etc. Mainly as a resistance wire for instrument shunts, field regulators and cinema arc resistances.

18.32. By far the most spectacular of the nickel–chromium-base alloys, however, are those of the "Nimonic" series. "Nimonic" is a trade name coined by Messrs. Henry Wiggin, the manufacturers of these alloys, which have played a leading part in the construction of jet engines. They are basically nickel–chromium alloys which will resist oxidation at elevated temperatures, but which have been further "stiffened" for use at high temperatures by the addition of small amounts of titanium, aluminium, cobalt, zirconium and carbon in suitable combinations. These elements help to raise the limiting creep stress at high temperatures by the formation of carbides or intermetallic compounds such as $NiAl_3$. The result is a series of alloys which will withstand oxidation and mechanical stresses at working temperatures in excess of 1 000° C. In addition to their use in jet engines, they find application as thermocouple sheaths, tubular furnaces for the continuous annealing of wire, jigs for supporting work during brazing or vitreous enamelling, gas retorts, extrusion dies and moulds for precision glass manufacture. Some of the current "Nimonic" alloys are shown in Table 18.5. Tensile strengths at various temperatures are shown, since they can be represented simply in tabular form. It should be appreciated, however, that the true criteria to usefulness will be high limiting creep stresses at elevated temperatures, available with these alloys.

18.33. Nickel-base Corrosion-resistant Alloys for use at ordinary temperatures are important to the chemical industry. "Corronel B" is a Henry Wiggin alloy containing 66% nickel, 28% molybdenum and 6% iron. It is particularly resistant to attack by mineral acids and acid chloride solutions under the most extreme conditions. Since it can be produced in the form of tubes and other wrought sections, it finds wide application in the chemical and petroleum industries for the construction of reaction vessels, pump and filter parts, valves, heat exchangers and immersion heaters.

"Ni-O-Nel" (40% nickel, 20% chromium, 3% molybdenum, 2% copper and balance iron) is also a Henry Wiggin alloy, similar in characteristics to some austenitic stainless steels but more resistant to general attack and also to stress–corrosion cracking in chloride solutions. Compared with 18–8 stainless steel, it is claimed to give better resistance to pitting attack in sea-water. "Hastelloy A" (58% nickel, 20% molybdenum and 20% iron) and "Hastelloy B" (65% nickel, 28% molybdenum and 5% iron) are used in the chemical industry for transporting and storing hydrochloric, phosphoric and other non-oxidising acids. "Hastelloy D" (85% nickel, 10% silicon, 3% copper and 1% aluminium) is a casting alloy which is strong, tough and very hard. It is difficult to machine and is usually finished by grinding. Its most important property is its resistance to corrosion by hot concentrated sulphuric acid.

The copper–nickel alloys are dealt with in 16.60.

Bearing Metals

18.40. The essential qualities of a bearing alloy are that it should be relatively tough and ductile to withstand mechanical shock, but at the same time be hard and abrasion-resistant, so that it will withstand wear and run with the lowest possible friction losses. Such properties would be difficult to obtain simultaneously in any single-phase alloy. Whilst solid solutions are tough, they lack hardness, and though intermetallic compounds are hard, they lack toughness. The problem is solved, therefore, by using an alloy which is a suitable combination of *both* phases. That is, it consists of hard particles of an intermetallic compound, firmly embedded in a matrix of solid solution, or sometimes a eutectic of two solid solutions. The soft matrix will yield to some extent

TABLE 18.5.—*"Nimonic" High-temperature Alloys*

"Nimonic" alloy	D.T.D. specn.	Approximate compositions %									Tensile strength at:							
		C	Fe	Cr	Ti	Al	Co	Mo	Nb + Ta	Ni	400° C		600° C		800° C		1 000° C	
											N/mm²	tonf/in²	N/mm²	tonf/in²	N/mm²	tonf/in²	N/mm²	tonf/in²
75	703B	0·1	2·5	20	0·4	—	—	—	—	Balance	680	44	587	38	247	16	92·7	6
80A	736A	0·05	1·5	20	2·2	1·4	—	—	—	Balance	1 140	74	1 080	70	541	35	77·2	5
90	747A 5027	0·06	1·5	20	2·4	1·4	18	—	—	Balance	1 140	74	1 080	70	695	45	77·2	5
105	5007	0·1	1·0	14·5	1·2	4·7	20	5	—	Balance	1 130	73	1 080	70	849	55	170	11
115	5017	0·1	0·5	15	4	5	15	4	—	Balance	1 080	70	1 080	70	1 020	66	432	28
PE 11	5037	0·05	Bal	18	2·3	0·8	—	5	—	38	973	63	958	62	494	32	92·7	6
PE 16	5047	0·05	Bal	16·5	1·2	1·2	—	3·2	—	43·5	788	51	711	46	371	24	77·2	5
PK 31	—	0·04	0·5	20	2·3	0·4	14	4·5	5·1	Balance	1 210	78	1 110	72	741	48	—	—
PK 33	5057	0·04	0·5	18	2·2	2·1	14	7	—	Balance	1 020	66	958	62	741	48	154	10

under pressure, and thus accommodate any slight misalignment of a shaft. At the same time it tends to wear, leaving particles of inter-metallic compound standing in relief, so that not only are friction losses reduced but also minute channels are provided which will assist the flow of lubricants.

18.41. Copper-base Bearing Alloys include the phosphor-bronzes (containing from 10 to 13% tin and 0·3 to 1·0% phosphorus) and the plain tin bronzes (containing from 10 to 15% tin). Both types of bronze satisfy the structural requirements of a bearing metal, since they contain particles of the hard intermetallic compound δ ($Cu_{31}Sn_8$) embedded in a tough matrix of the solid solution α. These alloys are very widely used for bearings of many types where the loading is heavy.

For small bearings and bushes of standardised sizes, sintered bronzes are often used. These are generally of the self-lubricating type, and are made by mixing copper powder and tin powder together, in the pro-portions of a 90–10 bronze, with the addition of graphite. The mixture is then compacted and sintered (6.53—Part I and 6.50—Part II) to produce a semi-porous, self-lubricating alloy suitable for low-duty bearings in sizes up to about 75 mm diameter shaft.

Leaded bronzes are useful in the manufacture of the main bearings in aero-engines, and for automobile and diesel crankshaft bearings. They have a high resistance to wear, and their good thermal conductivity enables them to keep cool when running. Since lead is insoluble in bronze, these bearings are generally centrifugally cast to prevent ex-cessive segregation of the lead. Alternatively, segregation is limited by adding small amounts of nickel to the alloy.

A low-tin bronze (2·5% tin), to which 1·5% iron and 1·2% chromium have been added, is useful in conditions involving both high-tempera-tures and excessive wear. The necessary hard particles are present in the form of an intermetallic compound based on tin, iron and chromium, and these are embedded in a tough, ductile matrix of copper containing a small amount of tin in solid solution. This alloy is suitable as a bearing metal for valve guides of internal-combustion engines and for the bearings of high-duty worm wheels. An increase in the tin content to 7·5% gives a cast chromium bronze which will withstand very severe abrasion, such as is encountered in various types of mining equipment.

Because of their lower cost, brasses are sometimes used, both as direct bearing alloys or, alternatively, to act as backings for thin bronze bearing shells. Bearing brasses are usually of the 60–40 type, with up to 1·0% each of aluminium, iron and manganese.

18.42. "White" Bearing Metals may be either tin-base or lead-base. The former, which represent the better-quality white bearing alloys, are often called "Babbitt metals", after Isaac Babbitt, their original patentee. All of these alloys contain between 3·5 and 15·0% antimony, much of which combines with some of the tin present to form an intermetallic compound SbSn. This compound forms cubic crystals, usually called "cuboids" (see Plate 18.1), which are hard and of low-friction properties.

Let us assume that we have a plain tin–antimony alloy containing 10% antimony, as being typical of a basic bearing-metal analysis. Fig. 18.2

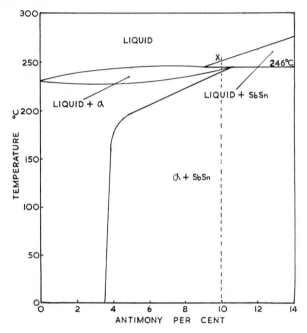

FIG. 18.2.—The Tin–Antimony Thermal-equilibrium Diagram.

shows the tin-rich end of the tin–antimony equilibrium diagram. Our 10% alloy will begin to solidify at X by depositing cuboids of the hard intermetallic compound SbSn. These SbSn crystals are of lower specific gravity than the liquid which remains, and they therefore tend to float to the surface of the melt. At 246° C a peritectic reaction takes place, and the solid solution α is formed as an in-filling between the cuboids. As the temperature falls, the solubility of antimony in tin decreases from about 10·3% at the peritectic temperature to 3·5% at room temperature, so that if cooling is slow enough, more cuboids of SbSn are precipitated uniformly within the solid solution α. Nevertheless, the segregation of the primary cuboids of SbSn at the surface of the cast bearing is a serious fault. It can, however, be minimised in one of the following ways:

(*a*) By rapidly cooling the cast alloy when possible. The metal then solidifies completely before the cuboids have had an opportunity to rise to the surface.

(*b*) By adding up to 3·5% copper to the alloy. This combines with some of the tin, forming a network of needle-shaped crystals of Cu_6Sn_5 which precipitates from the melt *before* the cuboids of SbSn begin to form. This network traps individual cuboids as they develop and prevents their movement towards the surface, thus ensuring a uniform cast structure. Moreover, the compound Cu_6Sn_5, being hard, improves the general bearing properties.

Lead is added in the interests of cheapness, and this forms solid solutions of limited solubility with both tin and antimony. These solid solutions form a eutectic structure. In some of the cheap bearing metals, e.g. "Magnolia metal", tin is omitted completely. The cuboids then consist of almost pure antimony, embedded in a eutectic matrix of antimony-rich and lead-rich solid solutions. These lead-rich alloys are intended only for low pressures. The more important white bearing metals are indicated in Table 18.6.

TABLE 18.6.—*White Bearing Metals*

B.S. 3332 designation	Composition (%)					Characteristics and uses
	Sn	Sb	Cu	Pb	Zn	
/1	90	7	3	—	—	High unit loads and high temperatures.
/2	87	9	4	—	—	Main bearings in automobile and aero engines.
/3	81	10	5	4	—	British Rail—moderately severe duties.
/4	75	12	3	10	—	Railway carriage work. Pumps, compressors and general machinery.
/5	75	7	3	15	—	
/6	60	10	3	27	—	
/7	12	13	1	74	—	Railway wagon bearings, mining machinery and electric motors.
/8	5	15	—	80	—	"Magnolia Metal"—Light loads and speeds at higher temperatures.
/9	68·5	—	1·5	—	30	Under-water bearings.

18.43. Other Lead-base Bearing Metals contain small amounts of the "alkaline-earth" metals calcium and barium, and sodium. These alloys are manufactured by electrolysing fused calcium, sodium or barium chloride, using a molten lead cathode in which the liberated metal dissolves. Such alloys are of the precipitation-hardening type, utilising the intermetallic compounds Pb_3Ca or Pb_3Ba for this effect.

18.44. Cadmium-base and Zinc-base Alloys are also used as bearing metals to a limited extent. The former have properties similar to the tin-base alloys, but the latter have been developed mainly as substitutes. They are used only for light work, since they lack plasticity and the ability to adjust themselves for any slight misalignment. Their anti-friction properties are also inferior to those of the tin-base alloys.

Type Metals

18.50. The alloys used for the manufacture of printing type are ternary alloys based on lead, tin and antimony. Antimony imparts hardness and resistance to wear, and also reduces contraction during solidification. This latter feature ensures that the alloy will take a good impression of the mould.

Tin increases toughness and fluidity and lowers the melting point. Ordinary type metal contains 10–25% tin, 20–30% antimony, up to 1·5% copper and the balance lead. "Linotype" machines use an alloy containing 2·5–5·0% tin, 10–13% antimony and the balance lead.

Fusible Alloys

18.60. These contain varying amounts of lead, tin, bismuth and cadmium, and are generally ternary or quaternary alloys of approximately eutectic composition. A number of such alloys will melt below

TABLE 18.7.—*Fusible Alloys*

Type of alloy	Composition (%)					Melting point (° C.)
	Bi	Pb	Sn	Cd	Sb	
Tinman's solder . .	—	38	62	—	—	183
Cerromatrix . . .	48	28·5	14·5	—	9	102–225
Rose's alloy . . .	50	28	22	—	—	100
Wood's alloy . . .	50	24	14	12	—	71
Lipowitz's alloy . .	50	27	13	10	—	65
Dental alloy . . .	53·5	17·5	19	—	Hg 10·5	60

the boiling point of water, and in addition to such obvious uses as the manufacture of teaspoons for practical jokers, they are used for metal pattern work, for the fusible plugs in automatic fire-extinguishing

18.1A.—Vertical Section Through a Small Ingot of Bearing Metal (8% Antimony; 32% Lead; 60% Tin).

A slow rate of cooling has enabled the cuboids of SbSn to float to the surface, where they have formed a layer. ×4. Etched in 2% Nital.

18.1B.—Vertical Section Through a Similar Alloy (with the Addition of 3% Copper).

Although cooled at the same rate as the ingot in 17.1A, segregation of the SbSn cuboids has been prevented by the needles of Cu_6Sn_5 which formed first. ×4. Etched in 2% Nital.

18.1C.—Bearing Metal Containing 10% Antimony; 60% Tin; 27% Lead; and 3% Copper.

Cuboids of SbSn (light) and needles of Cu_6Sn_5 (light) in a matrix consisting of a eutectic of tin-rich and lead-rich solid solutions. ×100. Etched in 2% Nital.

PLATE 18.1

sprinklers, for dental work, for bath media in the low-temperature heat-treatment of steels and for producing a temporary filling which will prevent collapse of the walls during the bending of a pipe. Bismuth-rich alloys such as "Cerromatrix" expand slightly on cooling, and are useful for setting press-tool punches in their holders.

Titanium

18.70. Mention of the use of small proportions of titanium in some steels (13.44) and aluminium alloys (17.57) has already been made. In recent years, however, a considerable amount of research has been carried out to assess the suitability of titanium for use as an engineering material in both the pure and the alloyed form.

Titanium is not a newly discovered element (it was discovered in Cornwall in 1791 by W. Gregor, an English priest), and is only rare in the sense that it is a little-used metal, and one which is expensive to produce. In Nature it is very widely distributed, being the tenth element in order of abundance in the Earth's crust. Of the metallurgically useful metals only aluminium, iron, sodium and magnesium occur more abundantly (Table 1.2).

18.71. Titanium is a bright silvery metal and, when polished, resembles steel in appearance. The high-purity metal has a relatively low tensile strength (216 N/mm^2) and a high ductility (50%), but commercial grades contain impurities which raise the tensile strength to as much as 700 N/mm^2 and reduce ductility to 20%. It is an allotropic element. The α-phase, which is hexagonal close-packed in structure, is stable up to 882.5° C, whilst above this temperature the body-centred cubic β-phase is stable. The allotropic change point is affected by alloying in a similar manner to the A_3 point in iron (13.11). Thus, alloying elements which have a greater solubility in the α-phase than in the β-phase tend to stabilise α over a greater temperature range (Fig. 18.3 (i)). Such elements include those which dissolve interstitially —oxygen, nitrogen and carbon—and also aluminium which dissolves substitutionally. Elements which tend to dissolve in β, and consequently stabilise it, are usually those which, like β, are body-centred cubic in structure. These are mainly the "transition elements" iron, chromium, molybdenum, etc., and the resulting equilibrium diagrams are of the types (ii) or (iii) (Fig. 18.3). Those alloys represented by an equilibrium diagram of type (iii) can be heat-treated by precipitation-hardening— as, indeed, could be expected with this type of diagram (9.82).

The β-phase in alloys tends to be hard, strong and less ductile than the relatively soft α-phase. Nevertheless, the β-phase forges well so that most of the commercial alloys, which are of the $\alpha + \beta$ type, are hot-

worked in the β range. The high-strength alloy I.M.I. Ti 680 (11%
tin; 2·25% aluminium; 4% molybdenum; 0·2% silicon) attains a
strength of 1 300 N/mm² when suitably heat treated and is used
in the form of structural forgings in the nacelle region of the Anglo-
French "Concorde" aircraft. Another alloy—I.M.I. Ex 700, containing
6% aluminium; 5% zirconium; 3% molybdenum and 1% copper—
is capable of a tensile strength of 1 540 N/mm².

Since the specific gravity of titanium is only 4·5, alloys based on the
metal have a high strength/weight ratio. Moreover, the creep properties
up to 500° C are very satisfactory and the fatigue limit is high.
Titanium therefore finds application in the compressors of jet aero-
engines, where the modern tendency is to increase the compression
ratio, with consequent increase in thermodynamical heating. Moreover,
titanium has a high melting point (1 725° C) coupled with a good
resistance to corrosion.

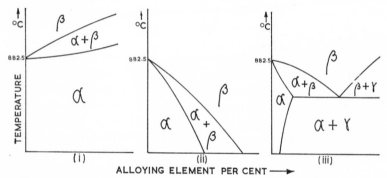

FIG. 18.3.—Effects of Alloying on the α ⟶ β Transition Temperature in
Titanium.

18.72. Despite the extravagance of many claims which have been
made in the popular Press concerning the future of titanium, it will undoubtedly play an increasingly important role in the design of aircraft
of the future. Thus, the design of the American Boeing 2707 super-
sonic transport plane is based largely upon titanium alloys. Although
the development of titanium technology in recent years has reduced the
price of billets to about £2·5 per kg, it may never be much cheaper to
produce since molten titanium combines rapidly with oxygen and
nitrogen, and must therefore be melted in a vacuum. It also reacts
readily with all known refractories and is, at present, melted by means
of an electric arc, in a water-cooled copper crucible, on to which molten
titanium does not weld.

Uranium

18.80. The mineral pitchblende was formerly thought to be an ore of tungsten, iron or zinc, until in 1789 M. H. Klaproth proved that it contained what he called "a half-metallic substance" different from the three elements named. He named this new element "Uranium" in honour of Herschel's discovery of the planet Uranus a few years earlier. It was not until 1842, however, that E. M. Péligot, by isolating the metal itself, showed that Klaproth's "element" was in fact the oxide.

18.81. In the solid form uranium is a lustrous, white metal with a bluish tint, and capable of taking a high polish. It is a very "heavy" metal, having a relative density of 18·7. Pure uranium is both malleable and ductile, and commercially it can be cast and then fabricated by rolling or extrusion, followed by drawing. It is a chemically reactive metal and oxidises at moderately high temperatures. Consequently, it must be protected during fabrication processes. For atomic energy purposes it is usually vacuum melted in high-frequency furnaces, and cast in 25 mm diameter rods.

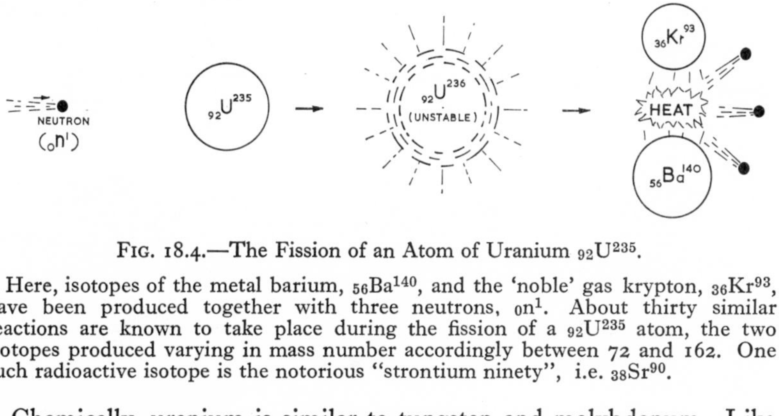

FIG. 18.4.—The Fission of an Atom of Uranium $_{92}U^{235}$.

Here, isotopes of the metal barium, $_{56}Ba^{140}$, and the 'noble' gas krypton, $_{36}Kr^{93}$, have been produced together with three neutrons, $_{0}n^{1}$. About thirty similar reactions are known to take place during the fission of a $_{92}U^{235}$ atom, the two isotopes produced varying in mass number accordingly between 72 and 162. One such radioactive isotope is the notorious "strontium ninety", i.e. $_{38}Sr^{90}$.

Chemically, uranium is similar to tungsten and molybdenum. Like them, it forms stable carbides which led to the experimental use of uranium in high-speed steels. These steels have not survived, since they showed no advantage over the orthodox alloys.

18.82. Until its development as a source of "fissionable material", uranium was little more than a chemical and metallurgical curiosity so that, before 1943, there was no commercial extraction process operating with the object of uranium as the sole product. Its industrial use, before the Second World War, was limited to the manufacture of electrodes for gas-discharge tubes; whilst its compounds were used in the manufacture of incandescent gas mantles.

18.83. The reader will, no doubt, associate uranium with the production of nuclear power. This element has two principal isotopes (1.90), the atoms of which contain 92 protons and 92 electrons in each case. However, one isotope has 146 neutrons giving a total mass of 238, whilst the other has 143 neutrons giving a total mass of 235. These isotopes are generally represented by the symbols $_{92}U^{238}$ and $_{92}U^{235}$ respectively. Here the lower (prefix) index represents the number of protons (92) in the nucleus (the *atomic number*, Z) and the upper (suffix) index represents the total number of protons and neutrons (the *atomic mass*, A, of the isotope).

Since a neutron has no electrical charge, and hence suffers neither attraction nor repulsion by either the negatively charged electrons or positively charged protons, it can be fired into the nucleus of an atom. When a neutron enters the nucleus of an atom of $_{92}U^{325}$ the latter becomes unstable and splits into two approximately equal portions (Fig. 18.4). At the same time a small reduction in the total mass occurs, for although the total number of particles (protons and neutrons) remains the same after fission, the sum of the masses of the resultant nuclei is slightly less. This is a measure of the *mass* equivalent of the nuclear binding *energy*. Einstein's Theory of Relativity states that mass and energy are not distinct entities but are really different manifestations of

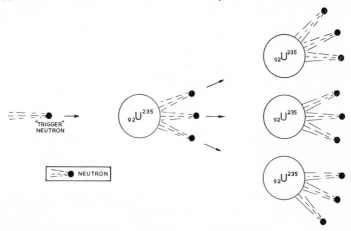

FIG. 18.5.—Chain Fission of $_{92}U^{235}$ Atoms.

the same thing. The two are related to the velocity of light (c) by the expression:

$$E = mc^2$$

Since the velocity of light is $2 \cdot 998 \times 10^8$ m/s it follows that even a small loss in mass can result in a great release of energy. As is often the case in engineering problems, this energy is emitted in the form of heat. It can be calculated that the energy produced by the loss of one gramme of matter in this way is sufficient to heat about 200 tonnes of water from 0 to 100° C.

Fission is also accompanied by the emission of neutrons and if the overall mass of $_{92}U^{235}$ is great enough, other atoms will absorb some of these neutrons so that a chain reaction occurs leading to an "atomic explosion". There is a "critical size" for a mass of $_{92}U^{235}$. Below this critical size neutrons will tend to escape rather than be absorbed by $_{92}U^{235}$ nuclei, and a chain reaction will not, therefore, be promoted.

18.84. The rate of production of nuclear energy from this source can be controlled by introducing into an "atomic pile" rods of some element which will absorb unwanted neutrons and thus control the rate of disintegration of other $_{92}U^{235}$ atoms. In natural uranium total disintegration will not occur since only 0·7% $_{92}U^{235}$ is present, mixed with the more common isotope $_{92}U^{235}$. Isotope $_{92}U^{235}$ will however, absorb a neutron to give an atom of nuclear mass 239. This undergoes further change to produce plutonium, $_{94}Pu^{239}$, which can be used as a concentrated fuel in such "fast reactors" as that at Dounreay.

Recent Examination Questions

1. Write a short essay on light-weight non-ferrous alloys, their properties and uses.   (University of Glasgow, B.Sc.(Eng.).)
2. (*a*) What is the main alloying addition made to zinc to produce zinc-based die-castings?

    (*b*) Indicate the characteristics that make these articles useful in engineering practice.

    (*c*) Describe ONE piece of equipment suitable for the manufacture of these products.

    (*d*) Indicate, with reasons, whether or not the equipment you have described is also suitable for making aluminium die-castings.

    (Institution of Production Engineers Associate Membership Examination, Part II, Group A.)
3. Discuss the factors which influence the choice of nickel-based alloys for high-temperature engineering.   (S.A.N.C.A.D.)
4. The development of materials to operate at elevated temperatures has been a most important aspect of metallurgical progress. Give a comprehensive account of the ferrous and non-ferrous alloys at present available for operation in the 600–1 000° C range, discussing such factors as resistance to oxidation, resistance to corrosion and retention of strength, and showing how these are obtained in particular alloys. Typical compositions and applications should be given.

    (Aston Technical College.)

5. Give the approximate composition of any commonly used metal in the group known as nickel-base alloys, outlining its main properties and giving two typical applications. (C. & G., Paper No. 293/23.)

6. Discuss the type of bearing materials produced both by powder metallurgy and by the more usual casting methods, with particular reference to the microstructure requirements in each case.

(West Bromwich Technical College.)

7. What are the requirements of a bearing metal? Give the standard composition and sketch the typical microstructure of one tin-base alloy and one lead-base alloy.

(Derby and District College of Technology, H.N.D. Course.)

8. Enumerate the principal properties which are to be desired in a bearing alloy. Illustrate how these properties are obtained by reference to typical metallic bearing materials.

(Constantine College of Technology, H.N.D. Course.)

9. List, and briefly describe, the essential requirements of a bearing metal. Show, by discussing the constitution and structure of the materials, how the requirements are realised in: (i) certain copper-based alloys; and (ii) white metals.

Compare the applications of the two groups and give typical compositions. (Aston Technical College, H.N.D. Course.)

10. In the field of aircraft design, certain components may be made from either aluminium-base alloys, iron-base alloys or titanium-base alloys.

What are the principal factors governing the selection of the material? Give a full description of the mechanism involved in the hardening of ONE of these alloys. (The Polytechnic, London, W.1., Poly.Dip.)

CHAPTER XIX

THE SURFACE-HARDENING OF STEELS

19.10. The service conditions of many steel components such as cams and gears, make it necessary for them to possess both hard, wear-resistant surfaces and, at the same time, tough, shock-resistant cores. In plain carbon steels these two different sets of properties exist only in alloys of different carbon content. A low-carbon steel, containing approximately 0·1% carbon, will be tough, whilst a high-carbon steel of 0·9% or more carbon will possess adequate hardness when suitably heat-treated.

19.11. The situation can best be met by employing a low-carbon steel with suitable core properties and then causing either carbon or nitrogen to penetrate to a regulated depth into the surface skin; as in the principal surface-hardening processes of carburising and nitriding. Alternatively, a steel of medium carbon content and in the normalised condition can be used, local hardness at the surface then being introduced by one or other of the flame-hardening processes. In the first case the hardenable material is localised, whilst in the second case it is the heat-treatment itself which is localised.

Case-hardening

19.20. The principles of case-hardening have been used for centuries in the conversion of wrought iron to steel by the Cementation Process. Both this ancient process and case-hardening make use of the fact that carbon will diffuse into iron provided that the latter is in the γ form which exists above 910° C. Thus case-hardening consists in surrounding the component with suitable carbonaceous material and heating it to above its upper critical temperature for long enough to produce a carbon-enriched layer of sufficient depth.

Solid, liquid and gaseous carburising media are used. The nature and scope of the work involved will govern which medium is best to employ.

19.21. Carburising in Solid Media. "Pack-carburising", as it is usually called, involves packing the components into cast-iron or steel boxes along with the carburising material so that a space of approximately 50 mm exists between the components. The lids are then luted on to the boxes, which are slowly heated to the carburising temperature, between 900 and 950° C. They are maintained at this temperature for up to five hours, according to the depth of case required. Fig. 19.1 indicates the relationship which exists between depth of case, carburising temperature and time of treatment.

Carburising media vary in composition, but consist essentially of some carbonaceous material, such as wood or bone charcoal or charred leather, together with an energiser which may account for up to 40% of the total

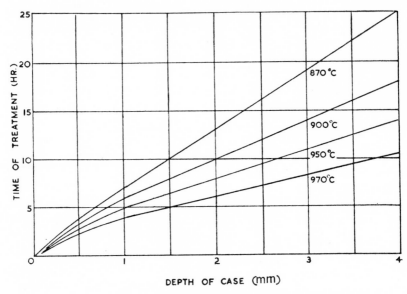

FIG. 19.1.—The Relationship Between Time and Temperature of the Carburising Treatment, and the Depth of Case Produced.

composition. This energiser is usually a mixture of sodium carbonate ("soda ash") and barium carbonate, and its purpose is to accelerate the solution of carbon by the surface layer of the steel.

It is thought that carburisation proceeds by dissociation of carbon monoxide which will be present in the hot box. When the gas comes into contact with the hot steel it dissociates thus:

$$2CO \rightleftarrows CO_2 + C$$

The atomic carbon deposited at the surface of the steel dissolves easily in the metal.

At the same time the barium carbonate reacts with some of the carbon from the charcoal, liberating more carbon monoxide thus:

$$BaCO_3 + C \rightleftarrows BaO + 2CO$$

The barium oxide then combines with carbon dioxide formed by the initial reaction:

$$BaO + CO_2 \rightleftarrows BaCO_3$$

The cycle of reactions is similar in the case of sodium carbonate, which also accelerates the absorption of carbon by the hot steel in the same manner as barium carbonate, though to a less degree.

If it is necessary to prevent any areas of the component from being carburised, this can be achieved by electro-plating these areas with copper to a thickness of 0·075–0·10 mm; carbon being insoluble in solid copper at the carburising temperature. An alternative method, which can be more conveniently applied in small-scale treatment, is to coat the area with a mixture of fireclay and ignited asbestos made into a paste with water. This is allowed to dry on the surface before the component is carburised.

When carburising is complete the components are quenched or cooled slowly in the box, according to the nature of the subsequent heat-treatment to be applied.

19.22. Carburising in a Liquid Bath. Liquid-carburising is carried out in baths containing from 20 to 50% sodium cyanide, together with up to 40% sodium carbonate and varying amounts of sodium or barium chloride. This cyanide-rich mixture is heated in "calorised" (22.64) pots to a temperature of 870–950° C, and the work, which is contained in wire baskets, is immersed for periods varying from about five minutes up to one hour, depending upon the depth of case required. One of the main advantages of cyanide-hardening is that pyrometric control is so much more satisfactory with a liquid bath. Moreover, after treatment the basket of work can be quenched. This not only produces the necessary hardness but also gives a clean surface to the components. The process is particularly useful in obtaining shallow cases of 0·10–0·25 mm.

The chemical reactions associated with cyaniding are not definitely known but are probably of quite a simple nature:

$$2NaCN + 2O_2 = Na_2CO_3 + 2N + CO$$

Dissociation of the carbon monoxide at the steel surface then takes place with the same result as in pack-carburising. The nitrogen, in atomic form, also dissolves in the surface and produces an increase in hardness by the formation of nitrides (Fig. 11.7) as it does in the nitriding process.

Cyanides are, of course, extremely poisonous, and every precaution *must* be taken to avoid inhaling the fumes from a pot. Every pot should be fitted with an efficient fume-extracting hood. Likewise the salts should in no circumstances be allowed to come into contact with an open wound. Needless to say, the consumption of food by operators whilst working in the shop containing the cyanide pots should be *absolutely forbidden.*

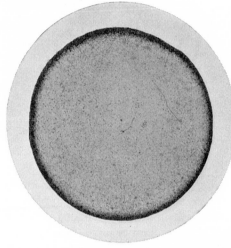

19.1A.—Cross-section Through a Carburised Bar with a Case Depth of Approx. 1 mm.

× 3. Etched in 5% Nital.

19.1B.—Structure in the Region of the Surface of a Carburised Bar, in the As-carburised Condition.

The original steel contains 0·15% carbon, but the carburised surface is now of eutectoid composition (0·83% carbon). × 35. Etched in 2% Nital.

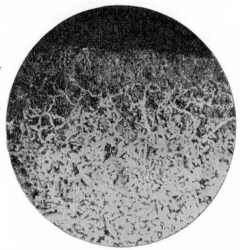

PLATE 19.1

19.23. Carburising by Gaseous Media. Gas-carburising is carried out in both batch-type and continuous furnaces. The components are heated at about 900° C for three or four hours in an atmosphere containing gases which will deposit carbon atoms at the surface of the components. The most important of these gases are the hydrocarbons methane, CH_4, and propane, C_3H_8. They should be of high purity in order to avoid the deposit of oily soot which impedes carburising. To facilitate better gas circulation—and hence, greater uniformity of treatment—the hydrocarbon is mixed with a "carrier" gas. This is generally an "endothermic" type of atmosphere made in a generator and consisting of a mixture containing mainly nitrogen, hydrogen and carbon monoxide.

The active agent in the atmosphere is possibly carbon monoxide, as is the case when carburising by so-called solid media:

$$2CO \rightleftharpoons C + CO_2$$

The atomic carbon thus released dissolves in the surface of the austenite (and on cooling to the lower critical temperature forms pearlitic cementite). The carbon dioxide formed in the above reaction probably reacts with methane as follows:

$$CO_2 + CH_4 \rightleftharpoons 2CO + 2H_2$$

The concentration of carbon monoxide is thus maintained so that carburising continues.

The relative proportions of hydrocarbon and carrier are adjusted to give the desired carburising rate. Thus, the concentration gradient of carbon in the surface can be "flattened" by prolonged treatment in a less rich carburising atmosphere. Control of this type is possible only with gaseous media.

Gas-carburising is becoming increasingly popular, particularly for the mass production of thin cases. Not only is it a neater and cleaner process but the necessary plant is more compact for a given output. Moreover, the carbon content of the surface layers can be controlled more accurately and easily, as was mentioned in the preceding paragraph.

19.24. Heat-treatment after Carburising. If carburising has been correctly carried out, the core will still be of low carbon content (0·1–0·2% carbon), whilst the case should have a maximum carbon content of 0·83% carbon (the eutectoid composition). If the carbon content of the case is higher than this, then a network of primary cementite will exist at the crystal boundaries, giving rise to intercrystalline brittleness and consequent exfoliation (or peeling) of the case during service.

The prolonged heating in the austenitic range will have introduced coarse grain, and further heat-treatment is desirable if the best properties are to be obtained.

19.25. Refining the Core. The component is first heat-treated with the object of refining the grain of the core, and consequently toughening it. This is effected by heating it to just above its upper critical temperature (about 870° C for the core) when the coarse ferrite–pearlite structure will be replaced by refined austenite crystals. The component is then water-quenched, so that a fine ferrite–martensite structure is obtained.

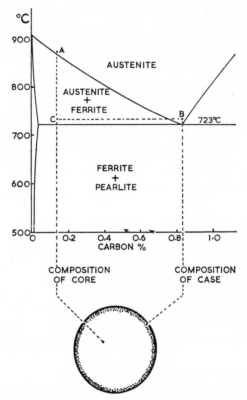

FIG. 19.2.—Heat-treatment After Carburising.
A Indicates the temperature of treatment for the core, and *B* the temperature of treatment for the case.

The core-refining temperature of 870° C is, however, still high above the upper critical temperature for the case, so that, at the quenching temperature, the case may consist of large austenite grains. On quenching these will result in the formation of coarse brittle martensite. Further treatment of the case is therefore necessary.

19.26. Refining the Case. The component is now heated to about 760° C, so that the coarse martensite of the case changes to fine-grained austenite. Quenching then gives a fine-grained martensite in the case.

At the same time the martensite produced in the core by the initial quench will be tempered somewhat, and much will be reconverted into fine-grained austenite embedded in the ferrite matrix (point C in Fig. 19.2). The second quench will produce a structure in the core consisting of martensite particles embedded in a matrix of ferrite grains surrounded by bainite. The amount of martensite in the core is reduced if the component is heated quickly through the range 650–760° C and then quenched without soaking. This produces a core structure consisting largely of ferrite and bainite, and having increased toughness and shock-resistance.

Finally, the component is tempered at about 200° C to relieve any quenching strains present in the case.

The above operations may be regarded as those which are highly desirable from a theoretical approach, but generally the heat-treatment of a case-hardened steel is modified for a number of reasons, not the least being the demands of economy. For example, the component may be "pot-quenched"; that is, quenched direct from the carburising temperature, followed by a low-temperature tempering process to remove quenching strains. This treatment will be satisfactory in the case of steels with slow grain growth at the carburising temperature.

In order to obtain maximum ductility in the core, the component can be slowly cooled from the carburising temperature and then reheated to 760° C and water-quenched. This quenching operation leaves the core quite soft and hardens the case, which will be fine-grained due to the low temperature of treatment.

Alternatively, the component can be cooled slowly from the carburising temperature, reheated to 820° C as a compromise between the upper and lower critical temperatures, water-quenched and then tempered at 200° C. The case will then be hard, though slightly coarse-grained, whilst the core will consist of a ferrite–bainite–martensite structure. The ferrite will be rather coarse-grained, remanent from the original carburising and will not be completely redissolved by heating to 820° C. Hence the core will not be of maximum toughness.

Case-hardening Steels

19.30. Both plain carbon and alloy steels are used for case-hardening, but the carbon content of either type should not be much above 0·3% if a tough core is to be maintained. Manganese may be present in amounts up to 0·9%, since it aids carburisation by stabilising the cementite. Manganese also increases the depth of hardening, but is liable to induce cracking during quenching. Silicon tends to cause graphitisation of the cementite, and, since it also retards carburisation, it is kept below 0·3%. Plain carbon steels used for case-hardening, then, contain up to 0·3% carbon and up to 0·9% manganese.

19.31. Alloy steels used for case-hardening contain up to 5·0% nickel and sometimes up to 1·0% chromium. Small amounts of vanadium and molybdenum are occasionally included. Nickel retards the

TABLE 19.1.—*Case-hardening Steels*

Typical composition (%)					Characteristics and uses
C	Mn	Ni	Cr	V	
0·10	0·8	—	—	—	Maximum toughness of core. Machine parts requiring shock-resisting properties.
0·17	0·7	—	—	—	Machine parts requiring a very hard surface and a tough core, such as gears, shafts and cams.
0·25	0·5	—	—	—	For high duty—ball- and roller-bearings in particular.
0·15	0·5	3·0	—	—	Used where combined hardness and toughness are required, as in gear-wheels, crank-pins, coupling pins, etc.
0·15	0·4	3·5	0·8	—	Parts requiring a glass-hard surface and toughness of core—high-duty gears, worm gears, crown wheels, clutch gears.
0·15	0·4	4·0	1·2	—	For a combination of surface hardness, stress-bearing and shock-resisting properties. Used for crown wheels, bevel pins, aero reduction gears.
0·40	1·5	—	1·0	0·15	Where maximum surface hardness is required and toughness of core is less essential.

grain growth during carburising, and so tends to give a fine-grained product. For this reason the core-refining treatment can be dispensed with in some applications of case-hardened nickel steels. The critical hardening speed is also reduced by the presence of nickel, so that an oil-quench may be used to harden the case. Strength and toughness are also increased, but the main disadvantages of nickel are that it retards carburisation and tends to promote graphitisation.

The main function of chromium is to increase the hardness and wear-resistance of the case and to improve the strength of the core without any serious loss in ductility. It must be used in limited amounts, however, because of its tendency to promote grain growth.

Nitriding

19.40. Nitriding resembles carburising in that the steel is heated for a considerable time in the hardening medium, but, whereas in carburising the medium is carbon, in nitriding it is gaseous nitrogen. Moreover, for the nitriding process special alloy steels are necessary, since hardening depends on the formation of the hard nitrides of such metals as aluminium, chromium and vanadium, at the surface of the component. Typical compositions of some of these special steels are given in Table 19.2.

19.41. Since it is conducted at a relatively low temperature, nitriding is made the final operation in the manufacture of the component, all

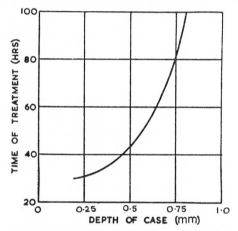

FIG. 19.3.—The Relationship Between Time of Treatment and Depth of Case Produced in the Nitriding Process.

machining and core-treatment processes having been carried out previously. The parts are maintained at 500° C for between 40 and 100 hours, according to the depth of case required, in a gas-tight chamber through which ammonia is allowed to circulate. Some of the ammonia dissociates according to the following equation:

$$NH_3 \rightleftharpoons 3H + N$$

Part of the "nascent" (or atomic) nitrogen thus formed is absorbed by the surface of the steel, forming nitrides both with the iron and with the metals mentioned above.

Attempts to nitride a plain carbon steel would fail because the iron nitrides formed would diffuse to quite a considerable depth, so that the surface hardness would be lost. Aluminium nitride, however, does not diffuse so easily and forms a shallow but extremely hard case of less than

TABLE 19.2.—*Nitriding Steels*

B.S. 970 designation	Proprietary trade names and codes	Composition (%)						Typical mechanical properties							Heat-treatment (for core)	Uses
		C	Mn	Cr	Mo	V	Al	Yield point		Tensile strength		Elongation (%)	Izod (J)	V.P.N. (case)		
								N/mm²	tonf/in²	N/mm²	tonf/in²					
—	Firth Brown LK1	0·5	0·65	1·6	0·2	—	1·1	1 000	65	1 240	80	13	39	1 050-1 100	Oil-quench from 900° C and temper 550-700° C.	Where maximum surface hardness is essential coupled with very high core strength.
905M39	Firth Brown LK3 Carr's P614	0·4	0·65	1·6	0·2	—	1·1	741	48	927	60	20	52	1 050-1 100	Oil-quench from 900° C and temper 600-700° C.	For maximum surface hardness and high core strength.
905M31	Firth Brown LK5 Carr's P612	0·3	0·65	1·6	0·2	—	1·1	556	36	741	48	24	65	1 050-1 100	Oil-quench from 900° C and temper 600-700° C.	For maximum surface hardness combined with reasonably high core strength.
—	Firth Brown LK7	0·2	0·65	1·6	0·2	—	1·1	463	30	618	40	30	72	1 050-1 100	Oil-quench from 900° C and temper 600-700° C.	For maximum surface hardness combined with ease of machining before hardening.
—	Firth Brown GK3	0·35	0·5	2·0	0·25	0·15	—	741	48	927	60	20	52	750-800	Oil-quench from 900° C and temper 600-700° C.	Moulds for plastics; other components requiring high hardness and good finish.
—	Firth Brown GK7	0·18	0·5	2·0	0·25	0·15	—	463	30	618	40	30	72	750-800	Oil-quench from 900° C and temper 600-700° C.	For ease of machinability and a high-class surface.
897M39	Firth Brown HCM3 Carr's P618	0·4	0·5	3·0	1·0	0·25	—	—	—	1 390	90	14	39	850-900	Oil-quench from 900° C and temper 550-650° C.	Ball races, etc, where high core strength is necessary.
722M24	Firth Brown HCM5 Carr's P615	0·3	0·45	3·0	0·4	—	—	772	50	1 000	65	20	52	800-850	Oil-quench from 900° C and temper 600-700° C.	Aero crankshafts, air-screw shafts, aero cylinders, crank-pins and journals.
—	Firth Brown HCM7	0·2	0·45	3·0	0·4	—	—	618	40	772	50	24	65	800-850	Oil-quench from 900° C and temper 600-700° C.	Aero-engine cylinders.

1·0 mm in depth. Chromium also helps to increase the case hardness, due to the formation of chromium nitride. It also diffuses to a greater depth than aluminium, and thus prevents a sudden change in composition from hardened skin to soft core. The latter situation would be likely to lead to spalling, or flaking, of the case. Molybdenum is added to steels for nitriding, mainly to toughen the core.

When core strength is important aluminium is usually omitted from the steel, and a somewhat lower case hardness is obtained, depending upon the formation of chromium and vanadium nitrides.

19.42. Before being nitrided, the components are heat-treated to produce the required properties in the core. The normal sequence of operations will be:

(a) oil-quenching from between 850 and 900° C, followed by tempering at between 600 and 700° C;

(b) rough-machining, followed by a stabilising anneal at 550° C for five hours to remove internal stresses;

(c) finish-machining, followed by nitriding.

Any parts of the component which are required soft are protected by coating with tin or solder, by nickel plating or by coating with a mixture of whiting and sodium silicate.

19.43. The Advantages of Nitriding are, briefly, as follows:

(a) Since no quenching is required after nitriding, cracking or distortion are unlikely, and components can be machine-finished before treatment.

(b) A very high surface hardness of up to 1 150 V.P.N. is obtained with the "Nitralloy" (aluminium-type) steels.

(c) Resistance to corrosion is good if the nitrided surface is left unpolished.

(d) Resistance to fatigue is good.

(e) The hardness is retained at elevated temperatures up to 500° C, whereas in a carburised component hardness begins to fall at about 200° C.

(f) The process is cheap when large numbers of components are to be treated.

19.44. The Disadvantages of Nitriding as compared with case-hardening are:

(a) The initial outlay for plant is higher than with case-hardening, so that the process is economical only when large numbers of components are to be treated.

(*b*) If the nitrided component is accidentally overheated the surface hardness will be lost completely, and the component must be nitrided again. A case-hardened component would only need to be heat-treated again (unless heating had been so excessive as to cause decarburisation).

19.45. The Satoh Process. This process, which was developed quite recently in Japan, provides a means of surface-hardening plain, low-carbon steels by nitriding in the presence of titanium. As was mentioned earlier (19.41), attempts to nitride a plain carbon steel are frustrated by the rapid diffusion of iron nitrides into the core. This is avoided in the Satoh process by plating a film of titanium on to the surface of the steel prior to the nitriding operation. During subsequent nitriding, titanium diffuses inwards along with the nitrogen and the two combine to form extremely hard and stable titanium nitride in the surface layers. It is claimed that a hard layer of adequate thickness is obtained after only 2 hours treatment, as against 40–100 hours with ordinary nitriding. Surface hardness is said to equal that of tungsten carbide.

Although the cost of the Satoh process is higher than that of ordinary nitriding because of the titanium-plating operation, it is pointed out that the increase in cost is more than offset by the savings in material costs arising from the possibility of using mild steel instead of the more costly alloy steels which are necessary for orthodox nitriding.

19.46. Carbonitriding. This is a surface-hardening process which makes use of a mixture of carburising gases and ammonia. It is sometimes known as "dry cyaniding"—a reference to the fact that a mixed carbide–nitride case is produced as in the ordinary liquid cyanide process (19.22). The relative proportions of carbon and nitrogen in the case can be varied by controlling the ratio of ammonia to hydrocarbons in the treatment atmosphere.

Flame-hardening

19.50. In this process both the core and the case are of the same composition, the surface being hardened solely by local heat-treatment. The surface is heated to above its upper critical temperature by a travelling oxy-acetylene torch, and immediately quenched by a water-jet attached to the torch. In order that heating and quenching shall proceed at the correct rate, machines are usually employed for the operation. Symmetrical components, such as gears, spindles and pins, are conveniently treated by this process, since they can be spun between centres.

19.51. Only steels with a carbon content above 0·4% can be hardened effectively, though small amounts of nickel (up to 4·0%) and chromium (up to 1·0%) can be added with advantage.

Before hardening, the components are generally normalised, so that the final structure will consist of a martensitic case about 3·75 mm thick, and a tough ferrite–pearlite core. Core and case will usually be separated by a layer of bainite, which reduces considerably the liability to spalling.

Induction-hardening

19.60. These processes are similar in principle to flame-hardening, except that the component is held stationary whilst the whole of its surface is heated simultaneously by electro-magnetic induction. The component is surrounded by an inductor block through which a high-frequency current in the region of 2 000 Hz, passes. This raises the temperature of the surface layer to above its upper critical in a few seconds. The surface is then quenched by pressure jets of water which pass through holes in the inductor block.

Thus, as in flame-hardening, the induction processes make use of the existing carbon content (which must be above 0·4%), whilst in both case-hardening and nitriding an alteration in the composition of the surface layer of the steel takes place.

RECENT EXAMINATION QUESTIONS

1. Explain why surface-hardening is carried out on steels.

   Describe in detail suitable surface-hardening procedures for plain carbon steels containing: (*a*) 0·2% carbon; (*b*) 0·6% carbon.

   (Lanchester College of Technology, Coventry, H.N.D. course.)

2. Explain the object of "case-carburising". Describe one method of case-carburising, commenting on the effects of any alloying elements present.

   (University of Glasgow, B.Sc.(Eng.).)

3. Select a mass-produced steel component which is subject to heavy surface wear but requires only moderate strength to withstand operating conditions. Suggest the type of steel and describe the treatment necessary to produce such a component, giving reasons for your choice of material and method.   (S.A.N.C.A.D.)

4. Write a short account of the process of surface-hardening by pack-carburising.

   Outline, with reasons, the heat-treatment necessary to develop optimum properties in a case-hardened component, indicating clearly the structural changes which occur.   (Carlisle Technical College.)

5. Describe a carburising process suitable for the surface-hardening of a mild steel. What heat-treatments are given after carburising before a component is in a suitable condition for service, and why are these necessary?   (The Polytechnic, London W.1., Poly.Dip.)

6. Describe the gas-carburising process for obtaining hard surfaces on steel, making reference to heat-treatment as well as to the introduction of carbon.

   Compare and contrast pack-carburising with the above process.
   (Constantine College of Technology, H.N.D. Course.)

7. Making full use of the Fe–C equilibrium diagram, describe the heat-treatment of a case-carburised steel.   (Brunel College, Dip. Tech.)

8. Draw the relevant portion of the iron–carbon diagram and use it to explain why the following sequence of treatments was specified for a case-hardened shaft 40 mm in diameter, made from plain carbon steel:

   (a) heat to 950° C. Oil-quench;
   (b) reheat to 780° C. Water-quench;
   (c) heat to 180° C. Air-cool.
   (University of Aston in Birmingham, Inter B.Sc.(Eng.).)

9. Give an account of the nitriding process for surface-hardening. What advantages does this method have over pack-carburising?
   (Derby and District College of Technology, H.N.D. Course.)

10. Discuss nitriding as a method of surface-hardening of steel and compare it with induction-hardening.   (S.A.N.C.A.D.)

11. Describe the processes of: (a) nitriding; and (b) gas-carburising, and discuss the factors which influence the case and core properties, mentioning the types of steel used. What are the advantages and disadvantages of each process?
    (Rutherford College of Technology, Newcastle upon Tyne.)

12. State the methods available for increasing the surface hardness of a steel component whilst still retaining a ductile core. Describe the methods briefly, emphasising their basic differences.
    (Nottingham Regional College of Technology.)

13. Outline the principles underlying the surface-hardening of steels by methods which involve:

    (a) changes in both the composition and structure of the surface relative to the main body of the component;
    (b) changes in the structure only of the surface layer, the composition remaining constant throughout.

    (U.L.C.I., Paper No. C111–1–4.)

14. Explain how you would produce a hard surface in the following conditions:

    (a) a shaft made from 1% Al–$1\frac{1}{2}\%$ Cr steel which has to operate at a temperature between 350 and 500° C;

    (*b*) a gear wheel made from a o·5% carbon steel which requires a hard surface to improve its wear-resistant properties.

Give reasons for the choice you make and a brief description of the process chosen.

(Lanchester College of Technology, Coventry, H.N.D. Course.)

15. Explain the differences in properties between carburised, nitrided and induction-hardened cases.

(University of Aston in Birmingham, Inter B.Sc.(Eng.).)

16. What are the principles underlying the case-hardening of gears?

(University of Aston in Birmingham, Inter B.Sc.(Eng.).)

17. Discuss the relative advantages and limitations of the hardened case as produced by the following processes:

    (i) gas-carburising;
    (ii) gas-nitriding;
    (iii) flame-hardening;
    (iv) induction-hardening.

(Derby and District College of Technology.)

18. Compare and contrast three *different types* of methods of surface-hardening metals.

(University of Aston in Birmingham, Inter B.Sc.(Eng.).)

CHAPTER XX

THE MACHINABILITY OF METALS AND ALLOYS

20.10. The ease with which a metal or alloy can be machined depends mainly upon two factors:

(*a*) the mechanical technique employed;
(*b*) the microstructure of the material being machined.

The first includes such features as the design of the tools and the method and type of lubrication employed. Clearly, these topics are outside the scope of a study of metallurgy, and we shall be concerned, therefore, with the second factor, namely, the microstructure of the material being cut.

Machining is really a cold-working operation in which the cutting edge of the tool forms chips or shavings of the material being machined, and the process will be facilitated if minute cracks form just in advance of the cutting edge, due to high stress concentrations set up by the latter. Very ductile alloys do not machine well, because local fracture does not occur easily under the pressure of the cutting tool. Instead, such alloys will spread under the pressure of the tool and "flow" around its edge, so that it becomes buried in the metal. Friction then plays its part, leading to the overheating and ultimate destruction of the cutting edge.

20.11. It follows, therefore, that a brittle alloy will be far more suitable for machining than would a ductile one. On the other hand, a brittle alloy will generally be unsuitable in ultimate service, particularly under conditions of shock. However, we can compromise by introducing what is, in effect, *local* brittleness in an alloy, whilst at the same time producing little or no deterioration in the impact toughness of the material *as a whole*. This can be done in three ways:

By the Presence of a Constituent which Exists as Isolated Particles in the Microstructure

20.20. Such particles, whether of hard or soft material, have the effect of setting up stress concentrations locally, as the cutting edge approaches them. A minute fracture will therefore travel from the cutting edge to the particle in question, thus reducing friction between the tool and the material being cut. In addition to the reduction of wear on the tool, there will also be a reduction in the overall power required. Moreover, due to the discontinuity introduced by the particles, the swarf produced will be in small, conveniently sized pieces, instead of the long, curly slivers obtained when a ductile material is machined. In fact, from some of the free-cutting materials, such as leaded brasses, the swarf produced is more in the nature of a powder.

20.21. Many alloys having a duplex structure will fall naturally into the free-cutting or semi-free-cutting class. The slag fibres in wrought iron, the graphite in cast iron and particles of the intermetallic compound $Cu_{31}Sn_8$ in a high-tin bronze are examples in which the presence of inclusions improves machinability. In many cases, however, a deliberate addition must be made in order to introduce such isolated particles. In the cheaper free-cutting steels the presence of sulphur is utilised by ensuring that sufficient manganese is also present to combine with the sulphur and form manganese sulphide. Manganese sulphide exists as isolated globules in the microstructure, whereas the ferrous sulphide, which would form if manganese were absent, would exist as brittle intercrystalline films. Thus, whilst in a good-quality steel of the ordinary type sulphur is usually present only in amounts less than 0·06%, in a free-cutting steel as much as 0·20% sulphur may be present, the manganese content being raised to between 0·90 and 1·20% to ensure that all the sulphur is combined with manganese.

20.22. Many alloys, whether in the liquid or solid states, will not dissolve lead. It therefore exists as isolated globules scattered at random in the microstructure instead of being distributed as intercrystalline films, as is often the case when *liquid solubility* followed by *solid insolubility* leads to coring. Particles distributed in this globular form cause very little deterioration in the mechanical properties.

Lead is used in this way to improve the machinability of steels as well as of non-ferrous alloys (in particular, brasses and bronzes), but whereas the amounts used are in the region of 1·0–3·0% in brass, in steels the amounts are only of the order of 0·15–0·35%, so that the particles of lead are so small that they are not generally visible using a simple microscope. These "Ledloy" steels can be either of the plain carbon or alloyed type and, as shown in Table 20.1, machinability is improved by up to 35% (in terms of relative machinability constants) for alloy steels. The use of lead is particularly advantageous in introducing free-cutting properties into such alloys as the 13% chromium stainless steels, because the alternative method of utilising the presence of sulphur and manganese may lead to a reduction in the impact value by half, and a general reduction in other mechanical properties.

20.23. Elements other than manganese are sometimes added in the

TABLE 20.1.—*The Effects of Lead on the Free-cutting and Mechanical Properties of Various Steels*

Type of steel	Condition	Composition (%)						Mechanical properties				Comparative machinability constants	Improvement in machinability when lead is added (%)
		C	Mn	S	Ni	Cr	Pb	Tensile strength		Elongation (%)	Izod (J)		
								N/mm²	tonf/in²				
Free-cutting	6·35 mm diameter drawn	0·13	0·95	0·23	—	—	—	568	36·8	24·0	—	12·5	24·0
Leaded free-cutting	6·35 mm diameter drawn	0·12	0·98	0·22	—	—	0·20	571	37·0	23·0	—	15·5	
0·40% carbon	39·7 mm hexagonal normalised	0·40	0·61	0·03	—	—	—	567	36·7	26·0	32	6·2	29·0
Leaded 0·4% carbon	39·7 mm hexagonal normalised	0·42	0·61	0·03	—	—	0·20	602	39·0	25·0	34	8·0	
Nickel-chromium	31·75 mm diameter heat-treated	0·41	0·74	0·03	1·73	0·65	—	1 043	67·5	18·0	75	4·75	35·0
Leaded nickel-chromium	31·75 mm diameter heat-treated	0·42	0·25	0·02	1·74	0·65	0·17	981	63·5	18·0	71	6·4	

presence of sulphur to improve machinability. Molybdenum or zirconium, together with 0·1 % sulphur, will produce non-abrasive sulphide
globules. Selenium is sometimes used to replace sulphur, particularly
in the austenitic stainless steels. Selenium is chemically very similar

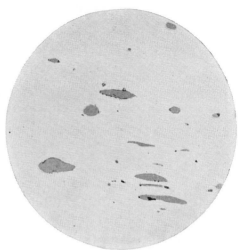

20.1A.—Free-cutting   Sulphur
   Steel.
   Longitudinal section showing
   inclusions of manganese sulphide. × 250. Unetched.

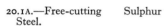

20.1B.—Free-cutting   Extruded
   Brass Rod (58% Copper; 2·5%
   Lead; Balance Zinc).
   Longitudinal section showing
   globules of undissolved lead.
   × 100. Unetched.

PLATE 20.1

to sulphur, and forms selenide globules which act as chip breakers in
the same way as sulphide globules. Some typical steels containing the
elements mentioned are shown in Table 20.2.

20.24. As has been mentioned, free-cutting properties are imparted to the copper alloys by the addition of about 2·0% lead. The machinability of brasses, bronzes and nickel silvers is improved in this way. The lead exists as tiny globules, plainly visible in the microstructure (Plate 20.1B). Since lead is insoluble in the liquid alloys, as well as in the solid, care has to be exercised to prevent segregation during the melting and casting of such alloys.

TABLE 20.2.—*Typical Free-cutting Steels*

Type of steel	B.S. 970 designation	Composition	
Free-cutting mild steel	220Mo7	Carbon *Manganese* Sulphur	0·1% 1·0% 0·25%
Semi-free cutting "35–45" carbon bright drawn	216M28	Carbon *Manganese* Sulphur	0·28% 1·3% 0·16%
Free-cutting "40" carbon	225M44	Carbon *Manganese* Sulphur	0·44% 1·5% 0·25%
Chromium rust-resisting steel (free-machining)	416S21 and 416S41	Carbon Chromium *Sulphur* *Manganese* *Selenium* *Molybdenum* Optional—*Lead*	0·12% 13·0% 0·3% max. 1·5% max. 0·3% max. 0·6% max. 0·35% max.
Free-cutting 18–8 stainless steel	303S21 303S41 325S21 and 326S26	Carbon Chromium Nickel *Manganese* *Sulphur* *Molybdenum* Titanium	0·1% 18·0% 10·0% 1·5% 0·25% 3·0% max. 0·6%

By Suitable Heat-treatment of the Material Before Carrying Out the Machining Operation

20.30. Low-carbon steels are most easily machined in the normalised condition. This treatment produces small patches of pearlite which break up the continuity of the ferrite, so that, during machining, they will behave in much the same way as the manganese sulphide globules in free-cutting steels. The effect will, of course, be less evident.

20.31. Fine-grained steels have poorer machining properties than coarse-grained steels of similar composition. Hence, machining of these fine-grained steels can be facilitated by heating to a temperature high enough to cause grain growth, and thus produce a coarse-grained austenite which will, in turn, give coarse ferrite and pearlite. Other mechanical properties will naturally be impaired by such treatment, so that after machining a normalising treatment should be given if the steel is not to be otherwise heat-treated.

20.32. High-carbon steels cannot easily be machined in their hard state. Machinability can, however, be improved by employing a spheroidising anneal (11.52). This involves annealing the steel at a temperature between 650 and 700° C, i.e. just *below* the lower critical temperature, so that the cementite "balls-up" into small spherical globules dispersed throughout the ferrite. These globules assist machining in the usual way, but annealing should not be prolonged, or large globules will result and lead to a rough machined surface due to local tearing.

By Cold-working the Material Prior to Machining

20.40. Material supplied in rod form for machining operations is usually cold-drawn or "bright-drawn". This treatment leads to some improvement in machinability by reducing ductility and causing some embrittlement of the surface of the rod.

RECENT EXAMINATION QUESTIONS

1. Discuss the metallurgical factors of materials that affect their machinability.   (Brunel College, Dip.Tech.)
2. Discuss methods by which the machinability of a steel may be improved by additions to its normal composition.
                    (U.L.C.I., Paper No. B111–2–4.)
3. Discuss the metallurgical factors which affect the machinability of: (i) ferrous alloys; (ii) non-ferrous alloys.
                    (Institution of Production Engineers Associate Membership
                    Examination, Part II, Group B.)
4. Discuss the factors which influence the machining properties of metals and alloys.
        Show how free-cutting properties are obtained in: (i) bright-drawn steel rod; (ii) extruded brass rod; (iii) grey cast iron; (iv) free-cutting quality 18–8 stainless steel.   (West Bromwich Technical College.)
5. (*a*) Discuss the relationship between microstructure and machinability.
        (*b*) What are the main metallurgical factors which account for the characteristic machining properties of: (i) grey cast iron; (ii) white cast iron; (iii) austenitic steel; (iv) pure copper?
                    (Institution of Production Engineers Associate Membership
                    Examination, Part II, Group B.)

CHAPTER XXI

METALLURGICAL PRINCIPLES OF THE JOINING OF METALS

21.10. Apart from purely mechanical methods, such as riveting, the chief methods available for the joining of metals are soldering, brazing and welding. Industrial applications of these processes are many and varied, and range from the soldering of sardine cans to the fabrication by welding of mass-produced ships by Henry J. Kaiser during the Second World War. In soldering, brazing and welding complete or incipient fusion takes place at the surfaces of the two pieces of metal being joined, so that a more or less continuous crystal structure exists as we pass across the region of the joint.

In the most ancient welding operation—that used by the blacksmith—the two pieces of metal are hammered together whilst at a high temperature, so that crystal growth occurs across the surfaces in contact, thus knitting the halves firmly together. A welding process of this type was probably used by Tubal Cain. By the time the Parthenon was being constructed in Ancient Greece, on the initiative of Pericles, between 447 and 438 B.C., welding played an important part in civil engineering. In the Parthenon the marble blocks used were not joined by mortar but by a system of iron dowels and double T-shaped clamps. Metallographic examination of surviving specimens has shown that the latter were made by welding the "feet" of two T-shaped pieces of iron together.

Soldering

21.20. A solder must be capable of "wetting", that is, alloying with, the metals to be joined, and at the same time have a freezing range appreciably lower, so that the work itself is in no danger of fusion. The mechanical strength of the solder must also be adequate.

21.21. Alloys based on tin or lead fulfil most of these requirements for a wide range of metallurgical materials which need to be joined, since tin will alloy readily with iron, with the alloys of copper and with lead. At the same time tin–lead alloys possess mechanical toughness, and melt at temperatures between 183 and 250° C, which is comfortably below the point at which deterioration in properties of the metals to be joined will take place.

Plain tin–lead solders are of two main types, depending upon whether they are to be used by the tinsmith or the plumber. Best-quality tin-man's solder contains 62% tin and, as indicated by the equilibrium diagram (Fig. 21.1), this is the eutectic composition. Such an alloy will possess distinct advantages in that it will pass quickly from complete liquid to complete solid without any intermediate pasty stage, during which a joint might, if disturbed, be broken. Nevertheless, since tin is an expensive metal compared with lead, the tin content may be reduced to 50% ("coarse" tinman's solder) or even less.

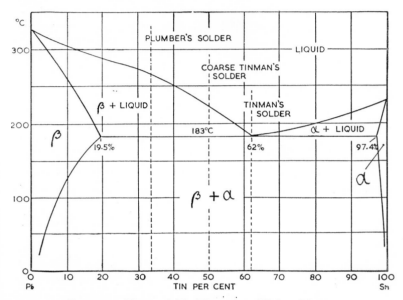

FIG. 21.1.—The Lead–Tin Thermal-equilibrium Diagram.

Plumber's solder contains about 67% lead, and will consequently contain a mixture of liquid and solid phases between 183° C and about 265° C. This extended range over which the alloy will be in a pasty state is of advantage to the plumber, since it enables him to "wipe" joints in lead piping, a feat which would be almost impossible to accomplish with an alloy melting or freezing over a small range of temperatures.

21.22. Solders are sometimes strengthened by the addition of small amounts of antimony which, within the limits in which it is added, remains in solid solution. When copper alloys are soldered it is always possible that at some point in the joint the concentrations of copper and tin will be such that one of the hard, brittle copper–tin intermetallic compounds, such as $Cu_{31}Sn_8$ or Cu_5Sn_6, will be formed in sufficient quantity to cause brittleness of the joint. This difficulty can be overcome by soldering copper with an alloy consisting of 97·5% lead and

2·5% silver, for, whilst lead and copper are insoluble in each other, silver alloys with each, and thus acts as a metallic bond between the two. A list of representative solders is given in Table 21.1.

TABLE 21.1.—*Tin–Lead-base Solders*

Composition (%)				Solidification range (° C)	Types and uses
Sn	Pb	Sb	Ag		
62	38	—	—	Solidification begins and ends at 183° C	Tinman's solder (eutectic alloy).
50	50	—	—	220–183	Coarse tinman's solder.
33	67	—	—	260–183	Plumber's solder.
31	67	2	—	235–188	Plumber's solder for wiping joints.
43·5	55	1·5	—	220–188	General-purpose solder.
30	69·35	—	0·65	250–180	Substitute solder for general purposes.
18	80·8	0·8	0·4	270–180	Substitute plumber's solder.
12	80	8	—	250–243	For soldering iron and steel.
—	97·5	—	2·5	Solidifies at 305° C	For soldering copper and its alloys (eutectic alloy).

21.23. In order that the solder shall "wet" the surfaces of the metals to be joined, the latter must be clean and free from oxide. (This renders very difficult the soldering of alloys containing aluminium.) To clean the metal surface a flux is used which will act as a solvent to the thin oxide layers liable to form before the surface can be wetted by the solder. Possibly the best-known flux is hydrochloric acid ("spirits of salts"), or the acid–zinc chloride solution which is obtained when a piece of zinc is dissolved in hydrochloric acid. Whilst being most effective in action, these fluxes have the disadvantage that the residue they leave behind is likely to lead to corrosion of the metal near to the joint. If it is inconvenient, therefore, to wash the finished joint, an organic type of flux is safer to use. Such fluxes usually have a rosin base and are almost completely non-corrosive, though only really effective on copper and tin-plate.

Brazing

21.30. Although the technique of this process may vary to some extent (17.80—Part II), metallurgically it is similar to soldering. Brazing is used when a tougher, stronger joint is required, particularly in alloys of higher melting point than those usually joined by soldering. Most ferrous materials and non-ferrous alloys of sufficiently high melting point can be joined by brazing. A borax-type flux is generally used, though for the lower temperatures involved in high-grade silver soldering a fluoride-type flux may be used.

21.31. Ordinary brazing solder contains about 50% copper and 50% zinc, and whilst this composition is liable to contain the brittle γ-compound, it must be remembered that much of the zinc volatilises during brazing, whilst some may be absorbed by the surrounding metal. The final joint will therefore most likely consist of a straightforward α-brass which is both tough and ductile.

Higher-grade brazing compounds, or silver solders, contain over 50% silver, which forms a eutectic with the copper present. Cheaper grades of silver solder contain between 10 and 20% silver and about 50% copper. Typical brazing alloys are shown in Table 21.2.

TABLE 21.2.—*Brazing Solders and Silver Solders*

Composition (%)					Freezing range (° C)	Type
Cu	Zn	Ag	Cd	Sn		
50	50	—	—	—	870–880	Ordinary brazing alloys for ferrous materials.
50	45	—	—	5	750	
16	4	80	—	—	740–795	High-grade silver solders for use on brass and light-gauge copper, monel and stainless steel.
20	15	65	—	—	695–720	
15·5	16·5	50	18	—	625–635	
45	30	20	5	—	775–815	Lower-grade silver soldering.
52	38	10	—	—	820–870	

Welding

21.40. Some of the welding processes resemble both soldering and brazing in so far as molten metal is applied to produce a joint between the two pieces. However, in welding, the added metal is, more often than not, of similar composition to the metals being joined, and a more positive state of fusion exists at the metal surfaces. Thus, differences between the weld metal and the pieces being joined are structural rather than compositional.

The method of production of a weld calls for rather different technique to that employed in brazing and soldering. The speed of working is particularly important if complete fusion of the metal near to the joint is to be avoided. Moreover, some welding processes rely on pressure to effect joining of the two halves. In such case no metal is added to form a joint, the weld metal being provided by the two halves being joined. Thus we have both fusion and pressure-welding processes, as indicated in Fig. 21.2.

Of recent years, welding has become one of the principal methods of fabricating and repairing metal products. It is an economical means of joining metals in almost all assembly processes and, assuming good welding technique, produces a very dependable result.

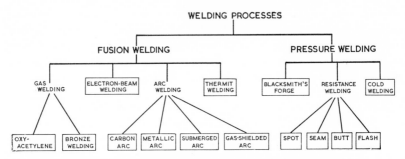

FIG. 21.2.—A Classification of the Chief Welding Process.

Fusion Welding Processes

21.50. Either gas flame, electric arc, the heating effect of a stream of electrons or (in specialised cases such as the Thermit process) chemical reaction can be used to effect fusion of the weld metal. These methods are dealt with in some detail in Part II, but are also mentioned briefly here.

21.51. Gas Methods. In these processes the surfaces to be joined are melted by the flame from a gas torch, the gases most commonly used being suitable mixtures of oxygen and acetylene, which produce temperatures up to 3 000° C. Such a high temperature will quickly melt all the ordinary metals and alloys, and is necessary in order to overcome the tendency of sheet metals to conduct heat away from the joint so quickly that fusion cannot occur. Moreover, rapid melting will reduce distortion, overheating and oxidation of the surrounding metal.

The atomic hydrogen arc is notable because of the high temperature (4 000° C) which it can produce. It is unsuitable for those metals which absorb hydrogen freely when in the molten state, since this will lead to the formation of a porous weld. For this—and other—reasons it has been replaced largely in recent years by gas-shielded arc processes.

In gas- and other fusion-welding processes a welding rod, or filler rod, is used to supply the necessary metal for the weld. The rod is held close to the work and melted by the flame, so that molten metal flows into the prepared joint between the pieces being welded. Since the edges of the pieces also fuse, a strong, metallurgically continuous joint is formed. Welding rods usually have a composition similar to that of the metals being welded, although in some cases they are richer in those elements which tend to volatilise during welding. The term "bronze welding" refers to the joining of metals with high melting points, such as steel, cast iron, nickel and copper, by the use of a copper-alloy filler rod.

Some form of flux is essential, particularly with alloys which oxidise easily and form an impervious oxide skin, as is the case with aluminium alloys. The flux should melt at a lower temperature than the metal being joined, and thus dissolve oxide films which form before fusion commences. The flux will also protect the metal from excessive oxidation.

21.52. Electric Methods. The oldest electric welding process is that which utilises a carbon arc. The arc is struck between a carbon electrode and the work itself, so that the heat generated melts the surfaces to be welded. Weld metal is supplied from a filler rod, the end of which is held in the arc.

The metallic arc process, however, is by far the most popular method of fusion welding. Basically it is similar to the carbon arc, but the electrode is a metal rod of suitable composition which serves also as the filler rod. The end of this rod melts and deposits on to the joint, whilst, at the same time, the heat generated melts the edges of the work, producing a continuous weld. The filler rod is usually coated with a suitable flux which not only supplies the weld but also assists in stabilising the arc. Although used chiefly for steel, the metallic-arc method also finds application in welding many of the non-ferrous alloys.

The submerged-arc process is essentially an automatic form of metallic-arc welding, which can be used in the straight-line joining of metals. Powdered flux is fed into the prepared joint so that, on melting, it envelops the melting end of the electrode and so covers the arc. The weld area is thus coated by protective slag, allowing a smooth, clean weld surface to be produced.

For those alloys in which oxidation is particularly troublesome, the gas-shielded arc method is used. In this process the electrode protrudes from the edge of a tube from which also issues an inert gas (argon or helium). The weld region is thus blanketed by a non-oxidising atmosphere making the use of a flux unnecessary (Fig. 21.3). Carbon dioxide, CO_2, is also used as a gas shield, and although it can have an

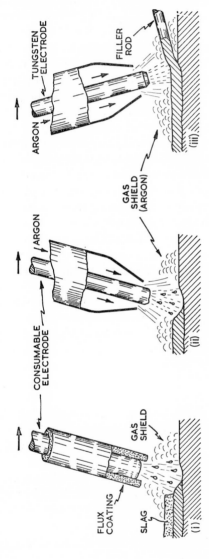

FIG. 21.3.—Electric-arc Welding Processes.

(i) The metallic arc—here molten flux affords protection of the weld region, assisted to a small extent by a gas shield provided by burning carbonaceous matter in the flux coating. (ii) The metallic–inert-gas arc (MIG) in which a very effective gas blanket is obtained by a flow of argon or helium around the weld region. (iii) The tungsten–inert-gas arc (TIG)—here a separate filler rod is used, the tungsten electrode being non-consumable.

oxidising action at high temperatures, this can be overcome by using filler rods containing deoxidants, such as manganese and silicon. It is a much cheaper gas than argon, and during recent years the CO_2 process has become the most popular semi-automatic method for welding steel.

An important feature, common to all gas-shielded arc methods, is that, since atmospheric nitrogen is kept away from the weld during fusion, nitrogen embrittlement is avoided. Gas-shielded arc welding will no doubt displace much of the gas- and metallic-arc welding now used, due to the superiority of the weld and the increased speed of manipulation which is possible.

21.53. Electron-beam Welding. In this process, fusion is achieved by focusing a beam of high-velocity electrons on the weld area. On striking the metal the kinetic energy of the electrons is converted into heat energy so that melting of the metal occurs. Welding must be carried out in a vacuum chamber, since the presence of oxygen and nitrogen molecules would lead to collisions with the electrons and their consequent scatter, as well as a loss of kinetic energy. Moreover, since the tungsten filament which emits the electrons is working at 2 000° C, it would oxidise rapidly if exposed to the atmosphere.

The process is particularly useful for joining refractory metals, such as tungsten, molybdenum, niobium and tantalum, and also metals which oxidise easily, such as titanium, beryllium and zirconium.

21.54. Thermit Welding. This is a process which now finds rather limited applications, chiefly in the repair of large iron and steel castings. A mould is constructed around the parts to be joined, and above this is placed sufficient thermit powder. The parts to be welded are pre-heated and the powder then fired.

Thermit powder consists of an intimate mixture of powdered aluminium and iron oxide in the correct molecular proportions. The affinity of aluminium for oxygen is much greater than the affinity of iron for oxygen. Therefore, aluminium reduces the iron oxide to metallic iron thus:

$$Fe_2O_3 + 2Al = Al_2O_3 + 2Fe$$

The heat of the reaction is so intense that the iron which is formed melts and runs down into the prepared mould, combining with the pre-heated surfaces and producing a weld.

Pressure-welding Processes

21.60. These processes include the oldest of all welding operations, that practised by the blacksmith at his anvil. The smith heats the two parts to be welded to a temperature short of actual fusion, and then applies pressure in the form of hammer blows to cause the surfaces to unite. The sand, which he often uses as a flux, combines with the iron oxide produced by heating to a high welding temperature, and forms a fusible slag which is scattered by the hammer blows.

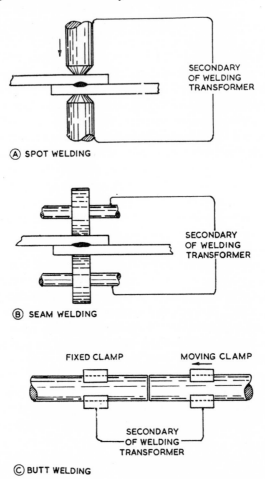

FIG. 21.4.—The Principles of Resistance Welding Processes.

Electrical-resistance welding is similar in principle, and is used in the following modifications:

21.61. Spot Welding, in which the parts to be joined overlap and are firmly gripped between heavy metal electrodes (Fig. 21.4A). When the current is passed, local heating of the sheets to a plastic condition occurs, resulting in welding at a spot. This method is used mainly for the joining of sheets and plates, and in particular, in the production of temporary joints.

21.62. Seam Welding is similar in principle to spot welding, but produces a continuous weld by employing wheels as electrodes (Fig. 21.4B). The parts to be joined are passed between the rotating wheels and heated to a state of plasticity by the current. The pressure applied by the wheels is sufficient to produce a continuous weld.

21.63. Butt Welding is used to join lengths of rod and wire. The ends are pressed together (Fig. 21.4C) and an electric current passed through the work so that the ends are heated to a plastic state due to the higher electrical resistance existing at the point of contact. The pressure is sufficient to form a weld.

21.64. Flash Welding is similar to butt welding in that ends of sheets and tubes can be welded together. The process differs, however, in that the ends are first heated by striking an arc between them. This not only produces the necessary high temperature but also melts away any irregularities at the ends, which are then brought into sudden contact under pressure so that a sound weld is produced.

21.65. Cold Welding. This process relies solely upon the application of high pressure between two surfaces in intimate contact, no heat source being used. Welding is generally made difficult by the presence of thin oxide films on the surfaces to be joined. In industrial processes, shearing forces are used to cause sliding between the surfaces in contact so that oxide films are ruptured and welding occurs.

The Microstructure of Welds

21.70. Since high temperatures are necessary during welding, a weld will exhibit a coarse crystal structure in contrast to the parent metal around it, which is often in the wrought condition. Some elements, such as chromium (in steel), accelerate grain growth, so that in these cases an enlarged grain may also be expected in that metal near to the weld.

A weld produced by one of the fusion methods will show an as-cast type of structure. Not only will the crystals be large but other as-cast features, such as coring, segregation and the like, may also be present, giving rise to intercrystalline weakness. The structures of some steels may be affected by the rate of cooling following welding. This is particularly true of steels of the oil- or air-hardening type, in which brittle martensite may be formed by the relatively rapid cooling of the metal near to the weld.

When possible it is of advantage to work a weld mechanically by hammering it while hot. This produces a fine grain as recrystallisation takes place; it also minimises the effects of coring and segregation.

The Inspection and Testing of Welds

21.80. Methods of examination vary according to whether the weld has been produced for service or for experimentation. In the former case methods of examination will obviously be limited in their scope, since the weld cannot be "dissected"; whilst in the latter case all methods of examination are available, since we are not interested in the weld as such, but only in its properties.

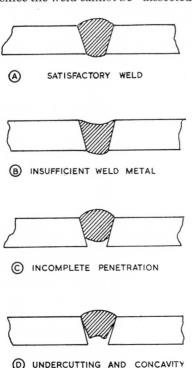

(A) SATISFACTORY WELD

(B) INSUFFICIENT WELD METAL

(C) INCOMPLETE PENETRATION

(D) UNDERCUTTING AND CONCAVITY OF WELD

FIG. 21.5.—Some Common Welding Defects Which are Visible on Inspection.

21.81. Visual Examination and Non-destructive Tests. The type of defect encountered may vary widely with the material being welded and the type of process used, but intelligent and careful examination of the resulting weld will often indicate whether it is likely to be satisfactory. Rough, burnt and blistered surfaces suggest overheating and gassing during welding, whilst spattering in arc welding indicates incorrect arc conditions. Insufficient fusion and fluxing are also defects which are relatively easy to detect.

In fusion welding the weld should be somewhat built-up above the parts being joined, to allow for subsequent trimming by grinding. Defects such as undercutting due to insufficient fusion, insufficient penetration and insufficient weld metal are shown in Fig. 21.5.

Cracks in or near welds can be detected by means of a magnetic crack detector. The part under test is placed across magnetic electrodes and more or less saturated with magnetic lines of force. If any crack exists, the field is broken so that each side of it becomes a magnetic pole, and when the surface is dusted with fine iron particles, these collect in a bunch in the region of the crack. Cracks may also be detected more simply by applying penetrating oil or paraffin to the surface, allowing it to remain for a few minutes and then carefully rubbing it off. Finely powdered chalk is then scattered over the surface and allowed to remain for some time. Any oil which seeps out of a crack will stain the layer of chalk.

The penetration method can be used to reveal cracks more sensitively by employing a liquid to which is added a small amount of a soluble, strongly fluorescent compound such as anthracene. This liquid is applied to the surface and the surplus washed off after allowing sufficient time for penetration into cracks to take place. The surface is then dried and examined under a quartz-tube mercury-vapour lamp screened by a filter of Wood's glass. This filter absorbs most of the light from the lamp, but transmits enough ultra-violet radiation to cause fluorescence of any penetrating liquid remaining in the cracks, which appear as bright lines upon the darker background of the metal.

When expense is justified, examination by X-rays or γ-rays (3.82) is useful for revealing porosity, inclusions, holes and cracks in welded joints; whilst pressure tests are applicable to vessels fabricated by welding.

Ultrasonic methods are used principally to detect internal flaws and discontinuities in steel plate destined for high-grade welded structures. Ultrasonic impulses are projected into the plate, and on meeting any fault, are reflected and shown as "traces" on a cathode-ray tube. The "probe" is contained in a small roller which traverses the surface of the plate under test.

21.82. Testing Specimen Welds. Welded joints can be produced in the form of test pieces and tested to destruction. The result will provide useful information regarding the general quality of the process and indicate whether similar results may be expected during subsequent production.

Bending tests are often applied, whilst tensile tests also provide useful information. Failure may occur in the weld, in the metal adjoining the weld or between the two. In the latter case, particularly when failure occurs at low stresses, inadequate fusion may be suspected and will often be clearly indicated by the appearance of the fracture, which, instead of being crystalline, is that of the smooth prepared surface. When failure occurs in the weld, visual examination will sometimes reveal such defects as porosity, slag globules and fissures.

Macro-examination of a suitably etched specimen will provide information on the soundness of a weld and also reveal the boundary between weld metal and base metal. A specimen should be cut from the welded joint so that it displays a complete transverse section of the weld. It can then be prepared as indicated in 10.40 and etched in one of the reagents mentioned in Table 10.6. The hydrochloric acid etch will reveal unsoundness by enlarging gas pockets and also by dissolving slag inclusions.

Microscopical examination of polished and etched sections will, in addition to revealing such defects as porosity, slag, cracks and oxide inclusions, also indicate such purely metallurgical defects as overheating, decarburisation and other microstructural defects associated with composition, temperature effects and rates of cooling. Special effects, such as carbide precipitation in stainless steels, can only be detected in this way.

The Weldability of Various Metals and Alloys

21.90. The suitability of an alloy for welding depends upon a number of factors, such as fluidity, the tendency towards oxidation and volatilisation of one or more of its constituent elements, and the behaviour of the weld metal and its surrounding parent metals during solidification and subsequent cooling. Nevertheless, by developing suitable techniques most metals and alloys can be successfully welded by one or other of the standard processes.

21.91. Wrought Iron and Plain Carbon Steels can be welded with varying degrees of success, depending upon the carbon content. Wrought iron and mild steel weld with ease when either gas or electric methods are used. For metallic-arc welding the electrode is usually coated with a mixture of lime, silica, ferromanganese and powdered aluminium. In addition to the fluxing action, this mixture acts as a deoxidant and also stabilises the arc.

Higher-carbon steels weld with difficulty due to decarburisation and the attendant porosity produced by gases which are a product of the oxidation of carbon. Structural difficulties may also arise, such as brittleness resulting from an increase in hardness brought about by rapid cooling. This can often be overcome by pre-heating (or post-heating) the work in order to induce a slower cooling rate. Carbon pick-up by the weld metal from the parent metal also causes brittleness, particularly since a coarse as-cast Widmanstätten structure can be expected to form, and this may lead to the formation of cracks in or near the weld during cooling. This is often overcome by a technique known as "buttering". A layer of weld metal is deposited on each half of the work piece before they are joined. Some carbon is absorbed by this layer but, since the parts are not joined, cracking will be unlikely to occur. The two "buttered" sections can then be welded with safety since little carbon is likely to reach the actual weld.

Where applicable, the hammering of welds whilst still hot brings about an improvement in properties due to the refinement in crystal structure which it produces. Normalising will also help to reduce the size of the as-cast type of crystals present in ferrous welds made by fusion processes.

21.92. Alloy Steels can generally be successfully welded, provided the correct type of filler rod is used. Low-alloy steels are often welded with plain steel welding rods containing 0·2% carbon, 1·4% manganese and 0·4% silicon, since there is generally sufficient pick-up of other alloying elements from the metals being welded.

The filler rods used to weld stainless steels are usually of a composition similar to the stainless steel being welded, thus limiting the possibility of electrolytic action at the weld during service. The 13% chromium air-hardening types present some difficulty on account of grain growth, which is accelerated by the chromium, and also due to the formation of brittle martensite in that part of the metal, near to the weld, which cools at a speed equivalent to that used in air-hardening.

The so-called weld-decay experienced in the 18–8 type of stainless steel has already been mentioned (13·44), and unless additions have been made to render it proof against weld-decay, the component must be heated, after welding, to about 1 050° C, and subsequently cooled in air or oil according to the thickness of the section, in order that carbides may once more be taken into solid solution. This process is not always practicable, particularly when a large or intricate section has been fabricated. In such cases a weld-decay-proofed steel must be used.

21.93. Cast Iron is readily welded when a suitable filler rod is used, that is, one relatively rich in silicon and manganese to allow for oxidation losses of these elements. Difficulties which are experienced include the hardening of the weld due to its absorbing carbon from the surrounding metal, followed by rapid cooling, which causes the retention of primary cementite and the formation of a white-iron structure. To overcome this, cast iron may be joined by bronze welding (21.51), though this introduces the danger of electrolytic action between the weld metal and the surrounding cast iron, and may lead to the accelerated corrosion of the latter, which will be anodic to the bronze weld (22.26).

21.94. Malleable Cast Iron of the whiteheart variety can be welded easily in the manner of mild steel. The temper carbon present in blackheart malleable castings, however, presents a problem similar to that arising from graphite in ordinary cast iron. This temper carbon is taken into solution on fusion; and, on cooling, will tend to be precipitated as cementite, thus embrittling the metal adjacent to the weld. The best results are obtained by welding at a low temperature, using a bronze filler rod.

21.95. Copper and its Alloys. In general, copper alloys are best welded using filler rods of a composition similar to that of the alloy being welded. Subsequent corrosion due to electrolytic action is thus minimised. For the arc welding of phosphor-bronzes and low-grade brasses, however, a silicon-bronze rod (3–4% silicon) is often used. The presence of silicon increases the fluidity of the weld metal.

Filler rods used for welding brasses sometimes contain small amounts of aluminium as a deoxidant. Welds in brass are improved by mechanical work which refines the grain. In the α + β phase brasses this work must be carried out whilst the weld is still hot, but in the ductile α-phase brasses it is best carried out in the cold and followed by a stress-relief annealing process which will lead to the formation of fine grain.

In the flame-welding of copper–nickel alloys care must be taken that sulphur is absent from the atmosphere, as this will cause a very serious deterioration in the properties of the metal around the weld. Nickel readily forms sulphides, for which reasons a sulphurous atmosphere always attacks alloys containing nickel.

Copper itself is usually welded with a filler rod containing a small amount of phosphorus. This acts as a deoxidant and minimises porosity, which may arise due to reactions between oxide globules and the welding atmosphere during solidification.

21.96. Aluminium Alloys oxidise easily, so that care has to be taken in the choice of a flux. The most suitable flux mixture contains the fluorides and chlorides of sodium, potassium and lithium, together with some potassium bisulphate. This flux must be washed off after use or it may cause corrosion in the region of the weld. For this reason the bulk of aluminium welding is now carried out using the gas-shielded arc process.

RECENT EXAMINATION QUESTIONS

1. Write an essay on "The Joining of Metals", indicating clearly the essential differences between soldered, brazed and welded joints.
   (Institution of Production Engineers Associate Membership Examination, Part II, Group B.)
2. Show that you understand the following terms: soldering; fusion welding; pressure welding.
   Give two examples of each process.
   What are the properties required of a flux in a joining process?
   (Rutherford College of Technology, Newcastle upon Tyne.)
3. What is the difference between soldering and welding?
   Describe the structure likely to be found in an arc butt weld formed by a single run between 0·3% carbon steel plates: (a) if the plates are pre-heated; (b) if the plates are not pre-heated.
   What effect would subsequent normalising have on the welds?
   (University of Aston in Birmingham, Inter B.Sc.(Eng.).)

4. Give an account of the principles of electrical resistance methods of welding. Outline TWO industrial welding processes.
(Derby and District College of Technology.)

5. Describe, with the aid of a sketch, an electric-arc welding process in which the work is protected by means of a gas shield.
For what metals and alloys is this process particularly suitable?
(U.L.C.I., Paper No. C111-2-4.)

6. With the aid of sketches, describe TWO of the following arc welding processes: (i) metallic arc; (ii) atomic hydrogen arc; (iii) argon arc.
Indicate the types of electrode and flux used and the main application of each process. (Nottingham Regional College of Technology.)

7. Write short notes on FOUR of the following: (a) stitch welding; (b) flash butt welding; (c) submerged arc welding; (d) MIG welding; (e) electro-slag welding; (f) electron beam welding.
(The Polytechnic, London, W.1., Poly. Dip.)

8. Discuss the arc welding of mild steel from the point of view of the heat-treatment which occurs during the process, and the structures which are obtained in the weld metal and the plate.
(Constantine College of Technology, H.N.D. Course.)

9. (a) Describe the basic differences in application and effect of the electric-arc and oxy-acetylene methods of fusion welding.
(b) Explain any difficulties or defects which may arise when the following materials are welded, and state any preventative measures that should be taken: (i) high carbon steel; (ii) aluminium alloys; (iii) austenitic stainless steel. (Aston Technical College, H.N.D. Course.)

10. Discuss the metallurgical and engineering problems associated with welding a 0·5% carbon steel. (S.A.N.C.A.D.)

11. Give an account of the structures which would result when a 6 mm mild-steel plate and a 6 mm low-alloy medium-carbon steel plate are joined by a filler weld. (University of Glasgow, B.Sc.(Eng.).)

12. What particular difficulties may be encountered in the fusion welding of: (i) 18–8 austenitic stainless steel; and (ii) ferritic stainless steel?
Explain how these difficulties are overcome in practice.
(Derby and District College of Technology, H.N.D. Course.)

13. What possible difficulties could be experienced in the gas welding of: (i) tough-pitch copper; (ii) high-carbon steel; (iii) cast iron; (iv) 13·0% chromium, 0·4% carbon steel; (v) aluminium alloy?
What precautions would be taken to minimise those difficulties and what other methods of fusion welding would be more suitable for the materials mentioned? (West Bromwich Technical College.)

CHAPTER XXII

METALLIC CORROSION AND ITS PREVENTION

22.10. Of all metallurgical problems which face civilisation, few can be economically more important than the prevention of metallic corrosion. In Great Britain alone it is estimated that some £500 million is spent annually in protecting iron and steel from rusting. This is perhaps less surprising when one considers, as an example, the team of painters permanently employed in protecting the Forth Railway Bridge from the ravages of air, rain and sea-water.

Other metals, in addition to iron and steel, corrode when exposed to the atmosphere. The green corrosion-product which covers a copper roof, or the white, powdery film formed on some unprotected aluminium alloys is clear evidence of this.

The Mechanism of Corrosion

22.20. Chemical Attack. Some corrosion is the result of direct chemical attack by gaseous media on the metallic surface. For example, sulphur gases will attack nickel and its alloys at high temperatures giving rise to intercrystalline corrosion. Atmospheric oxygen, however, is the gas responsible for the bulk of corrosion which arises from simple chemical attack. The rate at which such corrosion takes place is dependent largely upon the affinity of the metal for oxygen and upon the temperature prevailing. Oxidation occurs more rapidly at high temperatures as is demonstrated by the scaling of steel at red heat. Special materials are required to resist oxidation at high temperatures (13.70 and 18.31) and here we are concerned with corrosion which occurs at atmospheric temperatures.

In addition to affinity for oxygen, the nature of the oxide film produced also affects the extent to which oxidation can proceed. For example, the film of oxide which forms on the surface of pure aluminium (and some of its alloys) is dense and impervious to the passage of further oxygen. It therefore protects the metal beneath it from any further oxidation. The film of rust which is produced on the surface of iron or steel, however, is of a porous, loosely adherent nature, so that oxygen can penetrate it and cause further oxidation of the metal beneath. On examination, flakes of rust will be seen to have a laminated structure, suggestive of a corrosion process of this type. Thus, although aluminium has a much greater affinity for oxygen than has iron, it corrodes less quickly because of the protective nature of the oxide film which forms on its surface.

22.21. Electrolytic Action. Many readers will be familiar with the principles of the simple electric cell. Convention indicates electric current as flowing from copper to zinc in the external circuit. This often leads to confusion, since we now know that in fact a stream of electrons constitutes this "current" and that these electrons move in the *opposite* direction to that conventionally ascribed to the "current". With this in mind, we will now consider what happens in a simple cell when the circuit is closed (Fig. 22.1).

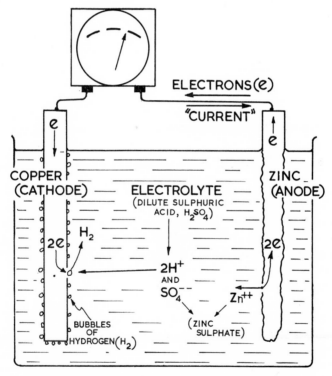

Fig. 22.1—The Chemical Reactions in the Simple Cell.

The conventional direction ascribed to the current in the early days of electrical technology is opposite to that in which electrons in fact flow.

A difference in "electrode potential" exists between the copper and zinc plates and, since zinc is anodic towards copper (Table 22.1), atoms of zinc at the surface of the zinc plate enter solution as zinc ions (Zn^{++}):

$$Zn \longrightarrow Zn^{++} + 2 \text{ electrons}$$

The electrons so released pass round the *external* circuit to the copper plate thus constituting the "current". On reaching the copper plate

TABLE 22.1.—*The Electro-chemical Series*

Metal	Normal electrode potential (volts, with respect to hydrogen)
Gold	+1·68 CATHODIC
Silver	+0·80
Copper	+0·52
(Hydrogen)	0·00
Lead	−0·13
Tin	−0·14
Nickel	−0·25
Iron	−0·44
Chromium	−0·56
Zinc	−0·76
Aluminium	−1·67 ANODIC

these electrons join hydrogen ions, present in the electrolyte, forming hydrogen molecules:

$$2H^+ + 2 \text{ electrons} \longrightarrow H_2 \text{ (forms bubbles at the copper plate)}$$

The zinc ions which entered the electrolyte "pair up" with SO_4^{--} ions from the sulphuric acid (in the electrolyte) to form zinc sulphate, which remains ionised whilst it is in solution. Thus we obtain electric current at the expense of loss of "chemical potential" of the zinc, which corrodes much more quickly than if the circuit were not closed.

22.22. The bulk of metallic corrosion is due to electrolytic action of this type. Electrolytic action is possible when two different metals or alloys are in electrical contact with each other and are also in common contact with an electrolyte. Action such as this occurs when a damaged "tin can" is left out in the rain. If some of the tin coating has been scratched away so that the mild steel beneath is exposed, electrolytic action takes place between the tin and the mild steel when the surface becomes wet with rain water or condensation (Fig. 22.2A). Impure water is ionised to the extent that it will act as an electrolyte, and electrons flow from the iron (mild steel) to the tin, leading to the release of ferrous ions into solution. Thus, once the coating is broken, the presence of tin accelerates the rusting of iron it was meant to protect, and the electrolytic action follows the same general pattern as that which prevails in the simple cell. The iron of the can corresponds to the zinc plate, the tin coating to the copper plate and the solution of oxygen in water to the dilute sulphuric acid used as the electrolyte in the simple cell. Industrial atmospheres accelerate corrosion because of the sulphur dioxide they contain. This gas is present due to the combustion of sulphur in coal and other fuels. Sulphur dioxide dissolves in atmospheric moisture forming sulphurous acid which has a much stronger electrolytic action than has dissolved oxygen.

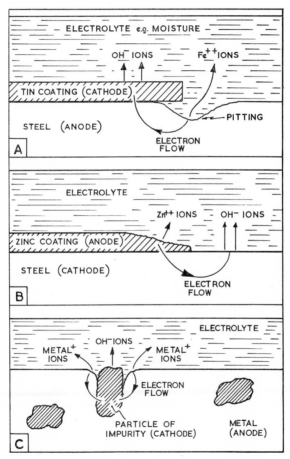

FIG. 22.2.—The Mechanisms of Electrolytic Corrosion.

22.23. Metallic coatings are frequently used to protect steel from corrosion, but it is necessary to consider in each case how corrosion will be affected if the coating becomes scratched or broken. Thus metals used to protect steel can be divided into two groups:

(*a*) Metals which are *cathodic* towards iron, such as copper, nickel or tin. These metals can be used only if a good-quality coating is assured, since protection offered is purely mechanical in that the coating isolates the surface of the steel from the corrosive medium. Suppose that at some point in a tin-coated iron sheet the tin film is broken (Fig. 22.2A). Here the tin film acts as a cathode and the steel the anode. Ferrous ions go into solution and electrons, so released, travel to the tin cathode where they react with water and dissolved oxygen:

$$H_2O + \tfrac{1}{2}O_2 + 2 \text{ electrons} \longrightarrow 2 \text{ OH}^-$$

Thus hydroxyl ions (OH $^-$) are released at the cathode and, away from the region of the cell, these combine with ferrous ions:

$$Fe^{++} + 2OH^- \longrightarrow Fe(OH)_2$$

Ferrous hydroxide, $Fe(OH)_2$, so formed quickly oxidises to ferric hydroxide, $Fe(OH)_3$, the basis of the reddish-brown substance we call "rust". Thus corrosion of the steel is actually accelerated when a coating of some substance which is cathodic to steel becomes damaged. Fortunately, in the case of both tin and nickel, metals commonly used to coat iron and steel, the difference in electrode potential in the cell so formed is small, so that the acceleration of attack is not great. The further apart metals are in the electro-chemical series (Table 22.1) the greater the rate of corrosion of the anode.

(*b*) Metals which are *anodic* towards iron, such as zinc and aluminium. These metals will go into solution in preference to steel (Fig. 22.2B) to which they will therefore offer some chemical, as well as mechanical, protection. Since they protect the steel by going into solution themselves, such protection is generally referred to as "sacrificial". Zinc is used in this way for galvanising steel products, whilst aluminium is effective as a protective paint partly for the same reason.

22.24. Having dealt briefly with the principles of electrolytic corrosion, we are now in a position to give an adequate explanation of the rusting of mild steel (Fig. 22.3). Here electrolytic action takes place between the iron and the oxide film which will be present on the surface —assuming of course that the oxide film is broken and that the gap is covered by an electrolyte (ordinary moisture containing dissolved atmospheric carbon dioxide and oxygen).

Since iron is anodic to its oxide, it will go into solution in the form of ferrous ions:

$$Fe \longrightarrow Fe^{++} + 2 \text{ electrons}$$

The electrons which are released travel to the cathode and take part in the reaction:

$$H_2O + \tfrac{1}{2}O_2 + 2 \text{ electrons} \longrightarrow 2OH^-$$
$$\text{(From atmosphere)}$$

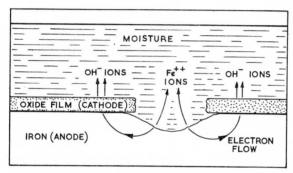

FIG. 22.3.—The Electrolytic Corrosion of Iron Due to the Presence of a Surface Film of Oxide.

As a result, ferrous ions go into solution at the anode and "hydroxyl" (OH^-) ions dissolve at the cathode. When these two different sorts of ions meet *away from the region of electrolytic action* the following reaction occurs:

$$Fe^{++} + 2OH^- \longrightarrow Fe(OH)_2$$

This ferrous hydroxide, $Fe(OH)_2$, is quickly oxidised by atmospheric oxygen to ferric hydroxide, $Fe(OH)_3$, which is precipitated as a reddish-brown substance, the main constituent of rust.

22.25. So far we have been dealing only with electrolytic action between some surface coating and the metal supporting it. The same type of corrosion may take place at the surface of a metal in which particles of an impurity are present as part of the microstructure. In Fig. 22.2C it is assumed that the particle of impurity is cathodic towards the metallic matrix, thus causing the latter to dissolve. Since impurities are often segregated at crystal boundaries, this will lead to gradual intercrystalline corrosion of the material as more particles of impurity become exposed by electrolytic action.

22.26. Those metallic alloys which have a crystal structure consisting of two different phases existing side by side are also prone to electrolytic corrosion. If one phase is anodic with respect to the other it will tend to dissolve when the surface is wetted by a suitable electrolyte. Patches of pearlite in steel will tend to corrode in this manner (Fig. 22.4).

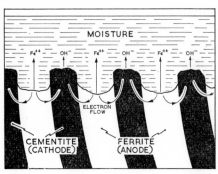

FIG. 22.4.—The Electrolytic Corrosion of Pearlite.

Ferrite is anodic to cementite, consequently ferrite goes into solution (as Fe^{++} ions) and the electrolytic action is similar to that which takes place between ferrite and an oxide film during the rusting of mild steel mentioned above. This leaves the brittle cementite platelets standing in relief, but these will ultimately break off as corrosion proceeds and increases their fragility. It is therefore apparent that in order to have a high corrosion-resistance, a metal or alloy should have a structure consisting of one type of phase only, so that no electrolytic action can take place. Similarly, as engineers will know, it is unwise to rivet or weld together two alloys of widely different electrode potentials if they are likely to come into contact with a substance which can act as an electrolyte. In such circumstances the alloy with the lower electrode potential may corrode heavily due to electrolytic action. An example of bad practice of this type is illustrated in Fig. 22.5. Here a steel pipe has been connected to a copper pipe. If the system carries water which is not chemically pure, then electrolytic corrosion of the steel, which is anodic to copper, can be expected. This

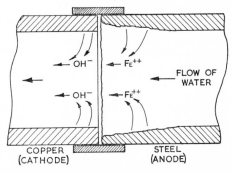

FIG. 22.5.—The Accelerated Corrosion of a Steel Pipe Due to the Presence of an Adjacent Copper Pipe.

corrosion will be much more rapid than if the whole pipe were of steel.

Those readers who are householders will probably be aware that the galvanised cold-water tank corrodes more quickly in that region adjacent to the *copper* inlet-valve mechanism, zinc being strongly anodic to copper. This form of plumbing is doubtless very good for trade, particularly as failure of the tank usually occurs when the room beneath has been newly decorated. The obvious remedy is to match the copper plumbing with a copper tank—which is less expensive than it sounds.

Corrosion Accelerated by Mechanical Stresses

22.30. We have seen that failure of a component may take place due to corrosion arising from electrolytic action between two different phases in a microstructure, or between two different materials in a fabricated structure. Failure of a component may also occur as a result of the complementary effects of chemical corrosion and mechanical stress. The methods of stress application may vary and this will affect the extent of corrosion which occurs. Forms of corrosion in which stress plays a part can be classified as follows:

22.31. Stress Corrosion. In a cold-worked metal the pile-up of dislocations at crystal boundaries and other points increases the energy in those regions so that they become anodic to the rest of the structure. Consequently, corrosion takes place in these regions of high energy and the locked-up stresses give rise to the formation of cracks which grow progressively with the continuance of corrosion. A similar process may take place in components in which unequal heating or cooling has given rise to the presence of locked-up stresses, as, for example, near to welded joints.

22.32. Corrosion Fatigue. As might be expected, any component which is subjected to alternating stresses and is working in conditions which promote corrosion may fail at a stress well below the normal fatigue limit (3.72). The action of the corrosive medium will tend to be concentrated at any surface flaw and behave as a focal point for the initiation of a fatigue crack. Once a crack has been formed it will spread more rapidly as a result of the corrosive action combined with alternating stress.

22.33. Fretting Corrosion is allied to corrosion fatigue and occurs particularly where closely fitting machine parts are subjected to vibrational stresses. In steel this form of corrosion appears as patches of finely divided ferric oxide (Fe_2O_3).

22.34. Impingement Corrosion refers to the combined effects of mechanical abrasion and chemical corrosion on a metallic surface. Mechanical wear can be caused by the impingement of entrained air bubbles or abrasive particles suspended in the liquid. The impingement of such media may lead to the perforation of any protective film existing on the surface. This film may be an oxide, which is cathodic to the exposed metal beneath. Corrosion will then take place in the manner indicated in Fig. 22.3. This type of corrosion is encountered in pump mechanisms, turbines and tubes carrying sea-water.

The Prevention of Corrosion

22.40. There are two principal methods by which corrosion may be prevented or minimised. First, the metallic surface can be insulated from the corrosive medium by some form of protective coating. Such coatings include various types of paints and varnishes, metallic films having good corrosion-resistance and artificially thickened oxide films. All of these are generally effective in protecting surfaces from atmospheric corrosion, though zinc coatings are used to protect iron from the rusting action of water, whilst tin coatings offer protection against most animal and vegetable juices encountered in the canning industry.

22.41. In circumstances where corrosive action is severe, or where mechanical abrasion is likely to damage a surface coating, it may be necessary to use a metal or alloy which has an inherent resistance to corrosion. Such corrosion-resistant alloys are relatively expensive, so that their use is limited generally to chemical-engineering plant, marine-engineering equipment and other special applications.

The Use of a Metal or Alloy Which Is Inherently Corrosion-resistant

22.50. The corrosion-resistance of a pure metal or a homogeneous solid solution is generally superior to that of an alloy in which two or more phases are present in the microstructure. As mentioned above, the existence of two phases leads to electrolytic action when the surface of the alloy comes into contact with an electrolyte. The phase with the lower electrode potential will behave anodically and dissolve, leading to pitting of the alloy surface; and the greater the difference in electrode potentials between the phases, the more rapid will be corrosion. Resistance to corrosion will generally be at a minimum when the second phase is segregated at the crystal boundaries, particularly if this phase is anodic to the matrix. In such circumstances serious intercrystalline corrosion will occur.

22.51. Most of the alloys which are used because of their high corrosion-resistance exhibit solid-solution structures. Aluminium–magnesium alloys (17.30) containing up to 7·0% magnesium fulfil these conditions and are particularly resistant to marine atmospheres. Stainless steel of the "18–8" type (13.44) is completely austenitic in structure when correctly heat-treated, but faulty heat-treatment may lead to the precipitation of carbides and, hence, to corrosion. Although such corrosion is partly due to the impoverishment of the austenitic matrix in chromium (since it is mainly chromium carbide which is precipitated), this corrosion is accelerated by electrolytic action between the carbide particles and the matrix. "Weld-decay" in steels of this type occurs for similar reasons.

22.52. As might be expected, coring in a cast alloy of the solid-solution type gives rise to a quicker rate of corrosion than that which obtains when the same solid solution has been cold-worked and

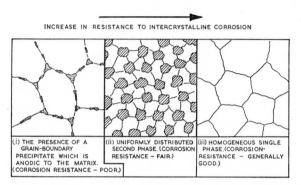

INCREASE IN RESISTANCE TO INTERCRYSTALLINE CORROSION

(i) THE PRESENCE OF A GRAIN-BOUNDARY PRECIPITATE WHICH IS ANODIC TO THE MATRIX. (CORROSION RESISTANCE – POOR.)

(ii) UNIFORMLY DISTRIBUTED SECOND PHASE. (CORROSION RESISTANCE – FAIR.)

(iii) HOMOGENEOUS SINGLE PHASE. (CORROSION-RESISTANCE – GENERALLY GOOD.)

FIG. 22.6.—The Relative Corrosion-resistance of Types of Microstructure.

annealed. This treatment produces greater homogeneity in the structure and removes the possibility of electrolytic action taking place between the core and outer fringes of individual crystals. Similarly, the presence of impurities segregated at the crystal boundaries of a metal used in either the alloyed or unalloyed state will accelerate corrosion by setting up electrolytic action with the metal when an electrolyte is present. The effects of cold-work in producing regions of high-energy value which accelerate corrosion have already been mentioned. Metals and alloys used in conditions promoting corrosion should, therefore, be of high purity and preferably free of locked-up stresses.

Protection by Metallic Coatings

22.60. Protection afforded by metallic coatings can be either "direct" or "sacrificial". Direct protection depends on an unbroken film of metal covering the article, and if the film becomes broken, corrosion may be accelerated by electrolytic action between the film and the metal beneath. In the case of sacrificial protection, however, the metallic film becomes the anode in the event of a break in the film, and thus dissolves in preference to the surface beneath. It follows that, when protection is limited to the direct type, as in the case of tin coatings on steel, the quality of the coating is most important, since acceleration, and not inhibition, of corrosion would follow a break in the film. In both cases, of course, protection of the direct type is the fundamental aim of the metal-coating process, and it is only in the possibility of the coating becoming broken that the effects of electrolytic action must be considered.

A number of methods are available for the production of metallic coatings. The most widely used are either electro-plating or dipping the articles to be coated into a bath of the molten metal. In some cases a successful coating can be produced by heating the articles to be coated in the finely powdered metal, whilst specialised use is made of the process known as "cladding".

22.61. Cladding. This process is applicable chiefly to the manufacture of "clad" sheet. The basis metal is sandwiched between pieces of the coating metal, and the sandwich is rolled to the required thickness. "Alclad", which is duralumin coated with pure aluminium, is possibly the best known of these products, whilst "Niclad" (nickel-coated steel) is also manufactured.

22.62. Hot-dip Metal Coating. Tin and zinc are the metals most often used to produce metallic coatings in this manner, though increasing use is being made of aluminium.

(*a*) *Hot-dip Tinning.* Tinplate is principally used in the manufacture of containers and boxes for packaging a wide variety of foodstuffs. The non-toxic properties of tin, combined with its good corrosion-resistance and the ease with which tin-coated articles can be soldered, make it a useful metal for coating articles which come into contact with all types of food.

The mild-steel sheet used for the manufacture of tinplate is usually about 0·25 mm thick, and, after being annealed, it is freed from all traces of oxide by pickling in either dilute sulphuric acid or dilute hydrochloric acid. The tinning unit comprises a thermostatically controlled vessel of molten tin in which is submerged a system of guides and rollers for conducting the pickled sheets down through a layer of molten flux into the tin and then upwards and out of the tin through a layer of palm oil. As they pass through the palm oil, the sheets are subjected to a squeegeeing action by tinned-steel rollers, which serve to control the final thickness of coating on the sheets.

After leaving the tinning machine, the tinned sheets are cooled and then cleaned, usually by washing them in an alkaline detergent. Finally, they are polished in either bran or wood meal.

Many other iron and steel articles used in food production are coated by hot-dip tinning. These articles include milk-cans, table-ware and domestic mincing-machines. After being cleaned, degreased and pickled, the component is immersed in a pot of molten tin, on the surface of which is a layer of flux. If the operations have been correctly carried out a brilliant finish is obtained which requires little in the way of after-work.

(b) *Hot-dip Galvanising.* The term "galvanising" is derived from the name of Luigi Galvani, the Italian physiologist, whose classic observations of dead frogs, in 1786, led to Volta's development of the electric cell. In 1829 Faraday observed that when zinc and iron, in contact with each other, were exposed to the air in the presence of a salt solution the iron did not rust; whereas if the zinc were removed the iron rusted rapidly. Faraday attributed this "to the effect of chemical action . . . in exciting electricity". In 1836 a Frenchman, Sorel, took out a patent for a process of coating steel with zinc by hot-dipping, and named the process "galvanising". The word was used to indicate the sacrificial electro-chemical protection afforded iron by the zinc.

Before being galvanised, the surface of steel must be thoroughly cleaned, either by shot blasting or, more generally, by pickling in dilute sulphuric acid. In order to prevent contamination of the molten zinc by iron, the work is then washed to remove all traces of iron salts formed by pickling. It is necessary, however, to protect the surface of the work from fresh oxidation, so it is immersed in a flux bath consisting of a solution of zinc chloride and ammonium chloride. This flux is usually dried on to the surface in an oven which is provided with an adequate draught and working at a low temperature, to prevent interaction between the flux and the iron surface.

The work, with its protective flux coating, is then immersed in the molten zinc bath. The flux coating peels away to form a layer on the surrounding bath surface, which in turn is protected from oxidation; and the surface of the work is left clean so that it is immediately wetted by the molten zinc. The thickness of the coating produced depends largely upon the temperature of the bath and the rate at which the work is withdrawn, but good-quality galvanised articles should carry between 0·3 and 0·6 kg/m^2 of zinc on the surface. Thinner and more flexible coatings are obtained by using a zinc bath containing about 0·2% aluminium. This limits the formation of intermetallic compounds of iron and zinc on the steel surface.

In addition to corrugated iron for roofing, the large number of galvanised products includes buckets, dust-bins, barbed wire, water tanks, marine fittings, agricultural equipment and window-frames.

(c) *Hot-dip Coating with Aluminium.* A process operating in Britain under the name of "Aludip" is used for the coating of steel sheets. These are first degreased and pickled and then treated in a solution containing copper ions to give a very thin copper film on the surface.

After being rinsed in an alcohol or glycerine solution, the sheets are immersed in molten aluminium. The copper film acts as a "key" between the steel sheet and the aluminium coating.

The Armco Iron Company (U.S.A.) also use an aluminium-dipping process, but which is followed by rolling. This gives a better finish and the product is used to replace tinplate for some food containers.

22.63. Coating by Means of a Spray of Molten Metal. Metal spraying consists in projecting "atomised" particles of molten metal from a special pistol by a stream of compressed air on to a suitably prepared surface. Surface preparation usually involves blasting the surface with an abrasive; steel grit having replaced sharp silica sand for this purpose on account of the silicosis hazards involved when the latter is used. The metal most commonly used for spraying is zinc, though coatings of aluminium, tin, lead, solder, cadmium, silver, copper and stainless steel can be deposited in this way.

A number of different types of pistol are in use for spraying zinc, but in all of them tiny particles of molten zinc are projected on to the surface to be coated by means of a blast of pre-heated compressed air. In one process molten zinc at 470° C is gravity-fed into the compressed-air stream, whilst in another, solid zinc powder is carried through a blow-pipe flame, and so melted. A further alternative method is to feed zinc wire axially into the centre of a blow-pipe flame. A stream of compressed air disintegrates the film of molten metal as it forms and sprays the metal from a nozzle.

In yet another process invented by Dr. Schoop, the originator of metal-spraying processes, electricity is used in place of gas for heating in those pistols which utilise zinc wire. In this pistol an arc is struck between two zinc wires so that their ends melt continuously as the wires are fed forward. The metal particles are carried forward by a stream of compressed air.

Metal spraying has wide application in view of its portability and flexibility; thus, large structures, such as storage tanks, pylons and bridges, can be sprayed on site. Notable recent examples include the Forth Road Bridge and the Volta River Bridge (Ghana), both of which were zinc coated using modern developments of the Schoop process.

22.64. Sherardising is a cementation process, similar in many respects to carburising, in that zinc is made to combine with an iron or steel surface by heating the work with zinc dust at a temperature *below* the melting point of zinc. The first patent for this process was taken out in London in 1901 by Sherard Cowper-Coles, after whom the process is named. Much thinner coatings can be obtained by sherardising than are possible with hot-dip galvanising, and generally about 0·15 kg/m² of zinc is used. Moreover, a more even film is obtained, thus making possible the treatment of such articles as nuts and bolts, the threaded portions of which would undoubtedly become clogged by hot-dipping.

As in all other coating processes, the surface of the work must first be properly prepared. This usually involves degreasing, followed by acid pickling in cold 50% hydrochloric acid or in hot 10% sulphuric acid. Shot blasting is also used, particularly in the case of iron castings, to remove graphite and core sand.

The work is packed into mild-steel drums along with some zinc powder. The drums are heated to 370° C and rotated slowly so that the tumbling action brings all the components into contact with the powdered zinc.

"Chromising" (chromium) and "calorising" (aluminium) are processes which are similar in principle to sherardising.

22.65. Electro-plating. The formation of metal coatings by electro-deposition is well known, and a wide variety of metals can be thus used, including copper, nickel, chromium, cadmium, gold and silver. Tin and zinc can also be electro-deposited, and a coating thus formed has advantages over one produced by hot-dipping in respect of flexibility, uniformity and control of thickness of film.

The usual degreasing and pickling processes must be applied to the surface before any attempt is made at electro-plating it. Most instances of peeling and blistering in the case of chromium-plating are due to inadequate preparation of the surface. The author has seen specimens in which the layer of chromium had been deposited on to oxide scale, with the inevitable peeling of the chromium as a result.

In the actual process of electro-plating the article to be plated is made the cathode in an electrolytic cell. Sometimes the metal to be deposited is contained, as a soluble salt, in the electrolyte, in which case the anode is a non-reactive conductor, such as stainless steel, lead or carbon. In most cases, however, the anode consists of a plate of the pure metal which is being deposited, whilst the electrolyte will contain a salt or salts of the same metal. Then, the anode gradually dissolves and maintains the concentration of the metal in the electrolyte as it is deposited on to the articles forming the cathode.

The conditions under which deposition takes place are very important, so that the cell voltage, the current density (measured in amperes per square metre of cathode surface), the ratio of anode area to cathode area and the time of deposition, as well as the composition and temperature of the electrolyte, must all be strictly controlled if a uniform adherent and non-porous film is to be obtained.

Protection by Oxide Coatings

22.70. In some instances the film of oxide which forms on the surface of a metal is very dense and closely adherent. It will then protect the metal surface beneath from oxidation. Stainless steels owe their resistance to corrosion to the presence of a high proportion of chromium, which is one of these elements that form oxide films impervious to oxygen. The "blueing" of ordinary carbon steel by heating it in air produces an oxide film of such a nature that it affords partial protection from corrosion.

22.71. Anodising. Reference has already been made (22.20) to the protection afforded aluminium by the natural film of oxide which forms on its surface. Anodic oxidation, or anodising, is an electrolytic process for thickening this oxide film. This process may be applied for several reasons, such as to provide a "key" for painting, to provide an insulating coating for an electrical conductor or to provide a surface which may be dyed, as well as to increase the resistance of aluminium to corrosion.

Before being anodised the surface of the article must be chemically clean. Preliminary treatment involves sand-blasting, scratch brushing or barrel polishing, according to the nature of the component. This is followed by degreasing in either the liquid or vapour of trichlorethylene; or by electrolytic cleaning. The latter process is used mainly for highly polished surfaces. The article is made the cathode in an electrolytic bath comprising a 1% solution of sodium hydroxide, and treatment is continued for about half a minute at a voltage of 14 and a current density of 550 A/m². The mechanism of this cleaning process is largely one of "flotation" of the film of grease by hydrogen bubbles given off at the surface of the article which comprises the cathode.

In the actual anodising operation which follows, the aluminium article to be treated is made the anode in an electrolyte containing either chromic, sulphuric or oxalic acid; the cathode being a plate of lead or stainless steel. When an electric current is passed, oxygen is formed at the anode and immediately combines with the aluminium surface of the article. The layer of oxide thus formed grows outwards from the surface of the aluminium. The normal thickness of a satisfactory anodic film produced commercially varies between 0·007 and 0·015 mm., and a film having a thickness within these limits would be formed by anodising a component in a 15% sulphuric acid solution at 20° C for about thirty minutes, using a current density of about 100 A/m² at a cell voltage of 15. A longer period of treatment produces a thicker, but soft and spongy, film which would be unsatisfactory in service. The thickness of the natural film produced on an aluminium surface by exposure to air at normal temperatures is of the order of 0·000 013 mm.

The surface produced by anodising can be dyed by immersion in either hot or cold baths of dyestuff, but whether or not the film is dyed, it should always be "sealed". The anodic coating is somewhat porous as it leaves the electrolytic tank and is easily stained, but sealing renders the coating impermeable. A simple, but quite effective, sealing process consists in treatment with hot or boiling water, which renders the coating non-absorptive without any visible change in appearance. Other sealing processes include treatment in solutions of 1% nickel or cobalt acetates at 100° C, followed by boiling in water, the application of linseed oil or lanolin to the surface or the application, while the article is warm, of a mixture containing equal parts of turpentine, oleic acid and stearic acid.

Protection by Other Non-metallic Coatings

22.80. Coatings of this type usually offer only a limited protection against corrosion and are, more often than not, used only as a base for painting.

22.81. Phosphating. A number of commercial processes fall under this heading, but in all of them a coating of phosphate is produced on the surface of steel or zinc-base alloys by treating them in or with a solution of acid phosphates. In order that the metal shall be made rust-proof a finishing treatment with varnish, paint, oil or lacquer is required.

The original phosphate process, named "Coslettising", was introduced in 1903; to-day a number of proprietary variations are available, the chief of which are "Bondorizing", "Granodising", "Lithoform", "Merlising", "Parkerising", "Rovalising" and "Walterising". Many readers may have used the preparation "Kurust", which is available for the small-scale treatment by phosphating of rusted steelwork such as the bodywork of decrepit motor-cars.

22.82. Chromating. Chromate coatings are produced on magnesium-base alloys, and on zinc and its alloys, by immersing the articles in a bath containing potassium bichromate along with various other additions. The colour of the films varies with the bath and alloy, from yellow to grey and black.

Cathodic Protection

22.90. This method of protection against corrosion can be used for buried or submerged pipe-lines and other structures. The pipe-line is made to act as a cathode by burying near it pieces of a metal which is much more electropositive than the iron of the pipe-line. These pieces of metal will therefore be anodic towards the iron of the pipe-line and will corrode sacrificially.

Alternatively, a current from D.C. mains can be passed through the soil or water on to the metallic surface concerned so as to keep it at a slightly negative potential with respect to its surroundings. When electric power is available this will be the cheaper method, since electricity can be obtained more cheaply from the mains than from any electro-chemical source. To protect the *whole* surface of a pipe-line by this means, however, would be expensive, but if the pipe has already been coated with paint or some other non-metallic substance, so that it is only necessary to protect any defective areas, the power cost is small, since very small currents only are necessary. In some parts of America the current is generated by dynamos driven by windmills.

RECENT EXAMINATION QUESTIONS

1. Explain the mechanism of metallic corrosion. Indicate the ways in which metals may be protected against corrosion.
    (Derby and District College of Technology, H.N.D. Course.)
2. Explain what is meant by the electro-chemical series of metals and discuss its significance in corrosion problems.
    Outline and illustrate with examples the general methods available to minimise electro-chemical corrosion.
    (Lanchester College of Technology, Coventry.)
3. Discuss the mechanism of electro-chemical corrosion in metallic structures. Comment briefly on some of the methods commonly used for its prevention. (University of Strathclyde, B.Sc.(Eng.).)
4. What is meant by the term "electrode potential" as applied to metals? What is the significance of a table of standard electrode potentials in relation to the corrosion of metals? Discuss the relative effectiveness of aluminium, cadmium, tin and zinc, as protective coatings for mild steel, and make particular reference to what occurs if the protective film becomes broken. (The Polytechnic, London W.1., Poly.Dip.)
5. Discuss the principles involved in "anodic protection".
    Outline ONE method of anodising a suitable component.
    (U.L.C.I., Paper No. B111-2-4.)
6. Outline the principles involved in the electrolytic corrosion of alloys.
    Explain the meaning of the term "sacrificial corrosion" and illustrate this principle in the protection of steels by metal coatings.
    (West Bromwich Technical College.)
7. Give an account of the principles of the galvanic corrosion of metals. Describe TWO methods for the protection of steel against corrosion which make use of these principles.
    (Derby and District College of Technology, H.N.D. Course.)

8. Discuss briefly the principal mechanisms of corrosion of metals and the basis of some common corrosion-prevention methods.

(S.A.N.C.A.D.)

9. "Mechanical isolation of the underlying metal and the galvanic relationship of the surface coating to the underlying metal are two major considerations of a metallic coating for corrosion-resistance."

Critically discuss this statement with reference to the protection of ferrous metals. (Aston Technical College, H.N.D. Course.)

10. Outline the conditions under which electrolytic corrosive attack can take place. Describe two methods used to protect metals from such attack.

(Rutherford College of Technology, Newcastle upon Tyne.)

11. Discuss the phenomena of corrosion fatigue in pure metals and alloys. Indicate its practical significance and how it may be minimised.

(Aston Technical College.)

12. What are the advantages of electro-galvanising over hot-dip galvanising? Give a brief outline of the latter process.

What other methods are available for applying a coating of zinc to steel? (U.L.C.I., Paper No. C111-2-4.)

13. (a) Describe briefly the following processes:

    (i) galvanising;
    (ii) electro-plating;
    (iii) anodising;
    (iv) phosphating.

(b) Explain the basic principles upon which each is based, and give one typical application of each process.

(Institution of Production Engineers Associate Membership Examination, Part II, Group B.)

14. Give an account of, and the principles involved in, the various methods of corrosion protection as applied to ferrous materials.

(Aston Technical College.)

15. Explain fully the following terms and processes, illustrating your answer with chemical formulae where appropriate:

    (i) cathodic protection;
    (ii) sherardising;
    (iii) anodising;
    (iv) corrosion fatigue and its effects on the S–N curves for steels.

(Institution of Production Engineers Associate Membership Examination, Part II, Group B.)

CHAPTER XXIII

METHODS OF MEASURING TEMPERATURE

23.10. Most metallurgical processes include the treatment of a metal or alloy at a high temperature. Moreover, in order to obtain the optimum properties with many of the modern alloys, a fairly accurate control of their heat-treatment temperatures is necessary. Pyrometry, or the measurement of high temperatures, is therefore one of the most important branches of metallurgical technology.

23.11. Whilst in America the Fahrenheit temperature scale is widely used, in Britain and Western Europe the Celsius scale is generally employed for temperature measurement in metallurgical operations. The fundamental, or "Absolute", scale of temperature, however, is based on the Second Law of Thermodynamics, as applied to a perfect gas. The absolute temperature unit, or "kelvin" (K)* covers the same temperature interval as does the degree Celsius, in that there are one hundred kelvin and one hundred degrees Celsius separating the melting point of ice and the boiling point of water (at standard pressure, i.e. 101 325 N/m²). The melting point of ice (0° C) corresponds to 273·15 K, whilst the boiling point of water corresponds to 373·15 K. Thus, "absolute zero" corresponds to −273·15° C.

23.12. The Gas Law states that

$$PV = RT$$

where $P =$ the pressure of the gas;
    $V =$ its volume;
    $R =$ a constant;
    $T =$ the *absolute* temperature of the gas.

The above gas law is true for a "perfect gas", and whilst none of the gases attain perfection in this sense, hydrogen and nitrogen can be used in practice, since their known deviations from perfection allow suitable corrections to be made in subsequent calculations. An instrument called the Constant Volume Gas Thermometer can be used to measure temperatures by an application of the Gas Law. The instrument consists essentially of a temperature-resistant bulb containing hydrogen or nitrogen. As the temperature rises the gas attempts to expand, but is

* Named after Lord Kelvin and now adopted as the unit of thermodynamic temperature in the Système International d'Unités.

prevented from so doing, and in consequence the pressure increases. This increase in pressure is measured, and since it is directly proportional to the rise in temperature, a means is thus afforded of measuring the latter.

23.13. Though it is a fundamental measuring instrument, the gas thermometer is cumbersome and fragile and quite unsuitable for the routine measurement of temperatures in a metallurgical process. It was used, however, to determine the melting points of many pure metals and chemical compounds. Since pure substances of this nature always melt at constant temperatures, these "recoverable temperatures" are used as standards against which a host of secondary pyrometers can conveniently be calibrated. The standard temperatures most often used are:

		° C
Freezing pure mercury .	. . . . .	−38·9
,, tin.	. . . . .	231·8
,, lead	. . . . .	327·3
,, zinc	. . . . .	419·4
,, antimony	. . . .	630·5
,, sodium chloride	. . . .	801·0
,, silver	. . . . .	960·5
,, copper .	. . . .	1 083·0
,, palladium	. . . .	1 555·0
Melting pure tungsten .	. . . . .	3 400·0

A fairly simple but robust type of secondary pyrometer, such as a thermocouple, can thus be chosen, and, in effect, be calibrated against the Gas Thermometer by using the intermediate non-variable values listed above.

Thermometers

23.20. The ordinary mercury-in-glass thermometer can be used only up to about 350° C, so that for metallurgical operations it is of restricted use. The range can be extended to 550° C by filling the upper part of the stem with nitrogen under pressure, but such a temperature range is still below that generally required. Moreover, the instrument would be more expensive than many of those dealt with later in this chapter.

Indicating Paints and Crayons

23.30. These materials are of particular use where small-scale heat-treatment of steel parts is carried out. Paints and crayons are supplied which will change in colour or appearance at a stated fixed temperature. A mark is made on the component with the substance, and it is then heated until the change in colour or appearance occurs.

The advantage of this type of indicator is that it ensures the attainment of the desired temperature *by the component itself*, and is not merely a reading of the furnace temperature, as is often the case with built-in pyrometers. Frequently, the temperature reached by a component on the hearth of a furnace bears little relationship to that recorded by the pyrometer in the roof. Moreover, the time for which the component is allowed to remain in the furnace will affect the temperature it reaches, whereas the indicating paint will not change colour until the component has reached the correct temperature. These paints are also useful in indicating the temperature of a component to be heated locally, e.g. when pre-heating for welding or pipe-bending, and in local stress-relieving.

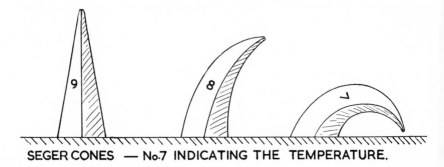

SEGER CONES — No.7 INDICATING THE TEMPERATURE.

FIG. 23.1.—Seger Cones.

Indicating Pellets and Cones

23.40. The well-known Seger cones are triangular pyramids made from various mixtures of kaolin, feldspar, quartz, magnesia, lime, iron oxide and boric acid; the composition being so adjusted that the cone wilts and finally collapses at a fixed temperature. A number of cones, of various melting points around the estimated temperature, are placed on a flat slab in a position convenient for observation. The end-point is judged to have been reached when the tip bends over and touches the base on which it is standing (Fig. 23.1).

Temperatures between 600 and 2 000° C can be indicated in this way, the main drawback in the use of Seger cones being that the end-point is affected by the rate of heating and by the nature of the furnace atmosphere.

Indicating pellets are used for temperatures up to 900° C, and melt sharply at the specified temperature, with an error stated to be not more than ±1%.

Electrical Resistance Pyrometers

23.50. These instruments make use of the fact that the electrical resistance of a metal wire varies as its temperature varies. The metal generally used is platinum in the form of a very fine wire wound on to a mica former and inserted into a refractory tube. This instrument is then used in the manner of a thermometer. The variation in resistance is determined by a Wheatstone-bridge type of circuit and the temperature then determined from a previously made calibration chart.

Although the platinum resistance thermometer gives accurate temperature readings, it is little used industrially because it is both bulky and fragile. Moreover, the characteristics of the coil are altered by the action of any reducing gases in the furnace atmosphere, which may dissolve in the wire. It is therefore used principally as a master instrument in the calibration of other pyrometers.

Thermo-electric Pyrometers

23.60. If two dissimilar wires are joined together at both pairs of free ends and one junction is heated to a higher temperature than the other, an electromotive force is set up in the circuit. This is known as the Peltier–Thomson effect. If the wires are now separated at the cold junction and connected instead to a sensitive millivoltmeter, the resultant E.M.F. in the circuit, due to the hot junction, can be measured. Moreover, it will be found that as the temperature of the hot junction increases, so will the resultant E.M.F. increase. This system, employing what is usually called a "thermocouple", is the basis for the most widely used industrial type of pyrometer.

23.61. The essential parts of a thermocouple installation are:

(*a*) two dissimilar wires of suitable composition which will produce a circuit E.M.F. large enough to be measured and able to function at the highest temperature required without melting, undue oxidation or fluctuation of electrical output;

(*b*) refractory insulation to prevent the wires from touching except at the hot junction;

(*c*) a refractory sheath to protect the couple from corrosive slags and injurious gases;

(*d*) a suitable millivoltmeter.

23.62. Most thermocouple wires are comparatively expensive, and since it is often necessary for the millivoltmeter to be at a considerable distance from the furnace, it follows that we shall need to connect the couple wires to the millivoltmeter by intermediate lengths of some cheaper wire. Supposing we do this as in Fig. 23.2A, using ordinary copper cable. A is the hot junction, which will record the temperature for us, and we will assume that the couple wires are connected to the copper cable at points B and C in a junction box on the wall of the furnace. B and C will also be points where dissimilar wires join and thus

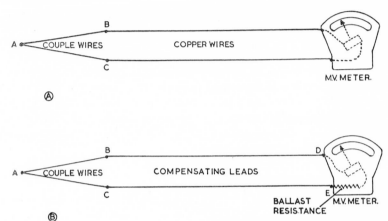

FIG. 23.2.—The Use of Compensating Leads.

constitute two more potential thermo-junctions. It should be realised, of course, that the millivoltmeter will record the resultant circuit E.M.F., and not the E.M.F. produced at the hot junction. Since the temperature of the junction box containing B and C is likely to change considerably, it follows that varying E.M.F.s will be set up at these points and so alter the total E.M.F. in the circuit. This will give rise to unknown errors in the recorded temperature.

The variable effects due to junctions B and C can be overcome in two ways:

(a) By using some means of keeping the temperature of B and C constant, both during calibration and in subsequent use of the thermocouple. In the laboratory this can be done conveniently by immersing the junctions in a Thermos flask containing pure melting ice (Fig. 23.3). Thus the term "cold junctions" is applied to B and C. This arrangement would be inconvenient in an industrial plant unless the expedient of burying the junctions about 3 m below ground were used. At this depth the temperature is fairly constant throughout the year. With expensive couple wires, however, this method is still rather costly, due to the extra 3 m required.

(b) By using "compensating leads". These are wires, which, taken as a pair, have the same thermo-electric constants as the couple wires themselves. One lead is usually of pure copper and the other of a suitable copper–nickel alloy, which, in conjunction with the copper lead, will produce similar thermo-electric E.M.F.s to the couple itself over the lower temperature range at which the cold junctions work. Thus, provided B and C are close together and hence at the same temperature in the junction box, the potential thermo-junctions are

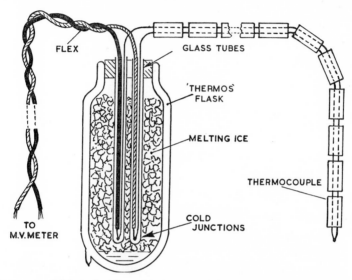

FLEX

GLASS TUBES

'THERMOS' FLASK

MELTING ICE

THERMOCOUPLE

TO M.V.METER

COLD JUNCTIONS

FIG. 23.3.—A Method of Ensuring Constant Cold-junction Temperature for a Laboratory Thermocouple.

now transferred to D and E (Fig. 23.2B), where the compensating leads are connected to the terminals of the millivoltmeter. Any errors here will be small, since variations in temperature of the millivoltmeter will be slight (it is sometimes installed in a works office). These slight variations can be compensated for by means of a bi-metallic spring attached to the pointer of the millivoltmeter.

23.63. Further error could be introduced by the variation in resistance of the compensating leads with their temperature, particularly if they were long. This is overcome by incorporating a ballast resistance in the millivoltmeter so that the resistance of the latter will be made large in comparison with that of the external circuit. The ballast resistance is usually of some alloy, such as manganin, which has a low-temperature coefficient of resistance.

23.64. The couple wires themselves may be of base or rare metal, according to their application. Base-metal wires are much cheaper,

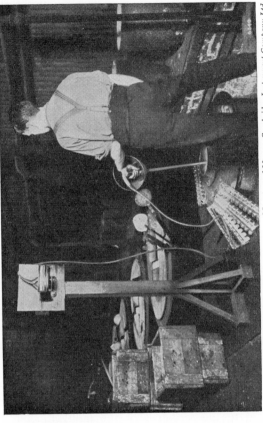

[Courtesy of Messrs. Cambridge Instrument Company Ltd.]

23.1B.—The Thermocouple in Use.
The millivoltmeter is carried on the wooden stanchion.

PLATE 23.1

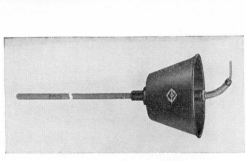

23.1A.—An Industrial Thermocouple for Immersion in Molten Metals.

so that thicker, stronger wires can be used. They are also more sensitive, but are limited in their use by the fact that they cannot generally be used above 1 100° C. For higher temperatures the platinum type of thermocouple must be used. This should be protected from injurious furnace

TABLE 23.1.—*Characteristics of the Common Thermocouples*

Thermocouple	Chemical composition		Sensitivity (millivolts/° C)	Temperature range (° C)	Max. temp.* (° C)
Copper–Constantan	Pure copper	Constantan 60% Cu 40% Ni	0·054	−200 to 300	600
Iron–Constantan	Pure iron	Constantan 60% Cu 40% Ni	0·054	−200 to 750	1 000
Chromel–Alumel	Chromel 90% Ni 10% Cr	Alumel 95% Ni 2% Al 3% Mn	0·041	−200 to 1 200	1 350
Platinum–platinum–rhodium	Pure platinum	Platinum–rhodium 90% Pt 10% Rh	0·0095	0 to 1 450	1 700

* For very *short* periods of exposure only.

atmospheres by a fireclay or fused-silica sheath, which can be, in turn, protected from possible mechanical damage by a metal sheath when the couple is for use at temperatures which the metal sheath will withstand.

The couple wires are usually welded together at the hot junction, but for the measurement of molten-metal temperatures they can be left unwelded, since there is a tendency for the couple wires to be dissolved

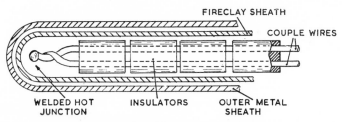

FIG. 23.4.—Insulation and Protection of the Couple Wires in an Industrial Installation.

in any case by the molten metal. The molten metal completes the circuit, and an almost instantaneous reading is obtained, indicating that actual contact with the metal has been made. If the couple wires were inserted only in the surface slag the circuit would not be completed, and no reading would be obtained.

23.65. The millivoltmeter may be of the plain indicating type, or it may also be made to produce a permanent record. In the latter type of instrument the millivoltmeter needle swings above a slowly rotating drum which carries a paper record-sheet, and between this and the needle is an inked ribbon. A motor-driven "chopper" bar falls every half-minute, producing a point on the record-sheet by pressing the needle on to the inked ribbon. This idea has been further developed to produce a temperature-control unit. In this device the needle, having reached a previously pre-set temperature, is pressed down on to a Bowden cable which tilts a mercury switch and switches off the furnace.

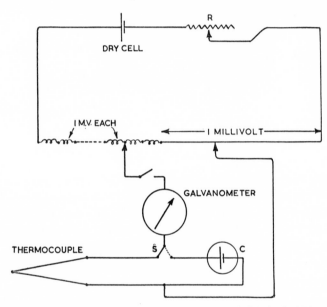

FIG. 23.5.—The Theoretical Circuit for a Potentiometric Method of Measuring the E.M.F. in a Thermocouple Circuit.

23.66. A more accurate measurement of the resultant E.M.F. produced in the pyrometer circuit can be made by employing a potentiometer system. Such an arrangement is particularly suitable for laboratory use, but a very excellent "workshop type" of potentiometer is also available in which all the variable resistances are of the rheostat type, and standardisation is effected by means of a standard cadmium cell built into the case. The principle of the instrument is shown in Fig. 23.5. The circuit is first standardised by adjusting the rheostat R until the circuit is balanced against the standard cell C, as shown by a zero reading on the galvanometer. The switch S is then operated to disconnect C and bring the pyrometer system into circuit so that the resultant E.M.F. it is producing can be read from the calibrated dials when the circuit is again balanced.

23.67. The Calibration and Checking of a Thermocouple. The calibration or checking of a thermocouple must be carried out with the couple connected up in the circuit in which it will ultimately be used, i.e. with the compensating leads in circuit, since, as already

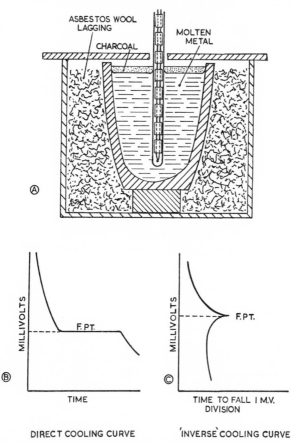

FIG. 23.6.—Calibrating a Thermocouple.
(A) Adequate lagging of the crucible to ensure slow cooling. (B) and (C) Types of cooling curve used.

mentioned, the E.M.F. generated by the thermocouple itself will not be the same as the E.M.F. measured by the millivoltmeter.

The couple can be checked against a special standard couple or, if one is not available, against the melting points of the pure metals as already outlined (23.13). This is done by melting the metal (under charcoal to prevent oxidation) in a small fireclay crucible and inserting the thermocouple as shown in Fig. 23.6A. It is advisable to prevent the metal from cooling too rapidly, or the temperature lag between the sheathed couple and the metal will be considerable. Cooling can conveniently be retarded by lightly lagging the crucible with asbestos wool.

Millivoltmeter readings are taken at intervals of, say, fifteen seconds as the metal cools; so that a cooling curve can be plotted as in Fig. 23.6B. The horizontal portion of the curve gives the millivoltmeter

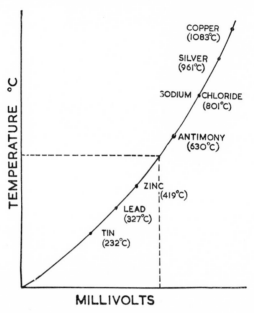

MILLIVOLTS

FIG. 23.7.—The Completed Calibration Curve for the Thermocouple, in Which the Freezing Points of the Metals Used Are Plotted Against the Readings (in Millivolts) Corresponding to the Freezing Points.

reading corresponding to the freezing point of the metal. Alternatively, an "inverse" cooling curve can be plotted (Fig. 23.6c) by taking the time interval necessary for the millivoltmeter needle to drop through one scale division. Whichever method is used, a series of millivoltmeter readings, corresponding to the melting points of the pure metals employed, will be obtained. A calibration curve for the pyrometer circuit can be plotted from these results, as shown in Fig. 23.7.

23.68. In addition to errors arising from inaccurate calibration, errors may also arise from the following faults:

(*a*) Bad connections or electrical leakages in the circuit.

(*b*) Lack of homogeneity in the couple wires leading to parasitic E.M.F.s being set up along their lengths.

(*c*) The use of thick couple wires, which will conduct heat away from the hot junction so rapidly that it never quite attains the surrounding temperature. This effect will be accentuated if the fireclay sheath enclosing the couple is so thick that it has a pronounced lagging action. Then not only will the thick couple wires conduct heat away from the hot junction rapidly but the fireclay sheath, having a low thermal conductivity, will also impede the transfer of heat from the medium outside the sheath. These two effects will therefore combine in making the pyrometer read low.

Radiation Pyrometers

23.70. It is often necessary to measure high temperatures which are beyond the range of the ordinary thermocouple, or to measure the temperature of a surface upon which, for some reason, it would be inconvenient or impossible to place an instrument of this type. In such cases some form of pyrometer is used which does not require to be in actual contact with the hot body, but measures the intensity of heat radiated from its surface.

23.71. In order that it should do this accurately, "black-body" conditions must prevail at the radiating surface. A black body can be defined as one which absorbs all radiations falling upon it without loss by reflection. Conversely, such a surface will radiate heat in the same way. (The reader may remember early physics experiments in which two cocoa-tins, one with a soot-coated surface and the other still bright, were both filled with hot water at the same temperature. Temperature readings taken at fixed intervals indicated that the blackened tin lost heat more quickly than did the bright one.)

The radiation received from the "muffled" interior of an almost totally enclosed furnace chamber closely approaches the ideal blackbody conditions, but the heat energy received from a body free to radiate in the open, where the surrounding air temperature is less than that of the hot body, is only a fraction of this. The fraction of heat energy radiated by a surface, as compared with an equivalent blackbody surface at the same temperature, is called the *emissivity* of the surface. This emissivity value depends upon the nature and temperature of the surface as well as its colour. Matt surfaces, for example, approach black-body conditions more closely than do smooth surfaces, irrespective of colour. A few emissivity values are given in Table 23.2.

23.72. "The total heat energy radiated by a body is proportional to the fourth power of its absolute temperature." This is the Stefan–Boltzman Law, which may be expressed thus:

$$E = Z(T^4 - T_0^4)$$

TABLE 23.2.—*Total Emissivity of Some Typical Surfaces*

Material	Emissivity
Copper (un-oxidised)	0·02
Copper (oxidised)	0·70
Copper (liquid)	0·15
Iron (un-oxidised)	0·05
Iron (oxidised)	0·85
Steel plate (rough and oxidised) . . .	0·97
Cast iron (strongly oxidised) . . .	0·95
Firebrick	0·75
Silica brick	0·85

N.B. The above values are approximate only, and depend largely upon the temperature of the surface and also upon its degree of roughness.

where E = the total heat energy radiated;

Z = a constant;

T = the temperature of the radiating surface (K);

T_0 = the temperature of the air surrounding the pyrometer (K).

When the temperature measurement to be made is concerned with a metallurgical process, T_0^4 is generally small compared with T^4, so that the expression may be simplified:

$$E = ZT^4$$

Since the energy emitted increases much more than proportionally to the temperature rise, it follows that radiation pyrometers are very sensitive at elevated temperatures. If the emissivity factor of the radiating surface is known, we can apply a correction for bodies radiating in the open. If the emissivity factor is e, then:

$$E = ZeT^4$$

where T is the true temperature. If the apparent temperature which we read is S, then:

$$E = ZS^4$$

$$ZeT^4 = ZS^4$$

$$eT^4 = S^4$$

By way of an example, let us assume that the observed temperature of a surface, coated with cuprous oxide and radiating in the open, was 800° C. If the emissivity factor of the cuprous oxide surface is taken as 0·70, the true temperature T can be found as follows:

$$eT^4 = S^4$$

$$\therefore \quad 0{\cdot}70 \times T^4 = (800 + 273)^4$$

$$\therefore \quad \text{Log } 0{\cdot}70 + 4 \log T = 4 \log (800 + 273)$$

$$T = 1\ 173 \text{ K or } 900° \text{ C}$$

23.73. Most radiation pyrometers are calibrated over the range 700–2 000° C and employ some form of optical focusing system to concentrate the heat radiation on to a thermocouple. Such pyrometers are independent of the distance between the instrument and the source of heat, provided the image of the latter is large enough to fill the field of view in the eyepiece. Errors may arise due to smoke, flame and water-vapour in the furnace atmosphere, so that, where possible, the fuel supply to the furnace should be turned off momentarily whilst the reading is taken.

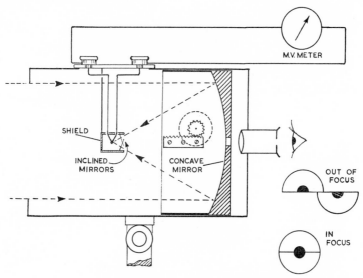

FIG. 23.8.—The Principle of the Féry Radiation Pyrometer.

23.74. A typical pyrometer of this group is the well-known Féry shown in Fig. 23.8. This consists essentially of a concave mirror which can be moved so that it focuses the heat rays on to a small thermocouple set in the tubular body of the instrument. Focusing is done by means of a split-image type of range-finder, the position of the mirror being adjusted until the two halves of the image coincide. The couple is generally protected from direct radiation by a small polished shield, whilst a black target is fixed to the hot junction so that it will readily absorb the heat focused upon it.

The modern tendency is to use a modified type of Féry pyrometer in which the concave mirror is replaced by a lens, as in the case of the solar-cell pyrometer (Fig. 23.10). Since no direct radiation will then fall upon the couple assembly, a shield is unnecessary. This makes a

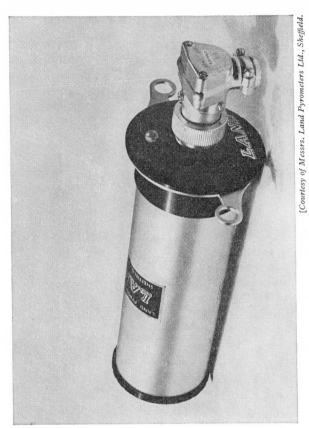

[Courtesy of Messrs. Land Pyrometers Ltd., Sheffield.

23.2A.—The "LAND" Silicon Solar-cell Pyrometer.

23.2B.—The Disappearing-filament
Pyrometer.

PLATE 23.2

stable zero easier to obtain than when a shield is present, but, at the same time, trouble may be experienced with "chromatic aberration" of the lens. Photographic lenses are expensive because they have to be of multiple construction to reduce such aberrations.

23.75. The chief advantage of the Féry pyrometer is that, unlike many optical pyrometers, it does not depend upon the human element in its operation, since no colour matching is involved. It can also be permanently installed and made to keep a continuous record automatically, by connecting it to a suitably calibrated recording milli-voltmeter.

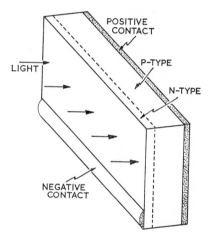

FIG. 23.9.—Construction of a Silicon Solar Cell.

The author has occasionally been asked by his students why, if the use of the Féry pyrometer is independent of its distance from the hot body, it cannot be used to measure the temperature of the sun. The answer, of course, is that the image produced by the sun would fill only a very small part of the field of view in the eyepiece of the instrument, instead of the complete disc, as is necessary for correct reading. This is perhaps fortunate, since were the Earth sufficiently near to the sun for the image of the latter to fill the field of view in a Féry pyrometer, the heat intensity would probably be sufficient to vaporise both the instrument and its operator.

The unit is approximately 6 mm long and 0·5 mm thick. Traces of impurity are added to the silicon which then forms a *p*-type crystal. A thin layer of *n*-type silicon is then deposited on the surface. A diode is thus formed with the *p–n* junction close to the surface which is then sight sensitive.

23.76. Photo-electric exposure meters are widely used in photography in order to determine the intensity of light. These are simple instruments usually based on the selenium "barrier layer" cell. When light falls upon such a cell an E.M.F. is generated which is proportional to the intensity of light falling upon it. The current arising from this E.M.F. is measured by a micro-ammeter. Similar cells can be used in photo-electric pyrometers. Being sensitive to *visible* radiation, they can be used in the range 800° C upwards, and within its temperature range the sensitivity of the instrument is high. A further advantage is that such a cell responds very quickly to temperature changes, but against this must be set the fact that if the temperature of the actual instrument exceeds 50° C, damage to the cell may result.

More recently, the silicon "solar" cell has been adapted for use in radiation pyrometers. This type of cell has been used widely in satellites and space craft to provide electricity for energising the radio transmitter and other electronic equipment. It consists of a small unit (Fig. 23.9) built up of separate layers of p- and n-forms of silicon. When light falls upon such a cell an E.M.F. is generated. Since the E.M.F. increases such that it is approximately doubled for every rise of 100° C, this means that the sensitivity of the cell increases with rise in temperature. An obvious drawback, associated with this increase in sensitivity, is that the range covered by such an instrument tends to be rather narrow unless a potentiometric type of recorder is used in conjunction with it. By using such a system to "tap off" only a fraction of the output of the cell at high temperatures, a range from say, 700 to 1 250° C can be covered on a *single scale*, though different scale ranges up to 4 000°C can be covered if necessary.

In addition to its sensitivity at high temperatures, the instrument has the advantage that, as with the Féry pyrometer, its reading is independent of the operator provided that the instrument has been correctly sighted. It was mentioned above that, when using a Féry pyrometer, the hot body must be near enough to fill the field of view as seen through the eyepiece. Since the sensitivity of the solar-cell pyrometer is so high, however, a lens system can be used to focus rays from a much greater distance and still fill the field of view. In this case the function of the optical system is similar to that of a telephoto lens in photography. Thus, with temperatures above 1 000° C., and a suitable optical system, the temperature of a target as little as 25 mm in diameter can be measured at a distance of 3 m. Moreover, since the time of response of the solar cell is of the order of 0·001 second, rapid temperature readings can be taken. The principles of the instrument are shown in Fig. 23.10.

Optical Pyrometers

23.80. The term "optical pyrometer" includes a number of ingenious devices which can be used to estimate temperature. The simpler types include graduated coloured filters, through which the hot body is viewed and the point on the scale where the colour becomes extinct is noted. These are, in principle, very similar to the cheaper forms of photographic exposure meter used to measure intensity of light. In point of fact we are only measuring the intensity of light, associated with a certain temperature, when we use an optical pyrometer. The principle

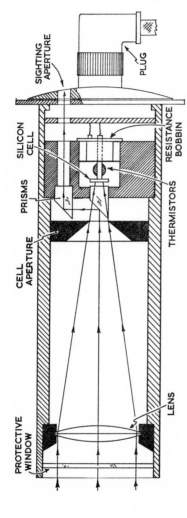

FIG. 23.10.—The "LAND" Silicon Solar-cell Radiation Pyrometer.

Light rays are focused on to the cell by means of the lens. The cell is shielded from reflections within the tube by the cell aperture. A small portion of the radiation is reflected through the eye aperture by the pair of unsilvered prisms in order to aid sighting of the instrument. Variations in ambient temperature are compensated for by means of the thermistor–resistance bobbin assembly which is in series with the cell.

of an optical pyrometer is based on Wien's Law, which can be expressed as follows:

$$I = C\lambda^{-5} \cdot e^{\frac{-k}{\lambda T}}$$

where I = the intensity of radiation given out by the body;
  λ = the wavelength of the radiation;
  T = the absolute temperature of the body;
C and k = constants;
  e = the base of Napierian logarithms.

23.81. The Wanner pyrometer makes use of a light polariser which compares the light of "red wavelength" from the hot body with the rays of similar wavelength from an electric lamp, the intensity of which has been previously calibrated. On looking through the eyepiece, the operator sees a circular field divided into half, one half being due to the hot body and the other half to the standard lamp. The polariser is then rotated until the two halves match in brightness.

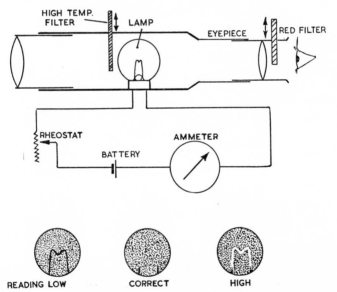

FIG. 23.11.—The Principle of the Disappearing-filament Pyrometer.

23.82. The best-known optical pyrometer, however, is of the so-called disappearing-filament type (Fig. 23.11). In this instrument a calibrated electric-bulb filament is sighted against the hot body and the current passing through the filament adjusted by means of a built-in rheostat until the colours of hot body and filament just match, i.e. the filament "disappears". The ammeter is calibrated in degrees Celsius with two temperature scales (700–1 300° C; and 1 000–1 800° C), a neutral filter being interposed between the hot body and the filament when using the instrument on the upper temperature scale. A red filter, which slides into the eyepiece, serves to remove the effects of any colour difference which exists between the hot body and the filament.

23.83. As with radiation pyrometers, the optical instruments also are affected by the emissivity of the surface viewed, and due allowances must accordingly be made. Moreover, false readings will be obtained if there is smoke or flame in the furnace atmosphere, or, on the other hand, if the surface whose temperature is sought is bright and can reflect light from other sources into the instrument.

RECENT EXAMINATION QUESTIONS

1. Outline the theory of the methods used for temperature measurement indicating their applications.

    (Constantine College of Technology, H.N.D. Course.)

2. Describe the experimental method used in the determination of the freezing range of a sample of a tin–lead alloy.

    (U.L.C.I., Paper No. B111–1–4.)

3. Outline the general considerations that must be taken into account when selecting a suitable pyrometric method for different applications.

    Without describing the methods, give an assessment of the thermo-electric and optical pyrometers in the light of these considerations.

    (Lanchester College of Technology, Coventry.)

4. Discuss, with reference to a thermocouple device:

    (a) the practical problems of measuring the temperature of a molten metal bath;

    (b) the relative merits of a potentiometer and a moving-coil milli-voltmeter as the measuring instrument;

    (c) the method of calibration;

    (d) the precise purpose of compensating leads.

    (Institution of Production Engineers Associate Membership Examination, Part II, Group A.)

5. Explain the term *black-body conditions* as applied to total radiation and optical pyrometers.

    Describe any radiation or optical pyrometer. State what precautions should be taken in use, in order to obtain a high accuracy.

    (U.L.C.I., Paper No. C111–1–4.)

6. Discuss the electrical methods available for temperature measurement in the range 300–1 600° C.

    (Lanchester College of Technology, Coventry, H.N.D. Course.)

ADDITIONAL MISCELLANEOUS QUESTIONS

1. Explain the following heat-treatments, stating the purpose of each and the effect on the properties of the metal:

    (a) the age-hardening of an aluminium–copper alloy;
    (b) the stress-relief annealing of a cold-worked mild steel;
    (c) normalising of a large steel casting.

    (C. & G., Paper No. 293/6.)

2. Describe briefly what is meant by any THREE of the following:

    (a) dendritic growth;
    (b) macro-examination;
    (c) temper-brittleness;
    (d) solution heat-treatment;
    (e) age-hardening.   (C. & G., Paper No. 293/21.)

3. With regard to physical and mechanical properties and relevant economic considerations, discuss the uses of the following metals and alloys to the engineer: pure copper; cast iron; mild steel; 1.2% carbon steel; low nickel–chromium steel; titanium-base alloys.

    (U.L.C.I., Paper No. C111–1–4.)

4. Sketch microstructures for FOUR of the following:

    (a) 0.15% carbon steel, normalised;
    (b) 95–5 aluminium bronze, cold-worked, annealed and then cold-worked again;
    (c) an alloy containing 88% aluminium and 12% copper in the cast condition;
    (d) a tin-lead solder containing 50% of each metal;
    (e) 60–40 brass in the extruded condition.

    Comment briefly on the structure you draw and label the phases in each case.   (U.L.C.I., Paper No. C237–2–8.)

5. Distinguish briefly but clearly between:

    (a) eutectic and eutectoid;
    (b) grey cast iron and white cast iron;
    (c) hardness and hardenability;
    (d) normal extrusion and inverted extrusion.   (S.A.N.C.A.D.)

6. Write notes on TWO of the following:

    (*a*) austenitic stainless steels;
    (*b*) age-hardening;
    (*c*) aluminium bronzes.   (S.A.N.C.A.D.)

7. Write an essay on one of the following topics:

    (*a*) bearing metals;
    (*b*) the surface-hardening of steels;
    (*c*) nickel-base engineering alloys.
        (Rutherford College of Technology, Newcastle upon Tyne.)

8. Give an account of the structure, properties and uses of FOUR of the following alloys:

    (*a*) 70–30 brass;
    (*b*) 60–40 brass;
    (*c*) spheroidal-graphite iron;
    (*d*) high-speed tool steel;
    (*e*) chromium–nickel austenitic steel.

        (Constantine College of Technology, Middlesbrough.)

9. Write notes on THREE of the following:

    (*a*) the structure and properties of 60–40 brass;
    (*b*) martensitic stainless steel;
    (*c*) sherardising and calorising;
    (*d*) anodising of aluminium;
    (*e*) weld-decay in austenitic stainless steels.

        (Constantine College of Technology, H.N.D. Course.)

10. (*a*) For a highly stressed machine component which has a cross-sectional area of about 2 in^2 a steel having a tensile strength of not less than 65 $tonf/in^2$ together with minimum elongation 16% is required. Why would a plain carbon steel be unsuitable for this purpose? Give the approximate chemical composition of a steel that would be suitable and describe the heat-treatment to be applied to the steel in order to obtain the required properties.

    (*b*) Describe *briefly* the most effective method of hardening (where possible) each of the following materials without altering its chemical composition:

    (i) pure lead;
    (ii) austenitic stainless steel;
    (iii) 0·4% carbon steel;
    (iv) an aluminium alloy containing 4% copper.

        (Aston Technical College.)

11. In any FOUR of the following cases, explain the reason for the addition of the element given:

   (a) sodium to certain aluminium–silicon alloys;
   (b) niobium or titanium to austenitic stainless steels;
   (c) chromium to copper;
   (d) iron to aluminium-bronze casting alloys;
   (e) copper to tin–antimony bearing alloys.

   For each case answered, give the approximate amount of the addition and the metallurgical effect.
   (Aston Technical College, H.N.D. Course.)

12. Give an account of any TWO of the following:

   (a) radiation pyrometers;
   (b) intermetallic compounds;
   (c) strain-ageing;
   (d) Allotropy in metals.

   (Derby and District College of Technology, H.N.D. Course.)

13. Write notes on FOUR of the following:

   (a) season-cracking;
   (b) temper-brittleness;
   (c) eutectic alloy;
   (d) allotropic change;
   (e) critical cooling rate;
   (f) weld-decay.   (Derby and District College of Technology.)

14. Sketch and describe the microstructure of any THREE of the following:

   (i) 0·5% carbon steel, oil-quenched from 850° C;
   (ii) 95–5 tin bronze, cold-worked and annealed;
   (iii) a bearing metal containing 10·0% antimony, 3·0% copper, 50·0% tin, balance lead;
   (iv) a modified aluminium casting alloy containing 11% silicon.

   (West Bromwich Technical College.)

15. State what material you would use to manufacture FOUR of the following:

   (a) marine condenser tubing;
   (b) non-sparking tools;
   (c) main bearings in an internal-combustion engine;
   (d) jaws in a jaw crusher;
   (e) canning material for nuclear reactor fuel.

   Give reasons for your choice.
   (Nottingham and District Technical College.)

16. Write an essay on ONE of the following topics:

   (a) the fatigue of metal;
   (b) the measurement of temperatures above 1 200° C;
   (c) stainless steels;
   (d) the hardenability of steel.

   (Institution of Production Engineers Associate Membership
   Examination, Part II, Group B.)

17. Discuss the compositions, uses and any points of metallurgical interest in
   any TWO of the following groups of non-ferrous alloys:

   (a) beryllium bronzes;
   (b) aluminium bronzes;
   (c) zinc-base alloys.

   (Institution of Production Engineers Associate Membership
   Examination, Part II, Group A.)

18. Explain the following terms:

   (a) Overageing;
   (b) stress-relief annealing;
   (c) M_s temperature;
   (d) Austempering.

   (University of Aston in Birmingham, Inter B.Sc.(Eng.).)

19. Write an account of the influence of residual stress, stress raisers and
   environment in determining the service life of stressed metal machine
   components.
   (University of Aston in Birmingham, Inter B.Sc.(Eng.).)

20. Discuss the various methods commonly used to increase the strength
   and hardness of metals.
   (University of Strathclyde, B.Sc.(Eng.).)

21. Write short explanatory notes on FOUR of the following terms:

   (a) twinning;
   (b) recrystallisation and grain growth;
   (c) the formation of dendritic structures;
   (d) intermetallic compounds;
   (e) temper-brittleness.

   (The Polytechnic, London W.1., Poly.Dip.)

22. Write short notes on any FOUR of the following:

   (a) weld-decay;
   (b) vanadium in steels;
   (c) mass effect;
   (d) deoxidation of steels;
   (e) nitriding.   (The Polytechnic, London W.1., Poly.Dip.)

23. Use the data in the table below to comment on the following statements:

    (a) Copper and its alloys will be replaced by aluminium and its alloys.
    (b) Plastics have replaced many metal components in the motor-car.
    (c) Steel production will not decrease substantially in the next decade.

(University of Aston in Birmingham, Inter B.Sc.(Eng.).)

	Tensile strength (tonf/in^2)	Density (g/cm^3)	Cost per ton £
STEEL			
Cast steel	24–80	7·8	140–220
Structural steel . . .	37–43	7·8	80 (works)– 100 (site)
Hot stampings and forgings .	35–45	7·8	140–200
Weldments	25	7·8	260–800
IRON			
Grey cast iron . . .	10–14	7·3	70–120
Malleable blackheart . .	20–33	7·3	112–140
Nodular ferritic . . .	26–32	7·4	120–200
COPPER ALLOYS (wrought) 60–40 Brass:			
Rod	25	8·36	240
Section	25		300–350
Hot stamping . . .	25		250–370
ALUMINIUM ALLOYS (wrought)			
HS. 30 sheet	13–19	2·7	260–600
HS. 30 plate	13–19	2·7	320–450
HS. 30 extrusions . . .	12–18	2·7	355–390
LIGHT ALLOY (castings)			
Aluminium castings LM 22 .	16	2·7	270–320
Magnesium castings . .	12	1·7	400
Zinc die-castings . . .	18	6·6	200–250
PLASTICS			
Polythene (moulded) . .	1	0·9	260
Polyvinyl chloride (pipe). .	4	1·4	200
Nylon (glass-filled moulded) .	10	1·38	950–1 150

24. (a) Discuss the mechanical and metallurgical effects of manganese as an alloying element in steel.

    (b) High-speed steel is widely used as a cutting-tool material. Explain why this is so with particular reference to the function of tungsten.

    Give a typical composition for this steel and briefly discuss any precautions that may be necessary during its heat-treatment.

    (c) Suggest a nickel steel suitable for use at temperatures in the range −100° to −200° C. Give reasons for your choice and suggest a practical application. (Aston Technical College.)

INDEX